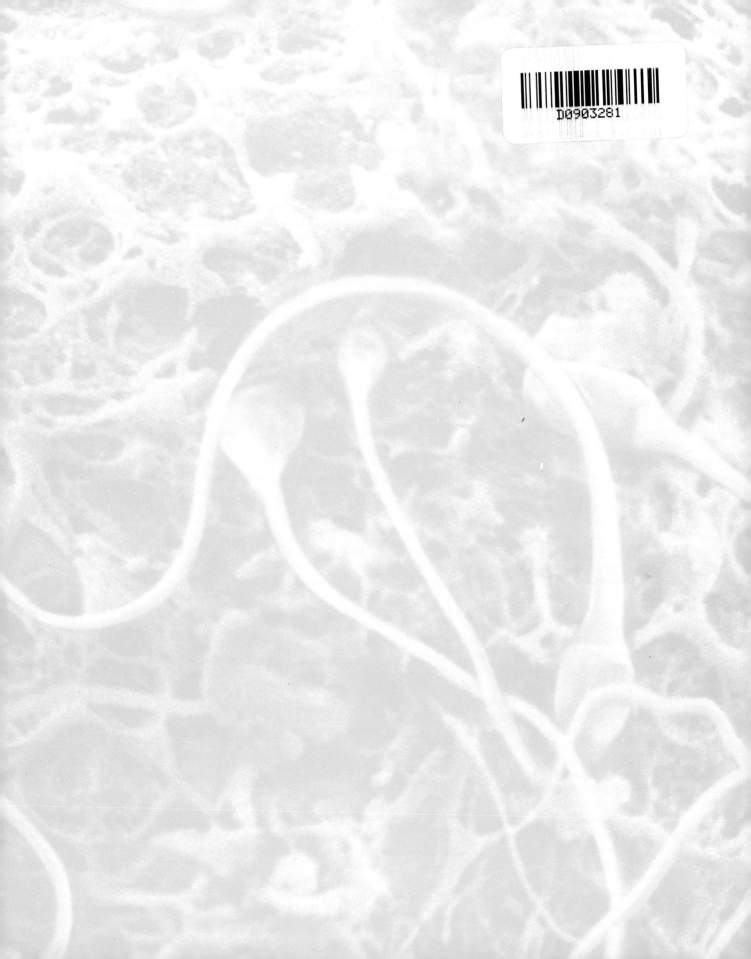

THE SCIENCE OF
PREGNANCY

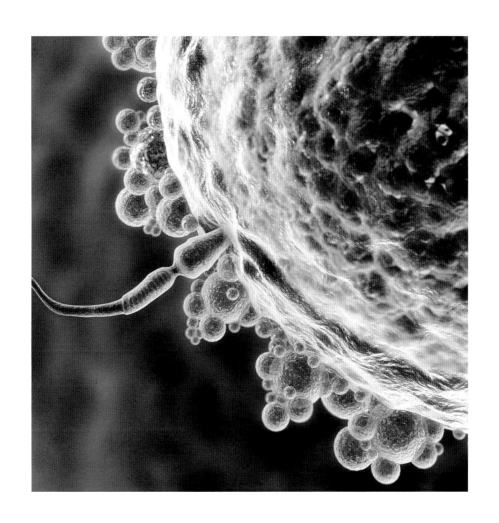

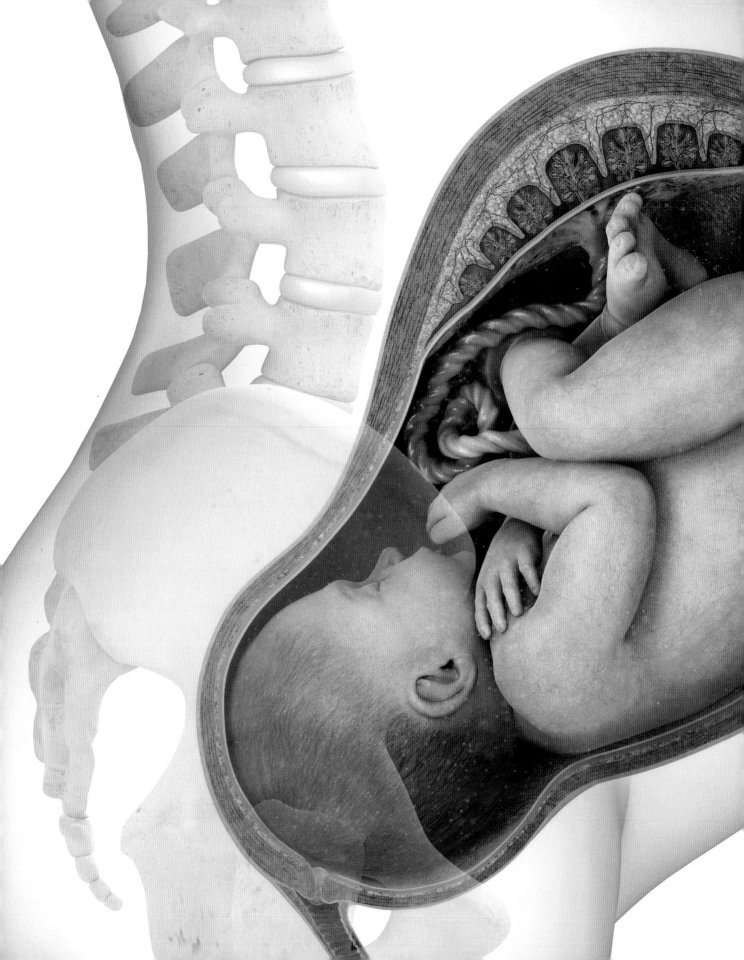

THE SCIENCE OF
PREGNANCY

DR. SARAH BREWER

SHAONI BHATTACHARYA

DR. JUSTINE DAVIES

DR. SHEENA MEREDITH

DR. PENNY PRESTON

Editorial consultant DR. PAUL MORAN

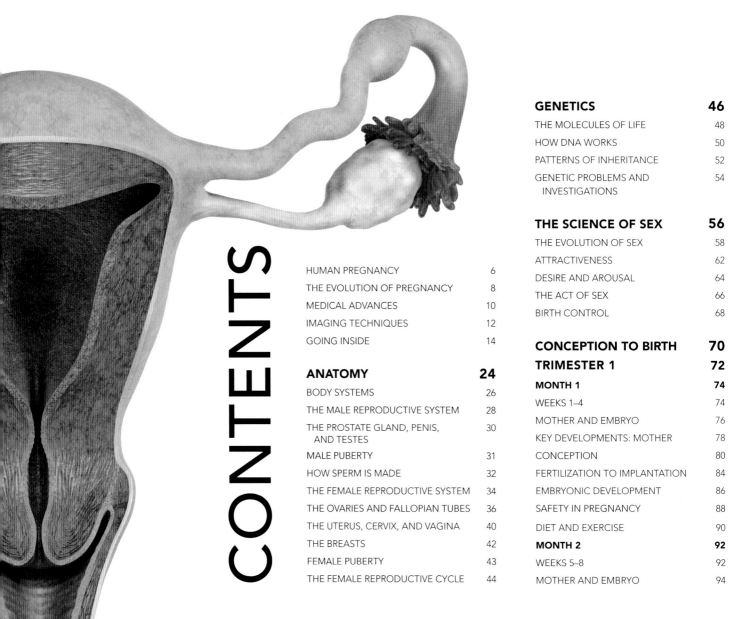

CONTENTS

 Penguin Random House

NEW EDITION

DK LONDON
SENIOR EDITOR Martyn Page
PROJECT EDITOR Shashwati Tia Sarkar
US EXECUTIVE EDITOR Lori Cates Hand
MANAGING EDITOR Angeles Gavira Guerrero
MANAGING ART EDITOR Michael Duffy
PRODUCER, PRE-PRODUCTION
Jacqueline Street-Elkayam
SENIOR PRODUCER Meskerem Berhane
JACKET DESIGNER Surabhi Wadhwa-Gandhi
JACKET EDITOR Emma Dawson
JACKET DESIGN DEVELOPMENT MANAGER Sophia MTT
ASSOCIATE PUBLISHER Liz Wheeler
ART DIRECTOR Karen Self
PUBLISHING DIRECTOR Jonathan Metcalf

DK DELHI
SENIOR EDITOR Janashree Singha
EDITOR Tanya Singhal
PROJECT ART EDITOR Rupanki Arora Kaushik
MANAGING EDITOR Soma B. Chowdhury
MANAGING ART EDITOR Sudakshina Basu
DTP DESIGNER Bimlesh Tiwary
PRE-PRODUCTION MANAGER Balwant Singh
JACKET DESIGNER Priyanka Bansal
SENIOR DTP DESIGNER Harish Aggarwal
JACKETS EDITORIAL COORDINATOR Priyanka Sharma
MANAGING JACKETS EDITOR Saloni Singh

FIRST EDITION

SENIOR EDITOR Peter Frances
SENIOR ART EDITOR Maxine Pedliham
PROJECT EDITORS Joanna Edwards, Nathan Joyce,
Lara Maiklem, Nikki Sims
EDITORS Salima Hirani, Mary Lindsay, Janine McCaffrey,
Martyn Page, Miezan van Zyl

RESEARCHER Dr. Rebecca Say
PROJECT ART EDITOR Alison Gardner
DESIGNERS Riccie Janus,
Clare Joyce, Duncan Turner
DESIGN ASSISTANT Fiona Macdonald
PRODUCTION CONTROLLER Erika Pepe
PRODUCTION EDITOR Tony Phipps
MANAGING EDITOR Sarah Larter
MANAGING ART EDITOR Michelle Baxter
ASSOCIATE PUBLISHER Liz Wheeler
ART DIRECTOR Phil Ormerod
PUBLISHER Jonathan Metcalf

ILLUSTRATORS

 Medi-Mation
Medical & Scientific Visualization

CREATIVE DIRECTOR Rajeev Doshi
SENIOR 3D ARTISTS Rajeev Doshi, Arran Lewis
3D ARTIST Gavin Whelan

ADDITIONAL ILLUSTRATORS

Peter Bull Art Studio, Antbits Ltd.

The Science of Pregnancy provides information on a wide range of medical topics, and every effort has been made to ensure that the information in this book is accurate and up-to-date (as of the date of publication). The book is not a substitute for expert medical advice, however, and is not to be relied on for medical, healthcare, pharmaceutical, or other professional advice in specific circumstances and in specific locations. You are advised always to consult a doctor or other health professionals for specific information on personal health matters. Please consult your doctor before changing, stopping, or starting any medical treatment. Never disregard expert medical advice or delay in seeking advice or treatment due to information obtained from this book. The naming of any product, treatment, or organization in this book does not imply endorsement by the editorial consultant, other consultants or contributors, editors, or publisher, nor does the omission of any such names indicate disapproval. The editorial consultant, other consultants or contributors, editors, or publisher do not accept any legal responsibility for any personal injury or other damage or loss arising directly or indirectly from any use or misuse of the information and advice in this book.

First American Edition published in June 2011 as
The Pregnant Body Book
This American Edition, 2019
Published in the United States by DK Publishing,
1450 Broadway, Suite 801, New York, New York 10018

A catalog record for this book is available
from the Library of Congress.
Published in Great Britain by Dorling Kindersley Limited.
ISBN 978-1-4654-8053-8

DK books are available at special discounts when purchased in bulk for sales promotions, premiums, fund-raising, or educational use. For details, contact: DK Publishing Special Markets, 1450 Broadway, Suite 801, New York, New York 10018
SpecialSales@dk.com

Printed and bound in China

A WORLD OF IDEAS:
SEE ALL THERE IS TO KNOW
www.dk.com

STATISTICAL SYMBOLS

- HEART RATE
- CROWN–RUMP LENGTH
- BLOOD PRESSURE
- CROWN–HEEL LENGTH
- BLOOD VOLUME
- WEIGHT

BODY SYSTEM SYMBOLS

- SKELETAL SYSTEM
- SKIN, HAIR, NAILS, AND TEETH
- MUSCULAR SYSTEM
- LYMPHATIC SYSTEM
- NERVOUS SYSTEM
- DIGESTIVE SYSTEM
- ENDOCRINE SYSTEM
- URINARY SYSTEM
- CARDIOVASCULAR SYSTEM
- REPRODUCTIVE SYSTEM
- RESPIRATORY SYSTEM

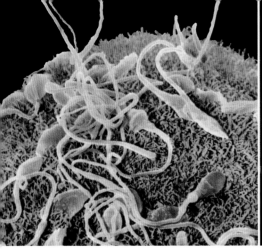

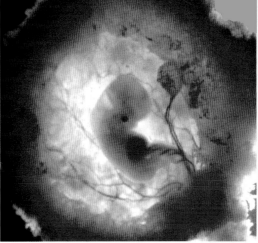

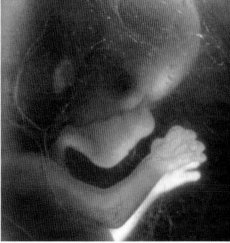

Just one of the millions of human sperm released will penetrate one egg to create a new life.

By seven weeks, most of the structures, organs, and limbs have already developed in the human fetus.

By 14 weeks, the fetus's facial features can be seen, although its head is disproportionately large.

HUMAN PREGNANCY

The growth of a new life inside a woman's uterus for the nine months of pregnancy is a truly amazing feat of biology. The creation of life is incredibly complex, and although each pregnancy is unique, some 130 million women worldwide experience its joys and risks each year.

The human body is capable of many astonishing things. But one of its most intricate, complex, and profound achievements is the ability to conceive, carry for nine months, and give birth to our helpless yet incredibly formed babies. As well as holding the promise of new life, pregnancy involves so many radical changes that it is little wonder that we marvel at and cherish the birth of children. Despite modern concerns about fertility, humans are remarkably fecund. By 2050 we will have reached a global population of 11 billion if we continue having children at the present rate.

A pregnant woman's body adapts in many amazing ways to accommodate and nurture the new life growing inside her. Her ligaments relax and stretch to allow space for her womb to grow, and her pelvic joints soften for birth. Her uterus expands from the size of a small pear to that of a watermelon by the end of pregnancy. She produces about 50 percent more blood so there is enough to pump around to the uterus and supply the growing fetus with a continuous supply of oxygen and nutrients, and her heart rate

increases by 20 percent by the third trimester—an extra 15 beats per minute. Even parts of her immune system will be suppressed so her body does not reject the fetus as "foreign."

Making babies

There is more than one way to have a baby. And all living organisms, including humans, have evolved to follow one of two strategies. One way to is to reproduce in great numbers, and have lots of offspring at the same time—this is called "big bang" reproduction. Having lots of babies is extremely energy consuming, and organisms that follow this strategy may breed just once and then die, such as Pacific salmon, some butterflies, and some spiders. Many of their offspring may perish, but because of their huge numbers, others will survive.

The second, less spectacular strategy, is to have only a few babies over a lifetime, but to invest more in each one so each individual is more likely to survive. This is the strategy that humans follow. It allows us to bear high-quality babies that thrive with parental care.

A male Emperor penguin incubates his egg and fasts while caring for his unborn offspring.

The marginated tortoise produces up to three clutches of between four and seven eggs a year.

A newborn Lemon shark emerges from its mother while remora fish break and eat the umbilical cord.

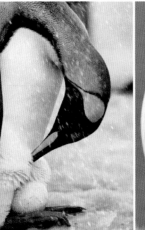

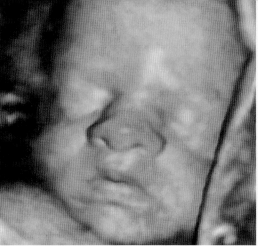
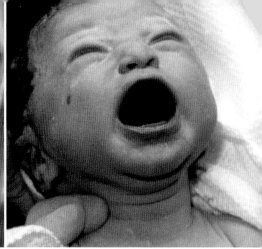

At 20 weeks, the baby is now growing rapidly. Eyebrows, eyelashes, and hair will have grown by this stage.

By 29 weeks, the baby's face is starting to fill out with fat as its rapid growth and weight gain continue.

A healthy baby girl cries moments after being born. Her skin is covered in vernix, which protects against infection.

How other animals reproduce

As humans we may take pregnancy for granted, but there are many weird and wonderful ways in which to produce the next generation. Some animals simply lay eggs, others carry eggs inside their bodies until they are ready to hatch, and many, like humans, go through pregnancy and bear live young. Although we might think that only birds and lower orders of animals lay eggs, there are even a few quirky mammals, such as the Duck-billed Platypus, that do so.

Animals that lay eggs follow ovipary; oviparous animals include all birds, most reptiles, and most fish. The egg comprises yolk, which contains all the embryo's nourishment, and its protective shell and layers keep the embryo safe inside. Often a parent has to keep eggs warm and protect them; many species incubate eggs until they hatch.

At the other end of the reproductive spectrum are those animals that house, protect, warm, and nourish developing embryos inside their own bodies. Humans, most other mammals, and a few rare reptiles, fish, amphibians, and scorpions, do this. This is known as vivipary. Humans and many other mammals are able to nurture young inside the uterus thanks to a special organ that develops during pregnancy: the placenta. Not all viviparous animals have this, and the placenta may have been pivotal in human evolution.

But there are some animals that fall between egg-layers and live-bearing animals—those whose embryos develop in eggs that remain within the animal's body, somewhat like a pregnancy. When the young are ready to hatch, the animal will "give birth" to a clutch of eggs, which will immediately spawn. Some fish and reptiles, such as sharks and anacondas, employ this strategy of ovivipary.

Parental duties

As soon as an embryo is conceived, the division of labor between mother and father begins. In many species the mother bears the burden of laying and guarding eggs, or pregnancy and birth, and even raising the offspring. But males can have a crucial role. In some species, the male becomes "pregnant." Male seahorses and pipefish nurture fertilized eggs in brood pouches. The female deposits her eggs in the male's pouch, where they are fertilized by sperm. And the male later "gives birth." Male Emperor penguins also make devoted fathers, painstakingly incubating a single egg on their feet for nine weeks in freezing temperatures, allowing their mates to go and feed after egg-laying. They, like many bird species, raise offspring together. Human children also thrive with both mother's and father's care, or other family support networks, because humans need a long, intense period of parenting.

Some animals, such as kangaroos, can stop their pregnancies by stalling the embryo from implanting in the womb. The pregnancy can then be started weeks, or even a year, later. These animals have evolved a way of bearing offspring when they can survive. Evolution has honed pregnancy to give offspring the best chances possible.

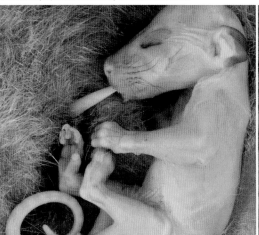

The common Japanese male seahorse becomes pregnant. The tiny seahorses are independent once born.

Common Brushtail Possums, unlike most mammals, are not nourished by a placenta but entirely on their mother's milk.

This four-day-old Japanese macaque reaches for its mother's nipple, and it may nurse for up to 18 months.

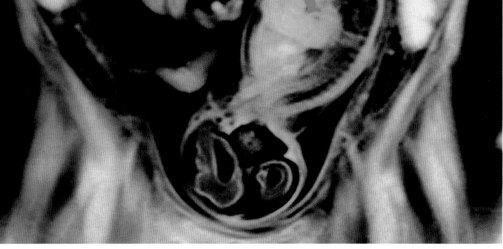

This color-enhanced MRI scan reveals the size and some of the anatomical features of the brain (shown in green) of a 36-week-old fetus.

This colored electron micrograph shows fetal tissues (villi) that protrude into the placenta, allowing for the exchange of vital gasses, nutrients, and wastes.

THE EVOLUTION OF PREGNANCY

Pregnancy evolved in humans to allow for extended care of the growing fetus and enable us to have large-brained babies with astonishing learning capabilities. The female body has evolved to cope with and adapt to the challenges of carrying a fetus for nine months.

Pregnancy may be an amazing condition, but it is not without perils. Why would humans evolve such a complex and risky way of reproducing when there are simpler methods available? The answer, quite simply, is that pregnancy's benefits outweigh the negatives.

Carrying a fetus in the uterus for nine months ensures each aspect of its environment is controlled: it is kept warm, safe, nourished, and supplied with oxygen. If we had evolved to lay eggs instead, as a handful of mammals do, the fetus would be limited to the supply of nutrients contained in the yolk. Pregnancy allows us to extend the period of care

and level of nourishment; and the longer this period lasts, the stronger the offspring are. Although a placenta is not essential for pregnancy (marsupials have a much simpler equivalent organ), it helps considerably in giving human babies a head start.

Crucially, a long pregnancy allows humans to bear large-brained babies. Large, complex brains, plus the ability to walk upright, makes humans special. Human brain volume is a massive 67–104 cubic inches (1,100–1,700 cubic cm) compared with the 18–31 cubic inches (300-500 cubic cm) of our closest living relative, the chimpanzee.

PREGNANCY FACTFILE

Pregnancy, birth, and newborns vary incredibly within the animal world. Human newborns are vulnerable compared with those of other mammals—wildebeest calves can run from predators within hours of birth, while bat babies can fly within two to four weeks of birth. Marsupials have short

pregnancies because they do not have a complex placenta, but then make up the difference with extended maternal care. Human babies require much parental care. In terms of motor, chemical, and brain development, a human baby displays the same levels at about nine months as those displayed by its primate cousins at birth.

	HUMAN	BLUE WILDEBEEST	ELEPHANT	RED KANGAROO	MOUSE	BAT
Gestation period	40 weeks	8 months	22 months	32–34 days	18–21 days	40 days to 8 months
Litter size	1 or 2 (very rarely more)	1	1 (rarely twins)	1	8–12	1 or 2 (3 or 4 in some species)
Average weight	6–9 lb (2.7–4.1 kg)	49 lb (22 kg)	198–265 lb (90–120 kg)	$\frac{1}{33}$ oz (0.75 g)	$\frac{1}{50}$–$\frac{1}{19}$ oz (0.5–1.5 g)	0–30 percent of mother's body weight
Ability at birth	Helpless: cannot hold up own head; can focus eyes to see only 17½ in (45 cm) ahead. Very long period of parental care required to reach adulthood	Can stand within 15 minutes; can eat grass within 10 days; weaned at nine months	Long period of maternal care and learning; weaned at 4–5 years	Climbs into mother's pouch unaided within 3 minutes; leaves the pouch at 240 days but suckles for another 3–4 months	Helpless; no pigment or hair; closed eyes and ears. By 3 weeks, have adult hair, open eyes, ears, and teeth and can be weaned	Completely dependent on mother for food and protection, but mature quickly and fly within 2–4 weeks; weaned shortly after
Time until next pregnancy	Can be within months, although many increase spacing	1 year	4–6 years, depending on female's age	Can get pregnant 1 day after birth but the pregnancy is paused until a suckling joey is 200 days old	Can conceive within hours of birth, but can delay pregnancy by up to 10 days if still nursing by halting implantation	Generally breed once a year but has various strategies to delay pregnancy

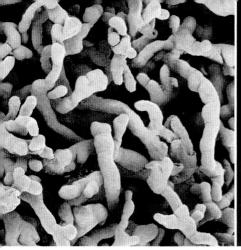

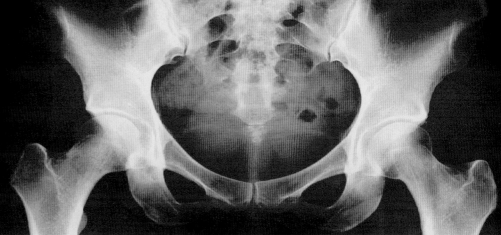

This colored x-ray shows a woman's pelvis is short and broad (an adaptation for child-bearing) and also has a narrow opening (adapted for walking upright).

Placenta
Provides fetus with nutrients and oxygen, removes wastes and carbon dioxide, and provides immunity

Pelvis
Narrow enough to allow upright walking, but with a large enough opening (pelvic inlet) for the head to pass through

Large head
Encases a large brain; must pass through pelvic inlet during birth

Pubic symphysis
Enlarges during pregnancy, allowing pelvis to be flexible during birth

A SPECIALLY DESIGNED PELVIS
Women have slightly shorter, broader pelvises than men to allow for the passage of babies' heads. Unlike other primates, human babies are about the same size as the birth canal, resulting in complicated and painful labors.

Human babies also have proportionately gigantic heads. A newborn's brain is already a quarter of the size of an adult's, making up about 10 percent of its body weight. In an adult, the brain makes up only about 2 percent of body weight.

The life-sustaining organ

Humans and other mammals may well owe their evolutionary and reproductive success to the placenta—a life-sustaining organ. Many scientists argue that we could not have developed large-brained young without it. The placenta enables a vital exchange between the blood of the mother and the fetus, passing nutrients and oxygen to the fetus, and passing wastes and carbon dioxide from the fetus's system to the mother's to be carried away. It also has an important immune function, because it acts as a barrier and allows some antibodies to pass from mother to fetus.

In humans, the placenta burrows deep into the uterine wall, and recent studies suggest that this depth may give better access to the nourishing maternal blood supply and, therefore, help humans have large-brained babies. Many mammals benefit from the placenta even after birth, by consuming the nutritious organ. Some human cultures have also been known to eat the "afterbirth."

Why women are special

Women's bodies have been sculpted to bear children, but evolution has had to accommodate two opposing challenges in order to do this. Humans are special because of their large, complex brains and their ability to walk upright. But these two massive evolutionary advantages are also in direct conflict.

A shorter, broader pelvis allows humans to walk upright. However, one side-effect of this is that the birth canal is no longer straight and wide, but curved and narrow. Although the birth canal is shorter, during the final stage of labor the mother must not only push the baby's head downward but also upward as it passes part of the vertebral column called the pelvic curve. This conundrum has meant that women have evolved special pelvises that are wide enough for a large-brained baby to pass through, but narrow enough for an upright lifestyle.

The many demands on our bodies have been delicately balanced by evolution. But amid these conflicts and compromises, child-bearing still has its dangers. Throughout the ages, humanity has sought the best ways to bring its young into the world, and now, in the modern era, medicine can give nature a helping hand in many ways.

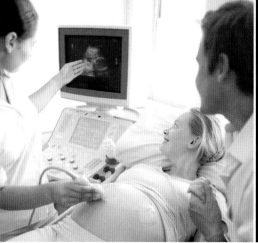
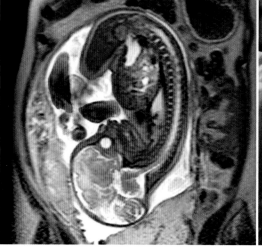
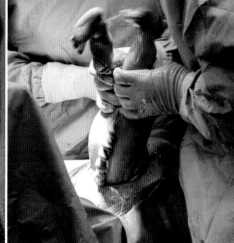

Ultrasound scanning of the abdomen offers expectant parents a glimpse of their baby.

A late MRI scan at 33 weeks reveals the placenta is blocking the cervix (placenta previa) in this woman.

A baby is extracted from its mother's womb by surgeons performing a cesarean section.

MEDICAL ADVANCES

Thanks to modern medicine there has never been a safer time to be pregnant. Advances in care mean that mother and child are cushioned from pregnancy's risks in most developed countries, and the situation is generally improving worldwide.

The care given to pregnant women during pregnancy and birth has improved unrecognizably, such that it is easy to take for granted and forget how hazardous pregnancy and birth once were. Even a century ago, it was not unusual to see maternal death rates of 500 in every 100,000 women giving birth in countries such as the US or the UK. Today, that figure is much lower, with between 4 and 17 women per 100,000 dying in developed nations.

This sea-change is a result of improvements in medicine and the quality of care, especially in the second half of the 20th century, alongside nutritional and socioeconomic improvements. Nevertheless, safety in pregnancy still needs to be improved internationally. In 2008 about 360,000 women died from pregnancy- or childbirth-related causes, mostly in the developing world. Globally, infant health has also massively improved, and the mortality rate in children under a year old is less than half of the mortality in 1960.

Preconception care

Because of improvements in our medical understanding, today many women may start preparing their bodies (eating a healthy diet and doing moderate exercise) before pregnancy to give their children the best possible start. Many women now take folic acid supplements before conception and in the first trimester, to protect against neural tube defects, such as spina bifida, in the fetus.

Couples planning a baby may adjust their lifestyles to improve their chances of conceiving. For example, in women, stopping smoking and cutting down on alcohol, caffeine, and even stress are recommended. Men may also be advised to cut down on alcohol and smoking because it can affect the quality of their sperm.

Advances mean that many women delay childbearing. A woman's age (too young or too old) and the spacing between children (too close together or too far apart) may impact on her and her child's health.

TIMELINE

Medical advances gathered pace in the second half of the 20th century. Notable advances before then include the first cesarean section—performed from ancient times in India, Rome, and Greece; the use of forceps to assist labor from the 17th century; the invention of the stethoscope in 1895; and the use of antibiotics from the 1930s, which massively cut maternal death rates.

1952 APGAR SCORING:
An examination carried out within five minutes of birth, this assesses the newborn's "Appearance, Pulse, Grimace, Activity, and Respiration", or its skin color, heart-rate, reflexes, muscle tone, and breathing. The score indicates any need for medical help.

1960 FEMALE "PILL":
The oral contraceptive pill gave women unprecedented control over their fertility, and has helped reduce unwanted pregnancies.

1966 REAL TIME ULTRASOUND:
This revolutionized scanning as the fetus's motion and life could be observed.

1973 SCANNING MEASUREMENTS:
Measuring certain aspects of the fetus in utero were used to give an indication of age, size, and weight.

1975 HOME PREGNANCY TEST INTRODUCED:
Available over the counter, this test gives instant results.

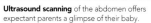

1950 · 1955 · 1960 · 1965 · 1970 · 1975

1959 FETAL ULTRASOUND SCANNING:
High-frequency sound waves were first used to measure a fetus's head, giving an idea of size and growth.

1962 HEEL PRICK TEST:
Also called "blood spot screening," this newborn test checks for rare disorders, such as phenylketonuria, which can benefit from early diagnosis and treatment.

1968 FETAL CARDIOTOCOGRAPH:
Now, fetal heart rates could be monitored electronically to tell if a baby was in distress during labor.

1975 SCANNING FOR SPINA BIFIDA:
The first case of ultrasound detection of this neural tube defect, leading to a termination of pregnancy.

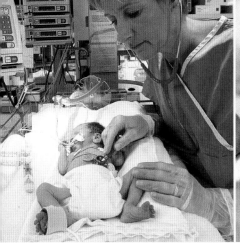
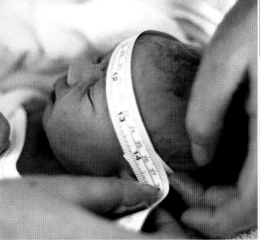
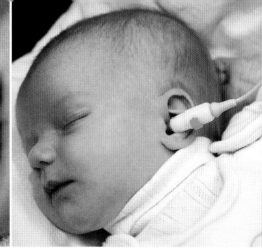

Premature babies have much better survival rates now, thanks to expert care in special baby units.

Newborn measurements allow health professionals to assess where a baby sits in the normal range.

Hearing tests can catch problems early, since hearing problems impact on speech and language development.

Advances in prenatal care

Care during pregnancy—the prenatal period—has improved incredibly in the modern age. Routine medical care is available in many countries. And leaps in technology, such as the invention of the stethoscope and, more recently, ultrasound, mean that we can now hear and see the fetus, which helps health professionals assess the care needed in any particular pregnancy.

The mother's health may be routinely monitored for conditions that may affect her unborn child. For example, urine will be tested regularly for urinary tract infections, which can lead to premature labor. And blood may be screened for sexually transmitted diseases, which, left untreated, could be transmitted to the baby either in utero or at birth, with harmful consequences. Blood tests may also detect conditions such as anemia or gestational diabetes in the mother, which can then be treated. Blood pressure monitoring can give warning of conditions such as preeclampsia.

Abnormalities may be spotted on an ultrasound scan or by tests such as amniocentesis (in which amniotic fluid from around the fetus is sampled and tested for a chromosomal disorder). In some cases where there is a high risk of an inherited disorder, genetic tests may be done. New techniques may also offer those facing genetic problems the option of selecting disease-free embryos for in-vitro fertilization.

Advances in perinatal care

The perinatal period runs from the 28th week of pregnancy to about four weeks after birth. This window is crucial to the well-being of mother and child. Advances such as the discovery of antibiotics and better hygiene have slashed death rates for mothers in the last century.

Now childbirth and its immediate aftermath can be much safer. Birth can be helped along—labor can be induced, assisted (for example, with forceps), or a cesarean section can be performed. Many types of pain relief are available to women in many countries, along with continuous monitoring of the fetus during labor, for signs of distress.

Advances in postnatal care

Immediately after birth, a newborn undergoes physical tests to assess whether it needs medical intervention. Newborn survival and health have been greatly improved by the availability of medicines and vaccines. Modern technology also gives premature babies a far greater chance of survival than they used to have.

Mothers and babies are often monitored for six weeks after birth. Health professionals will check both physical well-being (weigh the baby, give advice on feeding, and administer routine immunizations) and emotional health (looking for signs of postpartum depression and strong bonding, and offering advice and support as necessary).

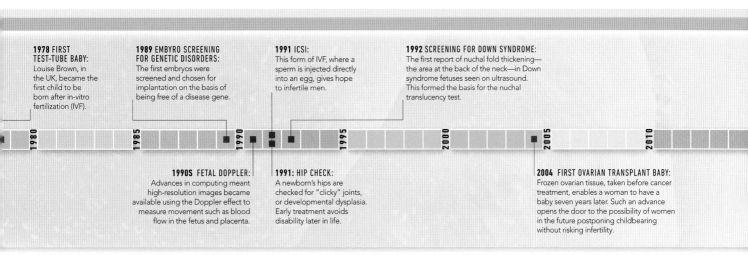

1978 FIRST TEST-TUBE BABY: Louise Brown, in the UK, became the first child to be born after in-vitro fertilization (IVF).

1989 EMBRYO SCREENING FOR GENETIC DISORDERS: The first embryos were screened and chosen for implantation on the basis of being free of a disease gene.

1991 ICSI: This form of IVF, where a sperm is injected directly into an egg, gives hope to infertile men.

1992 SCREENING FOR DOWN SYNDROME: The first report of nuchal fold thickening—the area at the back of the neck—in Down syndrome fetuses seen on ultrasound. This formed the basis for the nuchal translucency test.

1990S FETAL DOPPLER: Advances in computing meant high-resolution images became available using the Doppler effect to measure movement such as blood flow in the fetus and placenta.

1991: HIP CHECK: A newborn's hips are checked for "clicky" joints, or developmental dysplasia. Early treatment avoids disability later in life.

2004 FIRST OVARIAN TRANSPLANT BABY: Frozen ovarian tissue, taken before cancer treatment, enables a woman to have a baby seven years later. Such an advance opens the door to the possibility of women in the future postponing childbearing without risking infertility.

1980 1985 1990 1995 2000 2005 2010

IMAGING TECHNIQUES

The ability to see, hear, and monitor the fetus in the uterus has been one of the most profound medical advances of the 20th century. It has revolutionized prenatal care by allowing health professionals to check the health of a fetus and placenta and assess the progress of a pregnancy.

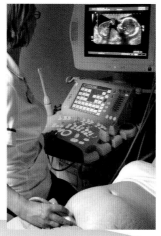

USING A TRANSDUCER
After gel has been rubbed over the woman's abdomen, the transducer is run with gentle pressure over the same area.

THE HISTORY OF ULTRASOUND

Until a few decades ago, the only way of checking a fetus's growth or position was by palpating the abdomen of a pregnant woman. Since the 1940s, scientists had been investigating the use of high-frequency sound waves to look inside the body, and World War II may have acted as a catalyst to their application to obstetrics. Ian Donald at Glasgow University was inspired by his experiences in the British Royal Air Force. He took the principles of sonar (which used sound waves to detect U-boats) and, with fellow obstetrician John McVicar and engineer Tom Brown, made the first ultrasound scanner to produce clinically useful 2D images. In 1958, the team published work describing how they used ultrasound to look at abdominal masses in 100 patients. They soon developed the technology to measure the fetus in the uterus, which became routine procedure.

HOW ULTRASOUND WORKS

Ultrasound harnesses high-frequency sound waves in the range of 2–18 megahertz. A hand-held probe called a transducer, which is pressed against the skin, contains a crystal that transmits sound waves. The transducer also contains a microphone to record returning echoes as the waves bounce off solid substances, such as organs or bone. The echoes are then processed by a computer to generate a real-time 2D image. This safe, painless procedure is widely used for routine prenatal checks. A similar technology, called Doppler ultrasound scanning, is used to look at moving substances, such as blood flow in the fetus or placenta. Recent technological advances make it possible to use ultrasound to build 3D images of fetuses too.

SONIC PICTURE
Sound waves passing through the mother's abdomen bounce off the fetus's body as well as other structures, such as the placenta and amniotic sac.

Microphone
This receives returning waves, whose pitch and direction may have been changed by internal structures.

Point of contact
Gel between the transducer and abdomen helps eliminate any air pockets.

Transducer
Applying electrical energy to a piezo-electric crystal inside the transducer distorts its mechanical structure. It expands and contracts, emitting ultrasound waves.

Sound waves
The frequencies used for imaging are inaudible to humans and have no known harmful effects on the fetus or mother.

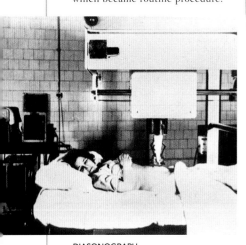

DIASONOGRAPH
Produced in 1963, this was one of the first commercial ultrasound machines. The patient lay beneath it while a probe moved horizontally and vertically above them.

Uterus
Ultrasound waves travel through this to give a picture of what lies inside

20-week-old fetus
Ultrasound scans can screen a fetus of this age for potential congenital abnormalities in an "anomaly" scan.

Cable to computer and monitor
The data is transmitted to a computer, where it is processed, and the resulting 2D scan image is displayed on a screen.

READING SCAN IMAGES

A 2D scan image shows contrasting black, white, and gray areas. These correspond to the type of structures that the sound waves encounter as they pass through the body, and how these structures create echoes. When ultrasound waves bounce off solid structures such as bone or muscle, they produce a white or light gray image. But soft or empty areas, such as the eyes or chambers of the heart, will appear black.

Seen as white
The fetus's bones are white on the scan as they cause the ultrasound waves to echo back.

Seen as black
Amniotic fluid shows as black because sound waves travel through it, so there is no echo.

Seen as gray
Muscle appears as gray, as it bounces sound waves back.

Nose
The soft parts of the nose cannot be seen, but the bone structure around it appears white.

FACIAL FEATURES
A fetus's face can be seen in an ultrasound scan image. Even 2D scans can give clues to the fetus's appearance by revealing some of its facial features—for example, the shape of its face.

Eye
The soft tissue of the eye appears black in the scan image, while the bone of the eye socket gives a white outline.

Mouth cavity
This is seen as black.

Two heads
The white outlines of the skulls indicate the two heads of twin fetuses. This scan image cannot reveal if they are identical or fraternal twins.

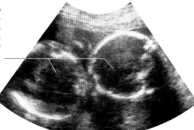

WHAT SCANS CAN TELL US
A scan reveals basic information about a pregnancy—the sex, size, and age of the fetus, its position (and that of the placenta) in the uterus, and if it is a multiple pregnancy. Scans can warn of potential problems, such as placenta previa (in which the placenta blocks the cervix, the fetus's exit route), or growth problems in the fetus or placenta. Screening for abnormalities is also an important function of scanning.

SEEING WITH SOUND WAVES
By moving the transducer, the sonographer can direct the ultrasound waves in order to reveal particular views that provide helpful information.

3D IMAGING

In recent years, striking, in-depth images of fetuses have been displayed by 3D scans. They are obtained by stitching together a series of successive 2D shots or "slices" into a 3D image using modern computer technology. Some parents acquire 3D scan images commercially as souvenirs, but many medical organizations advise against such "keepsake" scans because of the concern that, should the scan unexpectedly reveal abnormalities in the fetus, the parents, being in a nonmedical situation, may not have the appropriate support available.

MULTIPLE SCAN SLICES
A series of 2D "slices" or images are combined into a 3D image by a process called surface rendering.

Transducer

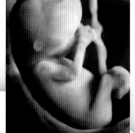

20-week-old fetus

A FETUS IN 3D
The third dimension, depth, enables us to see the shape of the fetus more clearly.

LOOKING INSIDE THE BODY

There are other imaging techniques that can be used to peer inside the body before or during pregnancy. Laparoscopy, a surgical procedure, can be used to investigate fertility by allowing doctors to examine the fallopian tubes, ovaries, and uterus. A fetoscopy may be performed to visualize the fetus, collect fetal tissue samples, and even to perform fetal surgery. To do this, a fiberoptic tube is inserted through the cervix or surgically through the abdomen. MRI scans may also be carried out on pregnant women to investigate suspected problems, although they are not advised in the first trimester.

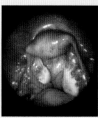 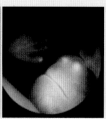 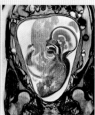

LAPAROSCOPIC VIEW
A flexible tube with a camera and light source is inserted through a cut in the abdomen. Shots of the reproductive system are then relayed to a screen.

FETOSCOPIC VIEW
An endoscope is inserted into the uterus to examine the fetus for diagnosis or to take skin samples—for example, to test for inherited diseases.

MRI SCAN
Powerful magnetic fields and radio waves produce a detailed image. Pregnant women are scanned only if the procedure is considered to be essential.

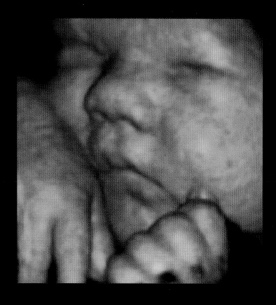

GOING INSIDE

Modern technology, especially the use of new imaging techniques, has given an incredible window into how a new human life develops in the uterus. It is now possible to see, photograph, and even film a fetus in unprecedented detail.

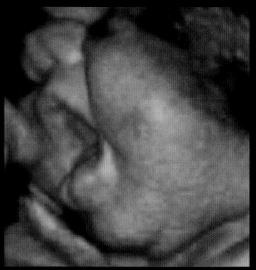

It is difficult to believe that only just over 50 years ago there was no way of checking the growth of a fetus except by feeling, or palpating, a pregnant woman's abdomen. The idea of actually being able to see a fetus rubbing its eye or sticking out its tongue was unimaginable. The development of obstetric ultrasound imaging in the late 1950s opened the door to a range of technological possibilities, and now not only is ultrasound imaging in pregnancy routine in many countries, but more detailed scanning is also possible. Ordinary two-dimensional ultrasound scans are often taken in the first trimester to date a pregnancy, and later, scans at around 20 weeks may be used to screen for various congenital problems, such as spina bifida or cleft palate. Even more detailed images can be obtained using three-dimensional ultrasound (including most of the images shown here) or MRI techniques, and movements such as blood flow in the placenta can be imaged using Doppler ultrasound. All of these techniques combine to offer powerful tools for monitoring and screening during pregnancy, and give the parents the chance to see their unborn baby.

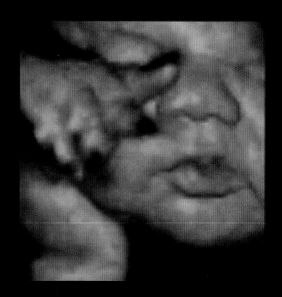

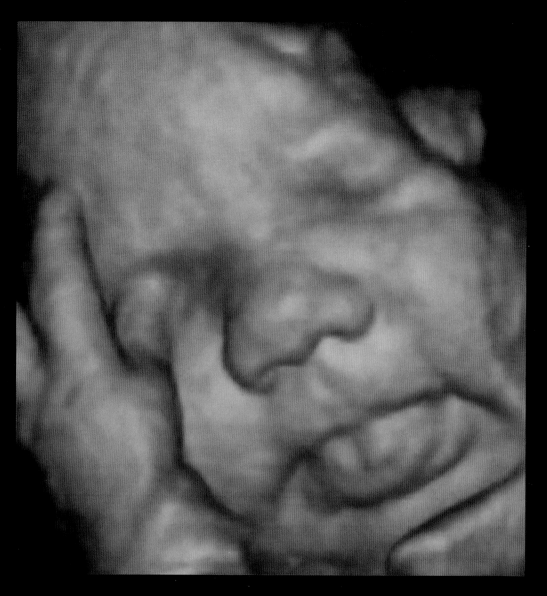

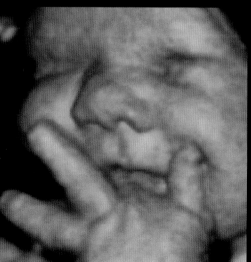

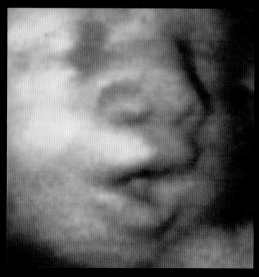

EXPRESSION

Three-dimensional ultrasound pictures reveal a range of expressions on the face of this 38-week-old fetus as it rubs its eyes and face, opens its mouth, and sticks out its tongue. Images like this are possible due to an explosion in computing power, which has meant that flat, two-dimensional scans can now be "sewn together" digitally to give three-dimensional pictures that can reveal amazing details such as fingernails and facial features. A fetus's face develops rapidly early in pregnancy, with tiny nostrils becoming visible and the lenses of the eyes forming by seven weeks, but it is not until the second trimester that the face takes on a humanlike appearance. By 16 weeks, the eyes have moved to the front of the face, and the ears are near their final positions. The fetus's facial muscles are also more developed, with the result that facial expressions such as frowning or smiling may also be seen.

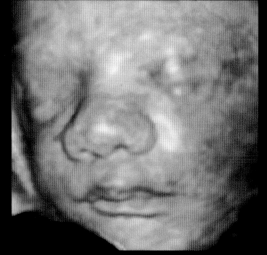

FRONT VIEW OF FACE AT EIGHT MONTHS

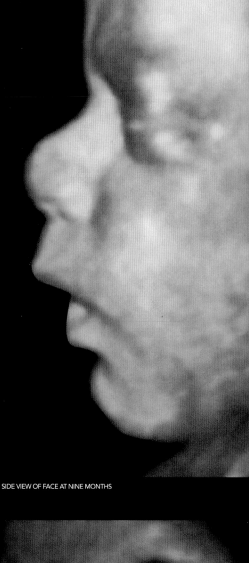

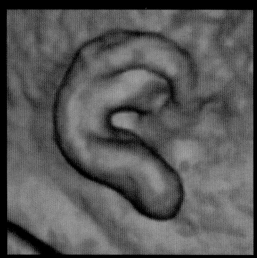

EAR AT ABOUT 39 WEEKS

SIDE VIEW OF FACE AT NINE MONTHS

THE HEAD AND FACE

The head and face start developing
early in pregnancy, although initially
development is relatively slow. Eye
buds and the passageways that will
become ears start developing on the
side of the head at about the sixth week.
By the tenth week, the head has become
rounder and the neck has started to
develop. In these early stages, the fetus is
very top-heavy: at 11 weeks, for example,
its head is half of its total body length.
The second trimester is a period of rapid
development for the head and face. This
is when the eyes move to the front of the
face (with the eyelids closed to protect
the eyes), the ears move to their final
positions, and the facial muscles develop.
By 22 weeks, the fetus's eyebrows may
be visible, and by 26 weeks it may have
eyelashes. By 27 weeks, the eyes open
and there is hair on the head. By the
time the baby is born, its head is more
in proportion to its body, although still

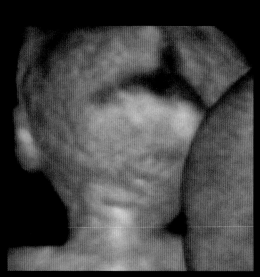

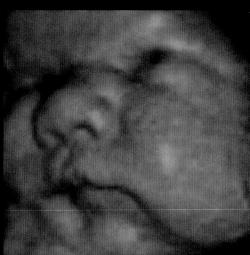

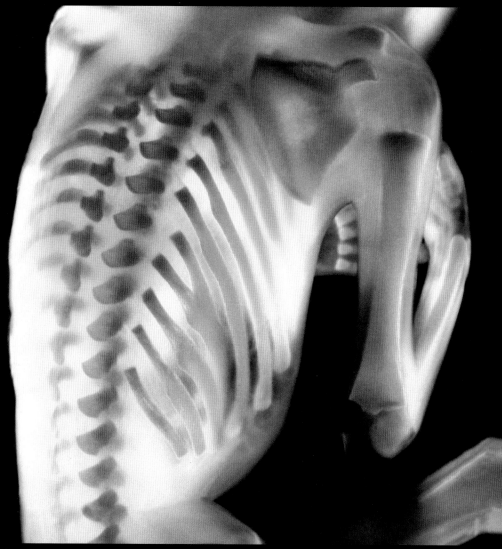

SKELETON AT 16 WEEKS

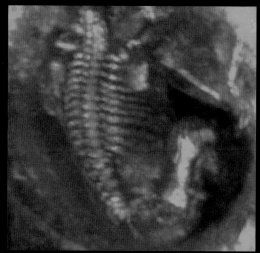

SKELETON AT 29 WEEKS

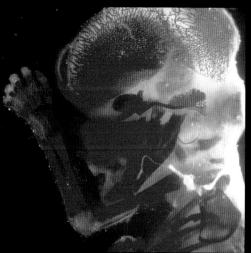

OSSIFICATION AT 12 WEEKS

THE SKELETON

The development of the fetus's skeleton begins in the first trimester, although the process is not complete until long after birth. The top image shows the fetus at 16 weeks. Before then, the tissues that will eventually become bone are laid down in the correct places—for example, around the head or in the arms, legs, and fingers—and these tissues are then ossified to make bone. This process of ossification can happen in two ways. Where there are membranes—as around the fetus's head—bone grows over the membranes to form bony plates. In other places, such as the limbs, ribs, and backbone, cartilage is gradually converted to bone from the middle outward. The image at the bottom right shows ossification in a 12-week-old fetus, with the partially ossified bones of the skull, arms, and ribcage shown in red. By 29 weeks (image at bottom left), the bones are fully developed, although they are still soft.

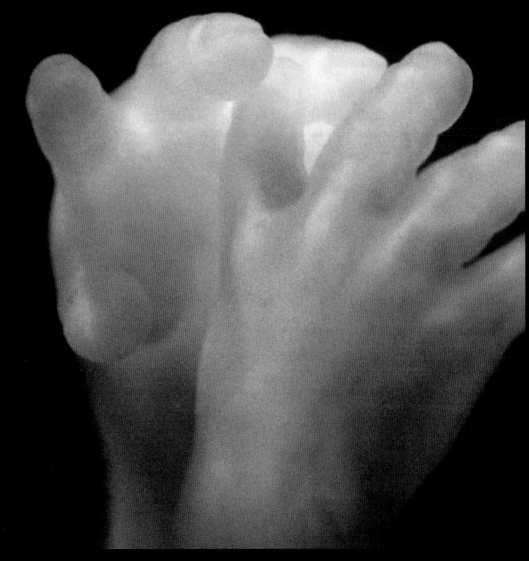

ARMS AND LEGS

The arms and legs grow from tiny limb buds that appear at about six weeks. Paddlelike at first, the limbs grow longer, and within a couple of weeks the fingers start to form. Toes appear at about nine weeks—the image at the bottom right shows the toes of a 10-week-old fetus. At nine weeks, the arms may develop bones and can bend at the elbow, and by 14 weeks the arms may already be the length that they will be when the baby is born. Finer details such as fingerprints and footprints start forming around 23 weeks. By 25 weeks, the hands are fully developed, and the fetus may use them to explore inside the uterus. Fingernails and toenails grow in the late second and early third trimesters; the main image on this page shows the well-developed hands of a 23-week-old fetus. As pregnancy progresses, the limbs develop further, and the fetus may deliver lively punches and kicks in the third trimester.

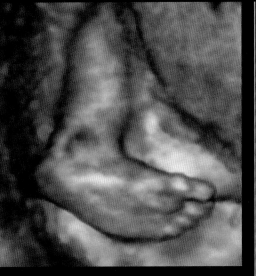

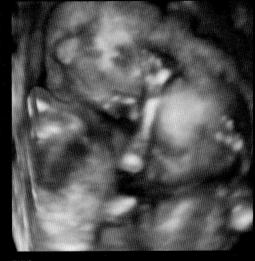

TWINS

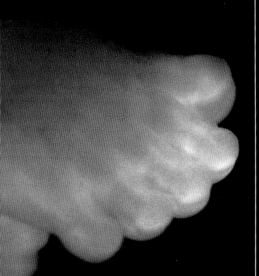

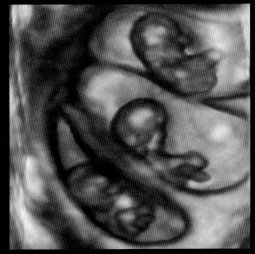

TRIPLETS

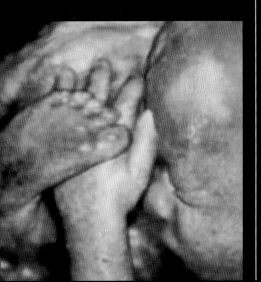

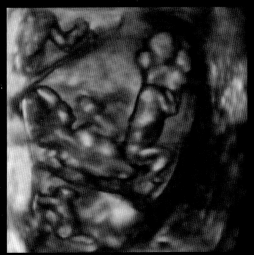

QUADRUPLETS

MULTIPLE BIRTHS

These three-dimensional ultrasound scans (near left) show twins, triplets, and quadruplets (from top to bottom). In the image of the triplets, a separate amniotic sac is clearly visible around each fetus. Between each amniotic sac, a small amount of placenta is seen to form a V-shape. This indicates that each of the triplets has a separate placenta. As a result of using such modern imaging techniques, medical professionals can not only discover whether a woman has a multiple pregnancy but can also gain valuable information about the state of the pregnancy. Multiple pregnancies are riskier than singleton ones, and scans can show, for example, whether fetuses share a placenta or amniotic sac, how each fetus is growing, and whether any of them is at particular risk. Such information can then be used to inform decisions, such as whether labor should be induced early.

FROM CELL TO FETUS

The journey from embryo to fetus to baby begins with rapid development in the first trimester, followed by massive growth in the second, and preparation for birth in the third. After conception, the embryo divides into a growing ball of cells, which implants in the uterine lining on about the sixth day. The cells differentiate into three layers, from which the fetus's major body systems will arise. By the fifth week of pregnancy, a spinal cord is forming, limb buds are sprouting, and the organs are developing. From the tenth week, the grape-sized embryo is termed a "fetus." And by 12 weeks, the fetus is fully formed. Its body grows rapidly in the second trimester, such that its head and body approach the proportions of an adult. By 14 weeks, its sex may be apparent. The brain grows rapidly in the last few weeks of the second trimester. By 30 weeks, in the third trimester, the fetus is becoming plump. In the run-up to birth, antibodies move into the fetus's blood from the mother, the fetus's eyes open, its sexual organs mature, and its lungs practice dilating.

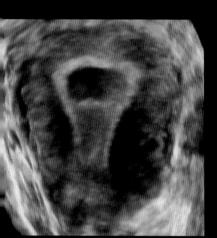

NON-PREGNANT UTERUS

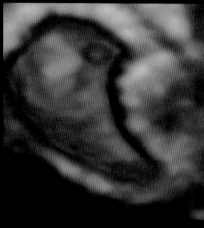

6 WEEKS

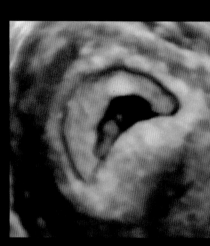

7 WEEKS

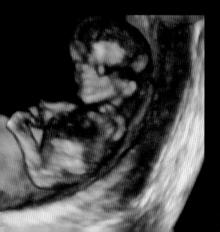

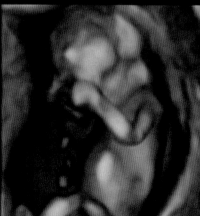

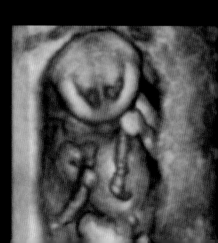

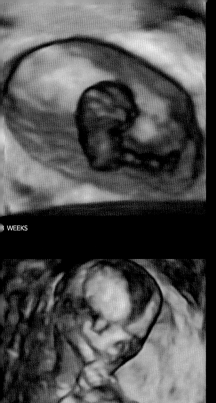

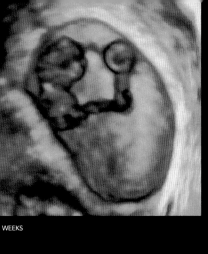

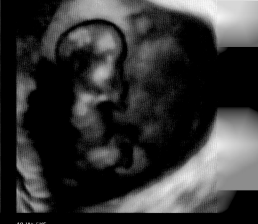

WEEKS

9 WEEKS

10 WEEKS

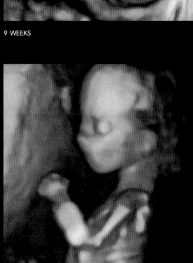

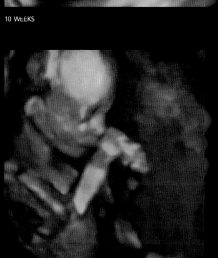

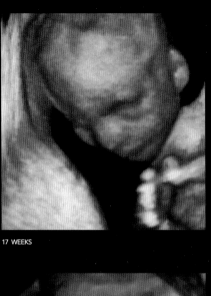

17 WEEKS

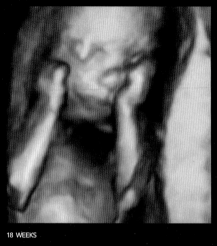

18 WEEKS

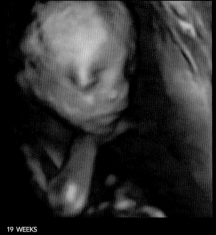

19 WEEKS

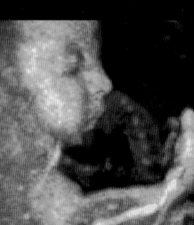

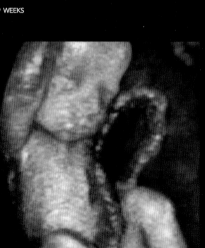

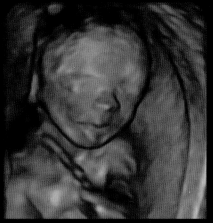

20 WEEKS

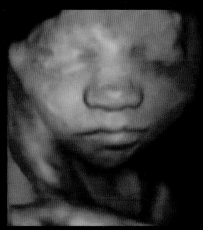

28 WEEKS

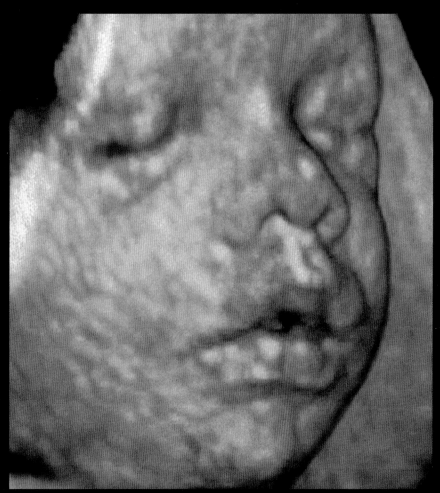

30 WEEKS

THE FEMALE AND MALE REPRODUCTIVE SYSTEMS CAN PRODUCE, STORE, AND BRING TOGETHER AN EGG AND A SPERM, GIVING THE POTENTIAL FOR A NEW LIFE. THE FEMALE SYSTEM IS ALSO ABLE TO NURTURE AND PROTECT THIS NEW INDIVIDUAL IN THE UTERUS FOR THE NINE MONTHS OF PREGNANCY, BEFORE DELIVERING IT INTO THE OUTSIDE WORLD AT BIRTH. FROM THEN ON, THE MOTHER CAN CONTINUE TO PROVIDE NOURISHMENT IN THE FORM OF BREAST MILK. ALL OF THESE PROCESSES TAKE PLACE AS A RESULT OF COMPLEX HORMONAL INTERACTIONS THAT TRIGGER THE BEGINNING OF THE REPRODUCTIVE PROCESS AT PUBERTY AND ENABLE IT TO CONTINUE THROUGHOUT THE FERTILE PART OF LIFE.

ANATOMY

BODY SYSTEMS

The human body can be divided into systems—groups of organs and tissues that work together to carry out a specific function or functions. During pregnancy, many of these systems alter their size, structure, and even their function to meet the needs of the growing fetus. Some of the changes are obvious, such as the rapidly expanding uterus and breasts. Other changes, such as the massive increase in blood volume, are more subtle yet essential for fetal well-being and a successful pregnancy.

REPRODUCTIVE SYSTEM

Female and male reproductive organs generate the egg and sperm to create new life. The ovaries produce the hormones needed to prepare the uterus for a fertilized egg. Once a woman becomes pregnant, her system undergoes dramatic changes: the uterus enlarges to fit the growing fetus; the placenta develops to connect fetal and maternal circulations; and the breasts prepare for lactation.

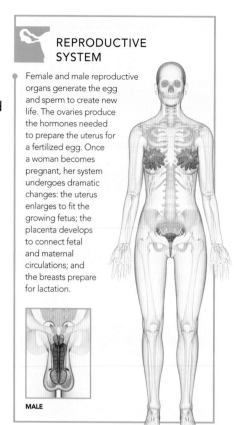

MALE

URINARY SYSTEM

This complex filtration system filters blood in the kidneys to eliminate waste products and to maintain the body's delicate equilibrium. The resulting waste is stored in the bladder as urine. Hormones control how much urine is made before it is excreted via the urethra. During pregnancy, the kidneys lengthen by ⅜ in (1 cm) and their blood flow increases massively, which causes frequent urination even before a developing fetus is large enough to press on the bladder.

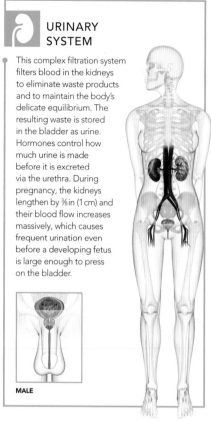

MALE

RESPIRATORY SYSTEM

The muscular diaphragm contracts and relaxes to bring air via the nose and trachea into the lungs and out again. Within the lungs, oxygen from the air diffuses into the blood, while carbon dioxide diffuses out of the blood and into the lungs, ready for exhalation. This gaseous exchange is vital for all body tissues. Oxygen consumption rises slowly in pregnancy, reaching an increase of 20 percent at full term. A woman's breathing rate rises to about 18 breaths a minute, up from 12–15. During labor, oxygen consumption may rise up to 60 percent, a reflection of the physical work involved.

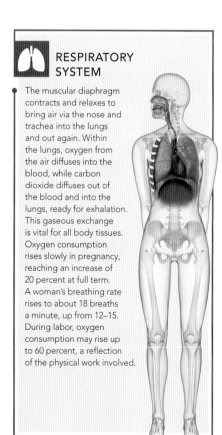

CARDIOVASCULAR SYSTEM

The heart works relentlessly to pump blood around the body in a complex system of blood vessels (arteries, arterioles, capillaries, venules, and veins) that supply every tissue and organ. During pregnancy, the volume of blood circulating increases by up to 50 percent to supply the growing fetus with everything it needs. Pumping more blood is extra work for the heart, so it contracts more forcefully and more frequently; the heart rate rises by up to 15 beats per minute.

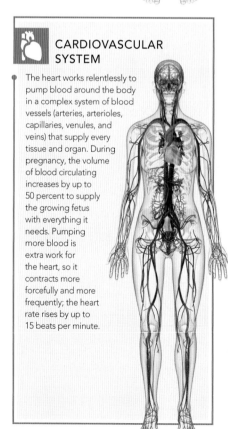

LYMPHATIC AND IMMUNE SYSTEM

The lymphatic system diverts excess tissue fluid back into the blood. The expanding uterus can press on blood vessels within the pelvis, resulting in a buildup of fluid in body tissues (edema), commonly those of the legs and feet. The immune system protects the body from infections and foreign invaders. Pregnant women appear to be susceptible to picking up colds and other common infections, but this may be due to the increased blood flow in mucus membranes.

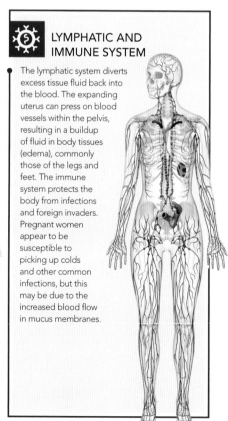

NERVOUS SYSTEM

The brain, spinal cord, and a network of nerves around the body continue to control the actions of the body and respond to what is happening. During pregnancy, the female sex hormone progesterone directly affects the brain's respiratory center to increase its sensitivity to carbon dioxide, thereby raising the breathing rate to "blow off" more carbon dioxide. Certain conditions affecting nerves, such as sciatica, may be more likely during pregnancy.

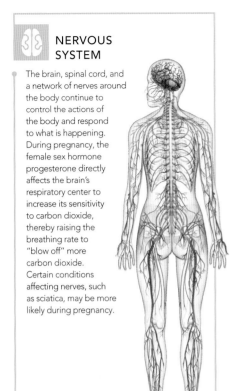

DIGESTIVE SYSTEM

Essentially, one long tube from mouth to anus (including the esophagus, stomach, and intestines), the digestive system breaks down food so that nutrients can be absorbed and waste products expelled. Accessory organs, such as the liver, pancreas, and gallbladder, provide biochemical help. During pregnancy, hormonal changes slow contractions that propel food and waste through the intestines, so constipation can occur. The valve between the esophagus and stomach may be more relaxed, resulting in heartburn.

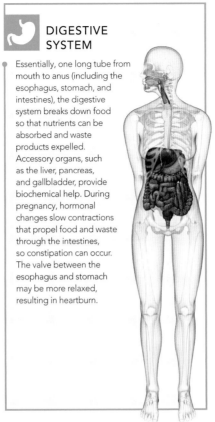

ENDOCRINE SYSTEM

This system of glands produces myriad hormones that maintain the body's equilibrium. Many hormonal changes occur at certain stages of a pregnancy. For example, one part of the pituitary gland releases oxytocin, needed to initiate labor, and another part releases prolactin, needed for milk production. The placenta not only forms a connection between fetal and maternal circulations, it also acts as an endocrine gland itself, producing estrogen and progesterone to sustain pregnancy.

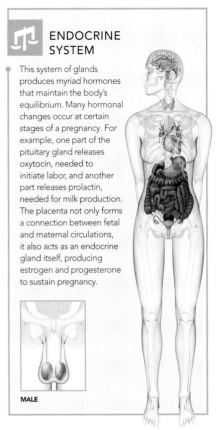

MALE

SKELETAL SYSTEM

The bones provide a moving framework for the body. During pregnancy, the hormones progesterone and relaxin increase the looseness of the joints, ultimately designed to allow a baby's relatively large head to pass through the pelvis during delivery. Intestinal absorption of calcium (to make the fetal skeleton) doubles during pregnancy. After birth, extra calcium for breast milk is temporarily "taken" from the mother's skeleton to meet the demands of a newborn.

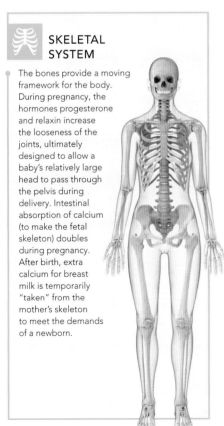

MUSCULAR SYSTEM

The muscles enable the bones of the skeleton to move. With the ligaments and tendons, they also work to maintain an upright posture. The increasing weight of the fetus causes the mother's posture to change during pregnancy, placing extra strain on the muscles, ligaments, and joints in the lower back. Also, many pregnant women notice a separation of the abdominal muscles, which allows the belly to grow too. The separated muscles usually rejoin in the weeks after childbirth.

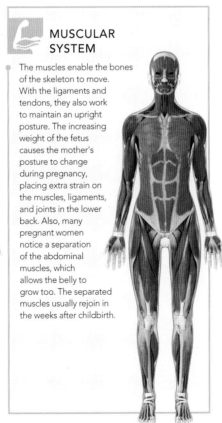

SKIN, HAIR, AND NAILS

The skin is the body's largest organ, measuring some 21½ sq ft (2 sq m), and helps regulate body temperature as well as forming a protective barrier. Skin, hair, and nails tend to look healthier during pregnancy; less hair is lost, so it looks thicker and more lustrous; and nails are smooth and not brittle. Pigmentation changes, such as the appearance of dark patches on the face (chloasma), and a dark vertical line (linea nigra) down the abdomen, may also develop.

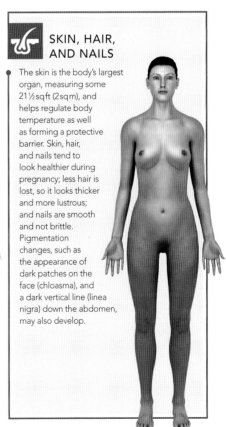

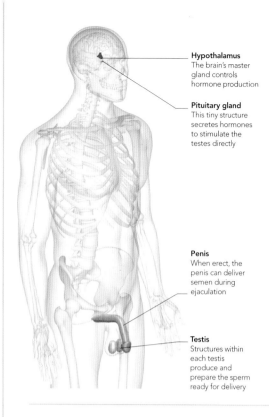

Hypothalamus
The brain's master gland controls hormone production

Pituitary gland
This tiny structure secretes hormones to stimulate the testes directly

Penis
When erect, the penis can deliver semen during ejaculation

Testis
Structures within each testis produce and prepare the sperm ready for delivery

THE MALE REPRODUCTIVE SYSTEM

The key parts of the male reproductive system, the penis and testes, work together with glands and other structures to produce and deliver sperm, which may combine with an egg to create a new life. The system begins to develop just six weeks after fertilization.

THE REPRODUCTIVE ORGANS

The male reproductive system is made up of the penis, a pair of testes that sit within the scrotum, a number of glands, and a system of tubes that connects them all. Once sperm have developed within each testis, they travel to each epididymis to mature and for temporary storage. They continue their journey along each vas deferens and then through the ejaculatory ducts to join the urethra, which runs the length of the penis. Columns of spongy tissue within the penis contain a rich network of blood vessels that fill with blood in response to sexual arousal (see pp.64–65). This engorgement causes the penis to become erect and able to deliver sperm to the top of the vagina (see pp.66–67).

LOCATING ORGANS OF THE MALE REPRODUCTIVE SYSTEM
The penis and testes are located outside the body cavity. The processes that occur in the testes are under hormonal control from the pituitary gland, which is regulated by the hypothalamus.

SPERM FACTORIES

Sperm are produced in abundance within the seminiferous tubules of the testes, a process called spermatogenesis (see pp.32–33). The developing sperm are protected and nourished by Sertoli cells, which extend inward from the walls of the tubules. Once a sperm leaves the testes, it moves on to the epididymis, where it matures and can be stored for up to four weeks. Semen is made up of sperm cells suspended in secretions—about 100 million sperm per 0.03 fl oz (1 ml) of fluid. About 0.1–0.17 fl oz (3–5 ml) of semen is delivered via the urethra of an erect penis at male orgasm.

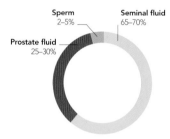

Sperm
2–5%

Seminal fluid
65–70%

Prostate fluid
25–30%

THE CONSTITUENTS OF SEMEN
Only a small percentage of semen is sperm; most is made up of milky white fluids, mainly produced by the prostate gland and the seminal vesicles.

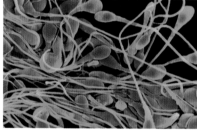

SPERM UP CLOSE
The basic structure of sperm can clearly be seen on this microscopic view of multiple sperm. Each sperm consists of a head, which carries half of a man's genetic information, and a long, thin tail.

TESTOSTERONE

The principal male hormone testosterone triggers development of the reproductive organs and the changes that occur at puberty, including deepening of the voice and a growth spurt (see p.31). Testosterone must be present for sperm production to take place. As with hormone production and egg development in women, testosterone and sperm production in men are controlled by hormones secreted by the pituitary gland (FSH and LH), which in turn are regulated by the brain's hypothalamus. Testosterone is produced by the Leydig cells located between the seminiferous tubules in the testes.

TESTOSTERONE CRYSTALS
Outside the body, testosterone can be crystallized and viewed under a microscope. Testosterone in the fetus causes the testes to descend into the scrotum before a baby boy is born. From birth until the surge at puberty, testosterone levels are very low.

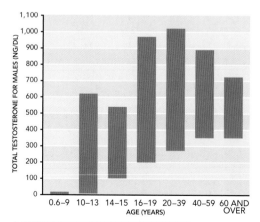

TOTAL TESTOSTERONE FOR MALES (NG/DL)

AGE (YEARS) — 0.6–9, 10–13, 14–15, 16–19, 20–39, 40–59, 60 AND OVER

A LIFETIME OF TESTOSTERONE PRODUCTION
Boys and men produce significant levels of testosterone throughout their lives, from puberty until well after the age of 60. Peak testosterone levels are present in young men between the ages of 20 and 40.

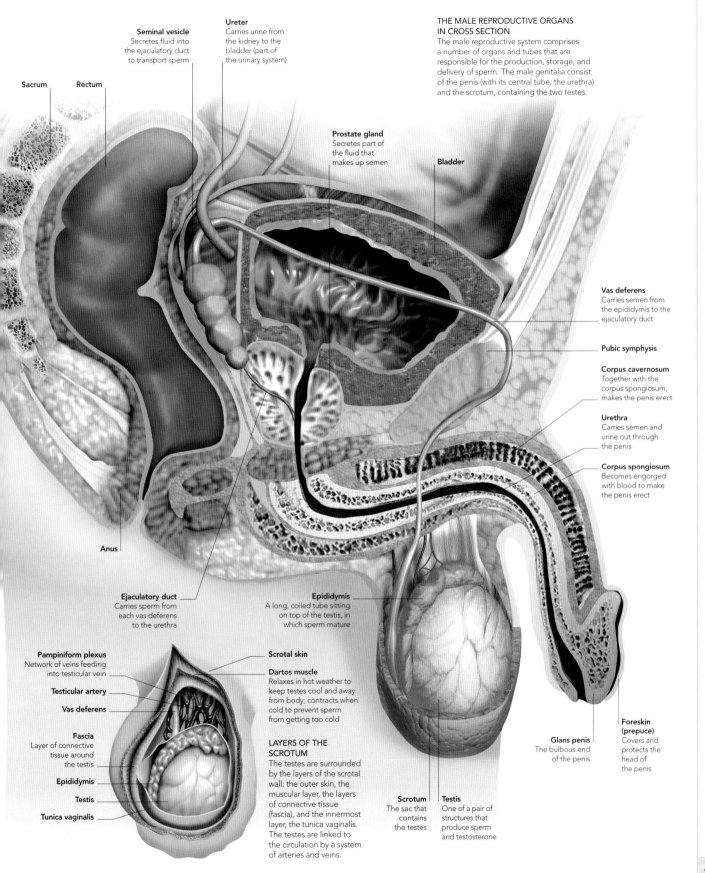

Sacrum

Rectum

Seminal vesicle
Secretes fluid into
the ejaculatory duct
to transport sperm

Ureter
Carries urine from
the kidney to the
bladder (part of
the urinary system)

Prostate gland
Secretes part of
the fluid that
makes up semen

Bladder

**THE MALE REPRODUCTIVE ORGANS
IN CROSS SECTION**
The male reproductive system comprises
a number of organs and tubes that are
responsible for the production, storage, and
delivery of sperm. The male genitalia consist
of the penis (with its central tube, the urethra)
and the scrotum, containing the two testes.

Vas deferens
Carries semen from
the epididymis to the
ejaculatory duct

Pubic symphysis

Corpus cavernosum
Together with the
corpus spongiosum,
makes the penis erect

Urethra
Carries semen and
urine out through
the penis

Corpus spongiosum
Becomes engorged
with blood to make
the penis erect

Anus

Ejaculatory duct
Carries sperm from
each vas deferens
to the urethra

Epididymis
A long, coiled tube sitting
on top of the testis, in
which sperm mature

Pampiniform plexus
Network of veins feeding
into testicular vein

Testicular artery

Vas deferens

Fascia
Layer of connective
tissue around
the testis

Epididymis

Testis

Tunica vaginalis

Scrotal skin

Dartos muscle
Relaxes in hot weather to
keep testes cool and away
from body; contracts when
cold to prevent sperm
from getting too cold

**LAYERS OF THE
SCROTUM**
The testes are surrounded
by the layers of the scrotal
wall: the outer skin, the
muscular layer, the layers
of connective tissue
(fascia), and the innermost
layer, the tunica vaginalis.
The testes are linked to
the circulation by a system
of arteries and veins.

Scrotum
The sac that
contains
the testes

Testis
One of a pair of
structures that
produce sperm
and testosterone

Glans penis
The bulbous end
of the penis

**Foreskin
(prepuce)**
Covers and
protects the
head of
the penis

29

THE PROSTATE GLAND, PENIS, AND TESTES

Sperm are developed and delivered by the prostate gland, penis, and testes. The prostate gland, located in the lower pelvis, and the penis and testes, which are outside the body cavity entirely, are connected by a system of incredibly long tubes.

THE PROSTATE GLAND

About 1½ in (4 cm) across, the prostate gland surrounds the urethra (the tube that carries urine from the bladder) as it emerges from the bladder. It produces a thick, milky, alkaline fluid that forms about 20 percent of semen volume and counteracts the acidity of other fluids in semen. The prostate gland is under the control of testosterone as well as nerves that, when arousal occurs, stimulate release of fluids by the prostate, seminal vesicles, and vasa deferentia. These fluids, together with the sperm, are released from the penis at ejaculation.

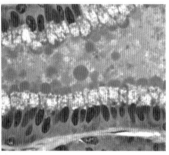

THE PROSTATE IN SECTION
This microscopic view of prostate tissue shows multiple secretory cells that release alkaline fluid, which neutralizes the acidity of semen, thereby improving sperm motility.

THE PENIS

The penis consists of a long shaft with a widened end, the glans. It has two functions: to deliver sperm and to expel urine. A penis contains three columns of erectile tissue: two corpus cavernosa, which lie alongside each other; and one corpus spongiosum, which encircles the urethra. When arousal occurs, blood vessels in these columns become engorged, making the penis erect (see pp.64–65). The average penis is about 3½ in (9 cm) long but can "reach" up to 7½ in (19 cm) when erect. Ejaculation is a reflex action.

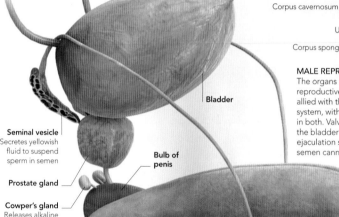

Veins

Arteries

Corpus cavernosum

Urethra

Corpus spongiosum

THE PENIS IN SECTION

Ureter

Bladder

MALE REPRODUCTIVE ORGANS
The organs and tubes of the male reproductive system are closely allied with those of the urinary system, with the penis featuring in both. Valves at the base of the bladder remain closed at ejaculation so that urine and semen cannot mix.

Seminal vesicle
Secretes yellowish fluid to suspend sperm in semen

Prostate gland

Cowper's gland
Releases alkaline fluid into the urethra during sexual arousal

Bulb of penis

Corpus cavernosum

THE TESTES

The paired testes are the powerhouses of the male reproductive system, producing sperm and the potent hormone testosterone. The testes are 1½–2 in (4–5 cm) long and comprise multiple conical sections (lobules), each containing tightly coiled tubes (seminiferous tubules) where sperm develop (see pp.32–33). The testes hang together in the scrotal sac. Within the scrotum, the temperature is 2–3.5° F (1–2° C) lower than body temperature—the optimal environment for sperm production. Leydig cells, clustered between the seminiferous tubules, secrete testosterone.

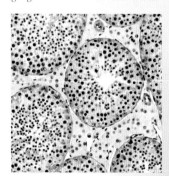

SEMINIFEROUS TUBULES IN SECTION
This magnified image shows seminiferous tubules packed with immature sperm and Sertoli cells; Leydig cells (stained green–brown) sit between the tubules.

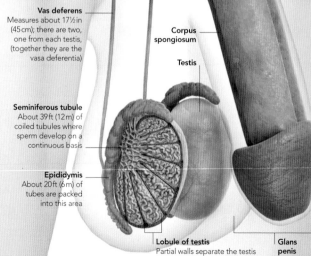

Vas deferens
Measures about 17½ in (45 cm); there are two, one from each testis, (together they are the vasa deferentia)

Corpus spongiosum

Testis

Seminiferous tubule
About 39 ft (12 m) of coiled tubules where sperm develop on a continuous basis

Epididymis
About 20 ft (6 m) of tubes are packed into this area

Lobule of testis
Partial walls separate the testis into about 250 compartments

Glans penis

MALE PUBERTY

The onset of puberty, brought about by the hormone testosterone, is a time of great physical and emotional changes. The body alters in shape and appearance, and within the body the sexual organs mature in readiness for sperm production.

PHYSICAL CHANGES

Puberty in boys (spermarche) tends to start between the ages of 12 and 15, on average two years later than it occurs in girls. The physical changes are very marked; some relate to the sexual organs themselves, the most obvious being the enlargement of the genitals; others appear unrelated, but all are the result of the dramatic increases in testosterone levels within the body. Puberty is accompanied by a final spurt of growth. Its later onset in boys than girls gives boys significantly more time to grow before they reach their final adult height.

WHY DOES A BOY'S VOICE BREAK?

Testosterone affects both the cartilage parts of the larynx (voice box) and the vocal cords themselves. The vocal cords grow 60 percent longer and thicker, and therefore start to vibrate at a lower frequency (making the voice sound deeper). At the same time, the larynx tilts and can start to stick out, forming the Adam's apple.

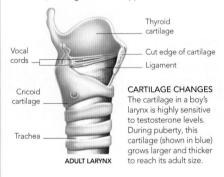

Thyroid cartilage

Vocal cords

Cut edge of cartilage

Ligament

Cricoid cartilage

Trachea

ADULT LARYNX

CARTILAGE CHANGES
The cartilage in a boy's larynx is highly sensitive to testosterone levels. During puberty, this cartilage (shown in blue) grows larger and thicker to reach its adult size.

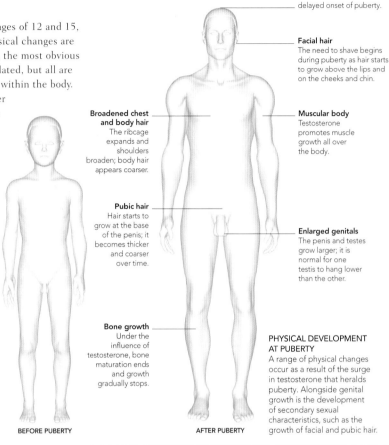

Height
Men are taller than women due to the delayed onset of puberty.

Facial hair
The need to shave begins during puberty as hair starts to grow above the lips and on the cheeks and chin.

Broadened chest and body hair
The ribcage expands and shoulders broaden; body hair appears coarser.

Muscular body
Testosterone promotes muscle growth all over the body.

Pubic hair
Hair starts to grow at the base of the penis; it becomes thicker and coarser over time.

Enlarged genitals
The penis and testes grow larger; it is normal for one testis to hang lower than the other.

Bone growth
Under the influence of testosterone, bone maturation ends and growth gradually stops.

BEFORE PUBERTY

AFTER PUBERTY

PHYSICAL DEVELOPMENT AT PUBERTY
A range of physical changes occur as a result of the surge in testosterone that heralds puberty. Alongside genital growth is the development of secondary sexual characteristics, such as the growth of facial and pubic hair.

HORMONAL CHANGES

From the age of about 10 years, the hypothalamus in boys begins to secrete a hormone (GnRH) that causes the pituitary gland to release hormones—FSH and LH—that control the testes. FSH, and to a lesser extent LH, promotes sperm production, but LH also stimulates the secretion of testosterone. High levels of testosterone cause the growth spurt and other pubertal changes. Once stabilized after puberty, testosterone levels in the body are regulated by a system of negative feedback.

TEENAGE BOYS AND AGGRESSION
It has been suggested that the surges of testosterone that occur in the teenage years in boys can be associated with increased levels of aggression.

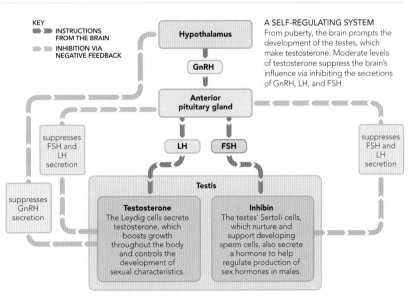

KEY
▶▶▶ INSTRUCTIONS FROM THE BRAIN
▶▶▶ INHIBITION VIA NEGATIVE FEEDBACK

Hypothalamus

GnRH

Anterior pituitary gland

suppresses FSH and LH secretion

suppresses GnRH secretion

LH

FSH

suppresses FSH and LH secretion

Testis

Testosterone
The Leydig cells secrete testosterone, which boosts growth throughout the body and controls the development of sexual characteristics.

Inhibin
The testes' Sertoli cells, which nurture and support developing sperm cells, also secrete a hormone to help regulate production of sex hormones in males.

A SELF-REGULATING SYSTEM
From puberty, the brain prompts the development of the testes, which make testosterone. Moderate levels of testosterone suppress the brain's influence via inhibiting the secretions of GnRH, LH, and FSH.

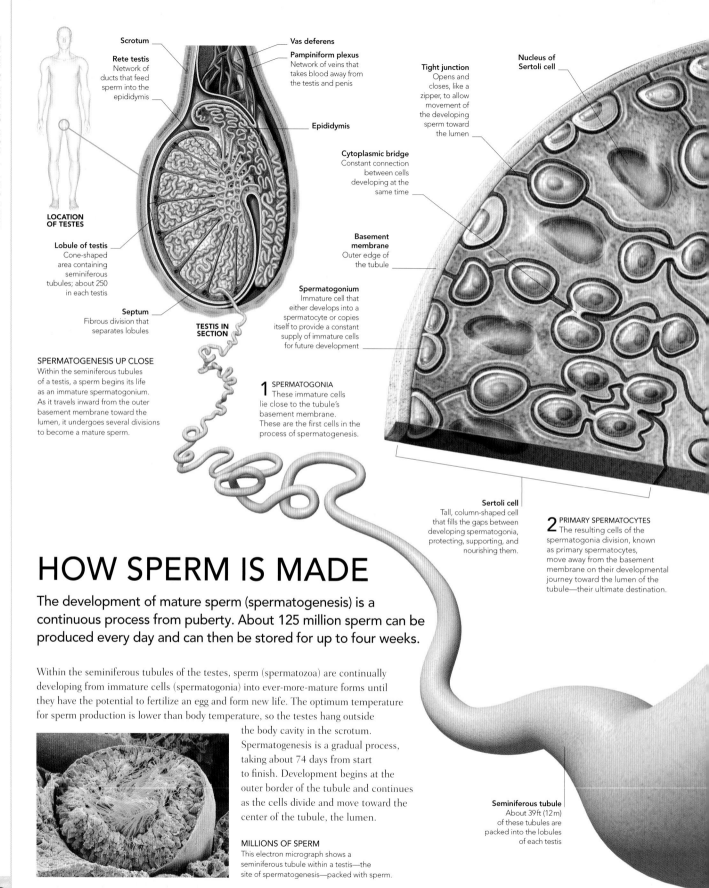

Scrotum

Rete testis
Network of ducts that feed sperm into the epididymis

Vas deferens

Pampiniform plexus
Network of veins that takes blood away from the testis and penis

Epididymis

LOCATION OF TESTES

Lobule of testis
Cone-shaped area containing seminiferous tubules; about 250 in each testis

Septum
Fibrous division that separates lobules

TESTIS IN SECTION

Tight junction
Opens and closes, like a zipper, to allow movement of the developing sperm toward the lumen

Nucleus of Sertoli cell

Cytoplasmic bridge
Constant connection between cells developing at the same time

Basement membrane
Outer edge of the tubule

Spermatogonium
Immature cell that either develops into a spermatocyte or copies itself to provide a constant supply of immature cells for future development

SPERMATOGENESIS UP CLOSE
Within the seminiferous tubules of a testis, a sperm begins its life as an immature spermatogonium. As it travels inward from the outer basement membrane toward the lumen, it undergoes several divisions to become a mature sperm.

1 SPERMATOGONIA
These immature cells lie close to the tubule's basement membrane. These are the first cells in the process of spermatogenesis.

Sertoli cell
Tall, column-shaped cell that fills the gaps between developing spermatogonia, protecting, supporting, and nourishing them.

2 PRIMARY SPERMATOCYTES
The resulting cells of the spermatogonia division, known as primary spermatocytes, move away from the basement membrane on their developmental journey toward the lumen of the tubule—their ultimate destination.

HOW SPERM IS MADE

The development of mature sperm (spermatogenesis) is a continuous process from puberty. About 125 million sperm can be produced every day and can then be stored for up to four weeks.

Within the seminiferous tubules of the testes, sperm (spermatozoa) are continually developing from immature cells (spermatogonia) into ever-more-mature forms until they have the potential to fertilize an egg and form new life. The optimum temperature for sperm production is lower than body temperature, so the testes hang outside the body cavity in the scrotum. Spermatogenesis is a gradual process, taking about 74 days from start to finish. Development begins at the outer border of the tubule and continues as the cells divide and move toward the center of the tubule, the lumen.

MILLIONS OF SPERM
This electron micrograph shows a seminiferous tubule within a testis—the site of spermatogenesis—packed with sperm.

Seminiferous tubule
About 39 ft (12 m) of these tubules are packed into the lobules of each testis

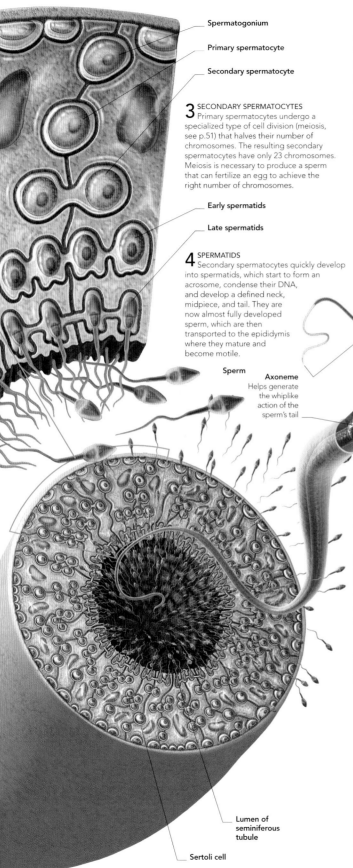

Spermatogonium

Primary spermatocyte

Secondary spermatocyte

3 SECONDARY SPERMATOCYTES
Primary spermatocytes undergo a specialized type of cell division (meiosis, see p.51) that halves their number of chromosomes. The resulting secondary spermatocytes have only 23 chromosomes. Meiosis is necessary to produce a sperm that can fertilize an egg to achieve the right number of chromosomes.

Early spermatids

Late spermatids

4 SPERMATIDS
Secondary spermatocytes quickly develop into spermatids, which start to form an acrosome, condense their DNA, and develop a defined neck, midpiece, and tail. They are now almost fully developed sperm, which are then transported to the epididymis where they mature and become motile.

Sperm

Axoneme
Helps generate the whiplike action of the sperm's tail

Lumen of seminiferous tubule

Sertoli cell

ANATOMY OF THE SPERM

Sperm are perhaps the tiniest cells in the body, yet they can propel themselves along and contain half the genetic information needed for a new individual to develop. The head contains the nucleus and at the front the acrosome, which contains enzymes that help it penetrate an egg. The midpiece contains the mitochondria, which provide all the energy a sperm needs on its long journey. Finally, the tail contains threads of tissue that slide next to each other enabling the whiplike action that propels the sperm forward.

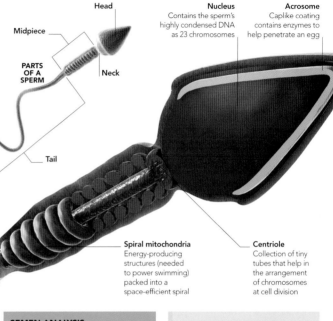

Head

Midpiece

PARTS OF A SPERM

Neck

Tail

Nucleus
Contains the sperm's highly condensed DNA as 23 chromosomes

Acrosome
Caplike coating contains enzymes to help penetrate an egg

Spiral mitochondria
Energy-producing structures (needed to power swimming) packed into a space-efficient spiral

Centriole
Collection of tiny tubes that help in the arrangement of chromosomes at cell division

SEMEN ANALYSIS

This test forms a crucial part of assessing couples with fertility problems. Several factors are routinely measured.

FEATURES OF THE SEMEN	NORMAL RANGE OF VALUES
Sperm count	More than 40 million per ejaculate
Semen volume	More than 0.07 fl oz (2 ml)
Sperm morphology (shape)	More than 70 percent with normal shape and structure
Sperm motility	More than 60 percent with normal forward movement
pH of semen	7.2–8.0
White blood cells	None (their presence may indicate infection)

ABNORMAL SPERM

Sperm can be abnormal in a variety of ways, such as having two heads, two tails, or a very short tail. Abnormally shaped sperm may not be able to move normally or to fertilize an egg. Some abnormal sperm are found in most normal semen samples. However, if the numbers are too high, fertility is likely to be affected.

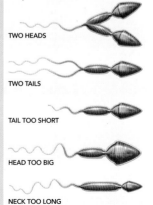

TWO HEADS

TWO TAILS

TAIL TOO SHORT

HEAD TOO BIG

NECK TOO LONG

THE FEMALE REPRODUCTIVE SYSTEM

The interconnected organs and tubes of the female reproductive system can provide everything needed to conceive and nurture a fetus. Once a baby is born, the system also provides it with the ultimate nourishment—breast milk.

Hypothalamus
The brain's "master gland" triggers and controls hormone secretion.

Pituitary gland
This tiny structure secretes hormones to stimulate the ovaries.

Breast
Made up of lobules, breasts produce milk in response to hormonal changes.

Ovary
Eggs develop here and are released every month.

Fallopian tube
This transport tube propels mature eggs from the ovary to the uterus.

Uterus
Every month its lining prepares for an embryo but is shed if fertilization does not occur.

Vagina
This elastic tube can stretch to allow a baby to be born.

LOCATING ORGANS OF THE FEMALE REPRODUCTIVE SYSTEM
The main reproductive organs lie within the pelvis. Their actions and those of the breasts are under the control of certain areas of the brain.

REPRODUCTIVE ORGANS

The uterus, vagina, ovaries, and fallopian tubes coordinate their actions to generate new life. The vagina receives an erect penis as it delivers sperm to the entrance of the uterus, the cervix. Eggs are stored and develop within the ovaries. Each month one egg (or, very rarely, two eggs) is released and moves along one fallopian tube to its ultimate destination, the uterus. If the egg has combined with a sperm en route, it will develop into an embryo (later called a fetus) and grow within the uterus, which stretches to many times its original size over the next nine months. The ovaries also produce hormones key to the reproductive process.

REPRODUCTIVE LIFE

At birth, the ovaries of a baby girl contain one to two million immature eggs, but the number dwindles over time; by puberty only about 400,000 remain. Usually, only one egg is released every month. The time available to women to have a baby is finite, although new technologies can prolong the window of reproductive opportunity for some women. Generally, the reproductive years, which start at puberty, end around the age of 50 when menopause occurs; men, meanwhile, can continue to father children to a much greater age.

IN THE FAMILY WAY
Mature eggs are released from the ovaries from puberty until menopause. A woman's fertility begins to decline gradually from about the age of 27, but starts to drop more rapidly from the age of 35.

PROGESTERONE CRYSTALS
This highly magnified and color-enhanced micrograph shows crystals of progesterone. This hormone helps prepare the uterine lining for pregnancy by causing it to thicken and its blood supply to be increased.

SEX HORMONES

Produced primarily by the ovaries, the female sex hormones estrogen and progesterone are responsible for the sexual development and physical changes that occur at puberty (see p.43), the monthly menstrual cycle (see pp.44–45), and fertility. Their production is under the control of two hormones—luteinizing hormone (LH) and follicle-stimulating hormone (FSH)—that are produced by the pituitary, the tiny gland at the base of the brain, which is in turn regulated by the hypothalamus. The sex hormones also influence emotions: many women experience mood changes during their menstrual cycle, which correspond with hormonal fluctuations. In addition, the male sex hormone testosterone also exerts effects within the female body, although it is present at relatively low levels.

EFFECTS OF SEX HORMONES ON THE FEMALE BODY

The female sex hormones estrogen and progesterone have key roles in the menstrual cycle, as well as more general physical effects. The male sex hormone testosterone is also present in women.

HORMONE	EFFECTS
Estrogen	Estrogen promotes the growth of the sex organs and the development of the physical changes that occur at puberty—secondary sexual characteristics. In the ovaries, it enhances the development of eggs, and it thins the mucus produced by the cervix so that it is easier for sperm to penetrate. Estrogen levels peak just before egg release (ovulation). It also stimulates growth of the uterine lining (endometrium).
Progesterone	Progesterone helps prepare the endometrium every month and maintains it if pregnancy occurs. If pregnancy doesn't occur, progesterone levels fall and menstruation results. Progesterone also prepares the breasts for milk production (lactation).
Testosterone	Despite circulating in relatively low levels, testosterone does affect the female body. It is responsible for the growth spurt of puberty and the closure of growth plates that signals the end of childhood growth.

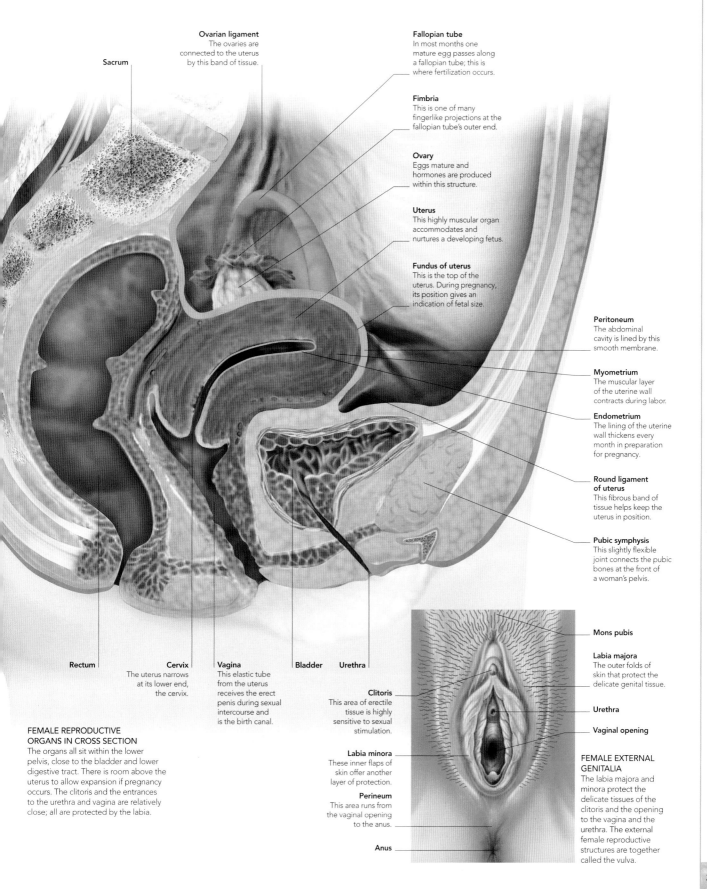

Sacrum

Ovarian ligament
The ovaries are connected to the uterus by this band of tissue.

Fallopian tube
In most months one mature egg passes along a fallopian tube; this is where fertilization occurs.

Fimbria
This is one of many fingerlike projections at the fallopian tube's outer end.

Ovary
Eggs mature and hormones are produced within this structure.

Uterus
This highly muscular organ accommodates and nurtures a developing fetus.

Fundus of uterus
This is the top of the uterus. During pregnancy, its position gives an indication of fetal size.

Peritoneum
The abdominal cavity is lined by this smooth membrane.

Myometrium
The muscular layer of the uterine wall contracts during labor.

Endometrium
The lining of the uterine wall thickens every month in preparation for pregnancy.

Round ligament of uterus
This fibrous band of tissue helps keep the uterus in position.

Pubic symphysis
This slightly flexible joint connects the pubic bones at the front of a woman's pelvis.

Rectum

Cervix
The uterus narrows at its lower end, the cervix.

Vagina
This elastic tube from the uterus receives the erect penis during sexual intercourse and is the birth canal.

Bladder

Urethra

FEMALE REPRODUCTIVE ORGANS IN CROSS SECTION
The organs all sit within the lower pelvis, close to the bladder and lower digestive tract. There is room above the uterus to allow expansion if pregnancy occurs. The clitoris and the entrances to the urethra and vagina are relatively close; all are protected by the labia.

Mons pubis

Labia majora
The outer folds of skin that protect the delicate genital tissue.

Urethra

Vaginal opening

Clitoris
This area of erectile tissue is highly sensitive to sexual stimulation.

Labia minora
These inner flaps of skin offer another layer of protection.

Perineum
This area runs from the vaginal opening to the anus.

Anus

FEMALE EXTERNAL GENITALIA
The labia majora and minora protect the delicate tissues of the clitoris and the opening to the vagina and the urethra. The external female reproductive structures are together called the vulva.

THE OVARIES AND FALLOPIAN TUBES

An egg starts its life in an ovary, where it is stored and then matures until ready for release at ovulation. The mature egg travels along a fallopian tube to the uterus where, if it has been fertilized en route, it embeds in the wall and pregnancy begins.

THE OVARIES

Lying on either side of the pelvis, the paired ovaries provide mature eggs (ova) that, if combined with a sperm, can form a new human being. They also produce estrogen and progesterone; these hormones control sexual development (see p.43) and the menstrual cycle (see pp.44–45). The ovaries are only the size of almonds, yet they contain tens of thousands of immature eggs. From puberty, eggs and their containing follicles begin a cycle of development and release from the ovary. When an egg is released, it enters a fallopian tube. The empty follicle remains in the ovary and produces hormones to sustain a pregnancy.

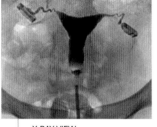

X-RAY VIEW
In this image, the uterus, ovaries, and fallopian tubes are highlighted by a contrast dye delivered by the probe seen in the vagina.

Ampulla
This long section is the most common site of fertilization.

Ovarian medulla
The central part of the ovary contains blood vessels and nerves.

Primordial follicle
This is the earliest immature follicle, present at birth.

Primary follicle
As a follicle's development gets underway, it is first called a primary follicle.

Secondary follicle
After further development, a primary follicle becomes a secondary follicle.

Ovarian ligament
This band of tissue connects the ovary to the uterus.

Uterus
This muscular organ accommodates the developing embryo, later called the fetus.

Blood vessels

Ovarian cortex
Follicles in various stages of development are found here.

Corpus luteum
Formed from the empty follicle, this produces both estrogen and progesterone.

Preovulatory follicle
This term is used for the mature follicle just before ovulation.

THE ESTROGEN FAMILY

The estrogens are a group of similar chemicals, three of which are produced in significant amounts: estradiol, estriol, and estrone. The levels of these hormones differ at various stages of a woman's life, but the main one—estradiol—predominates throughout her reproductive life, from menarche to menopause. Estrogen is mainly produced in the ovaries, but smaller amounts are manufactured in the adrenal glands, which lie on top of the kidneys, and in fat cells (adipose tissue). Being significantly overweight can be associated with higher levels of estrogen, which may affect the functioning of the ovaries and reduce fertility.

80   0 YEARS

12

16

Fat cells, or adipose tissue, produce a small amount of estrogen

Ovaries secrete estrone after menopause

KEY
■ ESTRADIOL
■ ESTRIOL
■ ESTRONE

Ovarian follicles produce estradiol from puberty to menopause

50

40

Placenta makes estriol during pregnancy

A LIFETIME OF ESTROGENS
Types of estrogen vary at different stages of a woman's life. Estradiol dominates the reproductive years.

INSIDE AN OVARY AND FALLOPIAN TUBE
Mature eggs are released from the surface of the ovary into the pelvis and are drawn into the nearby funnel-shaped end of the tube by the movement of fingerlike projections called fimbriae. The egg is propelled along the length of the tube (about 4½in/12cm) to the uterus.

Fallopian tube
The convoluted interior surface is made up of folds, and a layer of smooth muscle encircles the tube.

THE FALLOPIAN TUBES

Located on either side of the uterus, the fallopian tubes transport mature eggs from the ovaries to the uterus. Various features of the tubes facilitate an otherwise immobile egg to get to its destination—the fimbriae capture the egg initially, and the muscular wall and the beating cilia on the tube's interior propel the egg along. A fallopian tube has three main parts: the outermost infundibulum, the ampulla (the usual site of fertilization), and the innermost isthmus. Each region varies in diameter and microstructure; for example, the muscle in the isthmus wall is particularly thick to enable it to deliver the egg into the uterus. If fertilization occurs, the fertilized egg (zygote) divides as it passes along the tube ready for implantation in the uterus wall.

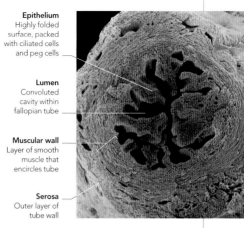

Epithelium
Highly folded surface, packed with ciliated cells and peg cells

Lumen
Convoluted cavity within fallopian tube

Muscular wall
Layer of smooth muscle that encircles tube

Serosa
Outer layer of tube wall

MICROSTRUCTURE OF A FALLOPIAN TUBE
This microscopic view shows a cross section through the ampulla region of a fallopian tube; the wall's different layers are clearly visible.

Fimbria
This delicate, fingerlike projection helps draw the egg into the fallopian tube.

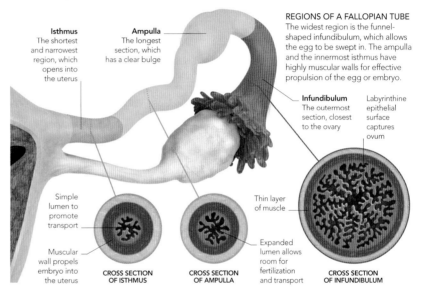

Isthmus
The shortest and narrowest region, which opens into the uterus

Ampulla
The longest section, which has a clear bulge

REGIONS OF A FALLOPIAN TUBE
The widest region is the funnel-shaped infundibulum, which allows the egg to be swept in. The ampulla and the innermost isthmus have highly muscular walls for effective propulsion of the egg or embryo.

Infundibulum
The outermost section, closest to the ovary

Labyrinthine epithelial surface captures ovum

Simple lumen to promote transport

Thin layer of muscle

Muscular wall propels embryo into the uterus

CROSS SECTION OF ISTHMUS

CROSS SECTION OF AMPULLA

Expanded lumen allows room for fertilization and transport

CROSS SECTION OF INFUNDIBULUM

HOW A FALLOPIAN TUBE PROPELS AN EGG

From the moment the egg (ovum) leaves the ovary, the fallopian tube is working to deliver it first to the middle third of the tube in preparation for penetration by a sperm (fertilization), and then on to the uterus. The movement of the fimbriae at the outer end of the tube combined with the beating of the cilia create a current that draws the egg into the flared end of the tube. Once inside, waves of muscular contraction and the action of cilia transport it to the uterus.

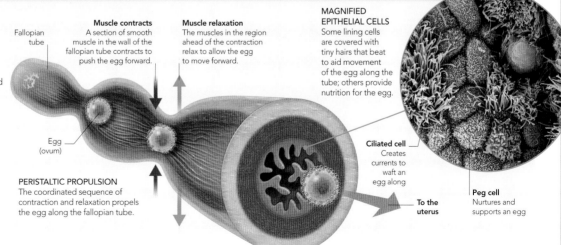

Fallopian tube

Muscle contracts
A section of smooth muscle in the wall of the fallopian tube contracts to push the egg forward.

Muscle relaxation
The muscles in the region ahead of the contraction relax to allow the egg to move forward.

Egg (ovum)

PERISTALTIC PROPULSION
The coordinated sequence of contraction and relaxation propels the egg along the fallopian tube.

MAGNIFIED EPITHELIAL CELLS
Some lining cells are covered with tiny hairs that beat to aid movement of the egg along the tube; others provide nutrition for the egg.

Ciliated cell
Creates currents to waft an egg along

To the uterus

Peg cell
Nurtures and supports an egg

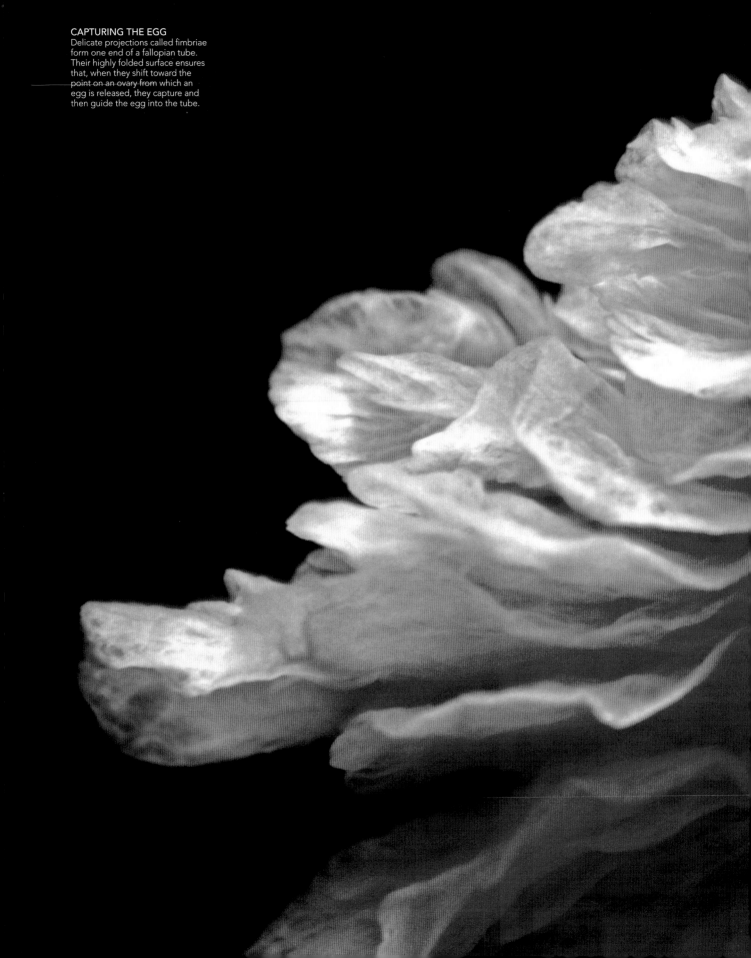

CAPTURING THE EGG
Delicate projections called fimbriae
form one end of a fallopian tube.
Their highly folded surface ensures
that, when they shift toward the
point on an ovary from which an
egg is released, they capture and
then guide the egg into the tube.

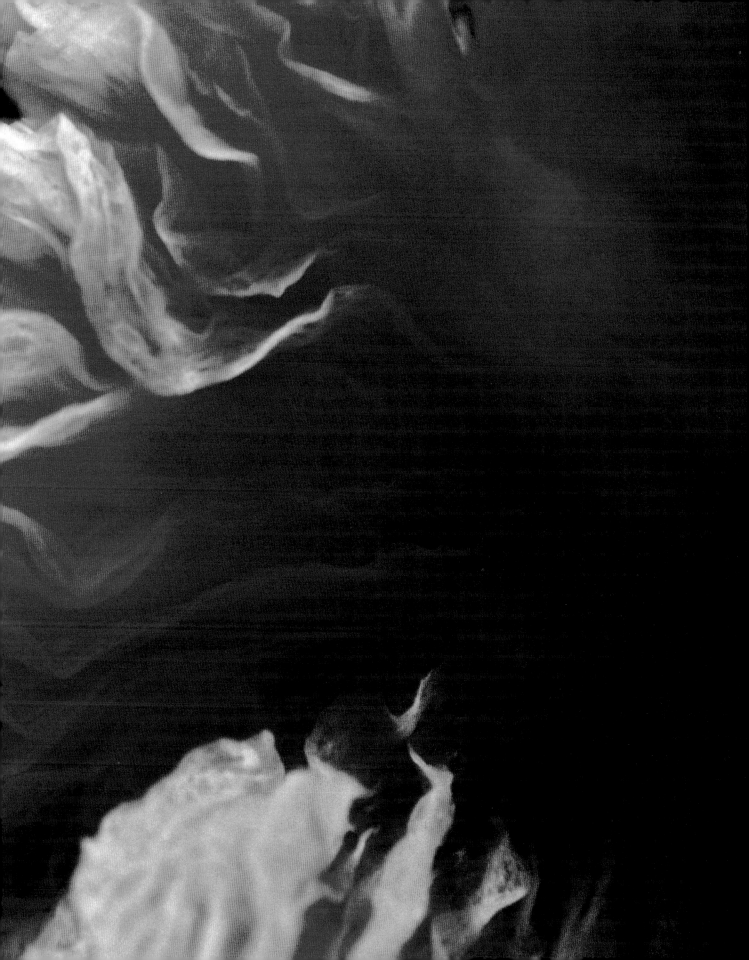

THE UTERUS, CERVIX, AND VAGINA

Every month, the lining of the uterus undergoes structural changes to prepare for the possible arrival of a fertilized egg. The uterus is the home for a developing fetus for the duration of a pregnancy; the cervix and vagina are its exit points to the outside world.

INSIDE THE FEMALE REPRODUCTIVE TRACT
The uterus, the central area of the female reproductive tract, is connected to the two fallopian tubes at its uppermost corners and to the vagina below, the exit being formed by the cervix.

THE UTERUS

A highly muscular organ, the uterus is the site of implantation for a fertilized egg. During pregnancy, it enlarges to many times its size as a fetus grows. The uterine wall is made up of three layers: the outer perimetrium, the middle muscular myometrium, and the inner endometrium. The endometrium builds up every month in preparation for a fertilized egg and then is shed if fertilization fails to occur. The uterus can be divided into sections: the upper dome-shaped fundus, the main body, and the neck, or cervix.

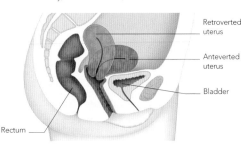

Retroverted uterus

Anteverted uterus

Bladder

Rectum

THE POSITION OF THE UTERUS
The angle of the uterus can vary, but in most women, it is tilted forward (anteverted); about 20 percent of women have a uterus that tilts backward (retroverted).

THE EXPANDABLE UTERUS

The wall of the uterus consists mainly of muscle, giving it an amazing capacity to enlarge to accommodate the growing fetus. The fundal height is monitored as a measure of fetal growth. Conveniently, the fundal height in centimeters usually corresponds to the length of the pregnancy in weeks.

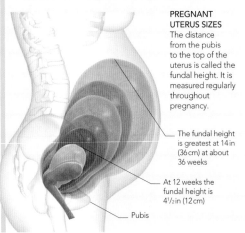

PREGNANT UTERUS SIZES
The distance from the pubis to the top of the uterus is called the fundal height. It is measured regularly throughout pregnancy.

The fundal height is greatest at 14 in (36 cm) at about 36 weeks

At 12 weeks the fundal height is 4 1/2 in (12 cm)

Pubis

THE LINING OF THE UTERUS

The uterine lining, the endometrium, is composed of a functional layer and a basal layer; the former thickens each month until a fall in hormones prompts it to be shed during menstruation. The basal layer remains to renew the functional layer once menstruation is over. The endometrium has a unique blood supply: straight arteries in the basal layer and spiral arteries in the functional layer. Most arteries in the body branch into arterioles and capillaries before rejoining to form venules and veins. Spiral arteries do this but they also have a shunt connecting them directly to veins. When hormone levels fall, the resultant shrinkage in the endometrium forces spiral arteries to coil until they restrict the blood flow, which diverts via the shunts until stopping. Tissue death sets in as cells of the functional layer do not have access to blood, and the capillary plexi and venous "lakes" rupture, all of which leads to the bleeding of menstruation.

Uterine cavity

Endometrium
Innermost layer of the uterus

Myometrium
Middle muscular layer of the uterus

Perimetrium
Outermost layer of the uterus

Functional layer
Highly regenerative layer with specialized blood vessels

Basal layer
Ever-present layer that helps rebuild the functional layer each cycle

Straight artery
Supplies only the basal layer

Shunt
Connection between spiral artery and venous "lake"; used as endometrium starts to shrink

Endometrial gland
Secretes mucus and other substances during the menstrual cycle

Venous "lake"
Blood pools here before rupturing at the start of menstruation

Capillary plexus
Network of single-cell-walled vessels that connect arterioles to venules

Spiral artery
Grows faster than the surrounding tissue, so coils tighter as functional layer nears completion

THE STRUCTURE OF THE ENDOTHELIUM
The thin layer of cells that lines the endometrium is called the endothelium; its detailed structure helps explain its ability to shed and renew itself every month. It has unique systems of blood vessels: straight arteries in the basal layer; and spiral arteries within the functional layer, which coil as the layer grows.

Fundus of uterus
Top part of the uterus

THE CERVIX

The neck of the uterus, more commonly referred to as the cervix, opens into the vagina at the external os, thereby forming a connection between the uterus and the vagina. The highly specialized epithelial cells that line the convoluted surface of the cervical canal present an obstacle course for sperm to navigate. They also secrete mucus, the nature and content of which varies during the menstrual cycle. The changes make mucus hostile to sperm for most of the cycle and then sperm-friendly around ovulation (see pp.44–45). If sperm-friendly mucus is present, it acts as a reservoir, prolonging the life of sperm past the usual 24 hours. A mucus plug seals the cervix during pregnancy, protecting it from the outside world.

Fallopian tube
Specialized transport tube that carries eggs from the ovary after ovulation to the uterus

Internal os
Inner boundary of the cervical canal, where it meets the uterus

Cervical canal
Has a vertical ridge front and back from which numerous folds (rugae) branch

SECRETORY EPITHELIUM OF THE CERVIX
The epithelium of the cervix contains columnar cells, which are responsible for secreting the cervical mucus. The production is affected by the hormonal changes of the menstrual cycle.

Convoluted surface
Folded surface of cervical canal presents an obstacle course to sperm after sexual intercourse

Columnar epithelium
Cells here secrete various chemicals as well as mucus

Vaginal fornix
Deepest portion of the vagina, extending into recesses created by the cervix

Cervical lumen
Space in the middle of the canal

CERVIX VIEWED FROM BELOW
This view of the cervix shows the external os. In a woman who has never had a vaginal delivery, this os is tightly closed; in a woman who has given birth vaginally it appears less tightly closed. Here the mucus appears whitish and watery.

FEATURES OF CERVICAL MUCUS

The amount of cervical mucus varies as it is affected by the hormones of the menstrual cycle. Mucus can be used as an indicator of the most fertile times in a cycle (see p.79).

SPERM–FRIENDLY MUCUS	HOSTILE MUCUS
Is produced in abundance	Is produced in small amounts
Is more stretchy and elastic	Is less stretchy and elastic
Contains more water and so is thinner	Contains less water and so is thicker
Is more alkaline (has a higher pH)	Is more acid (has a lower pH)
Has a strandlike structure	Has a globular structure
Has no anti-sperm antibodies	Contains anti-sperm antibodies

THE VAGINA

This elastic, muscular tube connects the uterus with the vulva. It receives the penis during sexual intercourse and expands greatly to provide the birth canal for the delivery of a baby. The vagina also allows menstrual blood and tissue to leave the body during menstruation. The vaginal wall is made up of an outer covering, a middle layer of muscle, and an inner layer of epithelium that is formed into ridges (rugae). The surface does not produce secretions itself but rather is lubricated by secretions from the cervix. The vagina contains natural bacteria, which creates a very acidic environment; this helps to protect against germs.

Ruga

External os
Outer boundary of the cervix, where it meets the vagina

Vagina
Elastic, muscular tube that connects the cervix to the vulva; ridges called rugae line the vagina

VAGINAL RUGAE
The ridges of the vaginal lining, known as the rugae, allow the highly elastic walls of the vagina to expand during sexual intercourse and childbirth.

41

THE BREASTS

Breast function is closely allied to the functions of the reproductive organs. Breast development occurs at puberty; further adaptations are made during and after pregnancy to produce breast milk for a newborn.

BREAST TISSUE

The breasts consist of glandular tissue, fat, and some supporting tissue that helps give the breasts their shape. The breast tissue is arranged in lobules, within which gland cells are formed into clusters called alveoli. Tiny tubes from the alveoli come together to form the main ducts that open onto the nipple. During pregnancy, high levels of estrogen and progesterone cause the glands and ducts to prepare for lactation (see pp.174–75). The shape of a woman's breasts is determined by her genes, the amount of fatty tissue the breasts contain, and muscle tone.

Fatty tissue

Lung

Blood vessel

Pectoral muscle

Lobule

Nipple
The center has tiny holes through which milk can flow

Milk duct

Rib

A BREAST IN CROSS SECTION
Breast tissue is arranged in a daisy pattern of 15–20 lobules. Ducts from these lobules drain milk directly to the nipple. The breasts are attached to the muscles beneath them by strong fibrous tissue.

Areola
Circular pigmented area around a nipple

Nipple
Lies at the center of the areola

Milk duct
Tube that transports milk from the lobule to the nipple

Lobule
Structure containing milk-producing cells

Ductule
Feeds into a milk duct

Alveolus
One of many glandular structures at the end of each lobule

Epithelial cell
Produces and releases milk during lactation

Adipose cell
One of multiple fat cells that make up the fatty tissue

FEATURES OF A BREAST
The breast is a highly glandular structure. The size and shape varies, but all contain a similar amount of milk-producing tissue. The nipple, which is surrounded by the areola, contains muscle that can cause it to become erect when stimulated. The nipple receives milk via the milk ducts that drain the lobules.

MICROSTRUCTURE OF THE BREAST
This enlarged view of the breast tissue shows the alveoli, which contain the milk-producing cells, embedded in fatty tissue. They are drained by tiny ductules.

FEMALE PUBERTY

This important stage in life, when sexual organs develop and marked physical changes occur, tends to begin at the age of 10–14 years in girls and usually lasts for three or four years.

PHYSICAL CHANGES OF PUBERTY

The changes of puberty take place in a particular order. Early breast development, known as thelarche, is the first physical change of puberty to occur, with the appearance of the so-called breast bud, when the nipple and a small area around it start to protrude from the chest wall (see panel, right) This is followed within about six months by the growth of hair in the pubic area and, soon after that, armpit (axillary) hair. Gradually, the breasts swell, more pubic and axillary hair grows, and the genitals develop. The uterus also enlarges leading up to the first menstrual period (menarche). While these changes are taking place, a girl grows taller and her body outline changes, with the hips and pelvis widening. The onset of puberty is about two years later in boys than in girls.

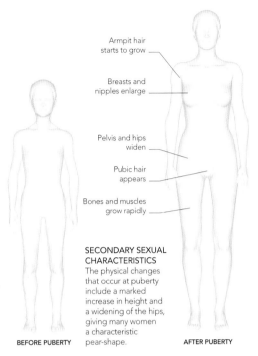

Armpit hair starts to grow

Breasts and nipples enlarge

Pelvis and hips widen

Pubic hair appears

Bones and muscles grow rapidly

BEFORE PUBERTY

AFTER PUBERTY

SECONDARY SEXUAL CHARACTERISTICS
The physical changes that occur at puberty include a marked increase in height and a widening of the hips, giving many women a characteristic pear-shape.

BREAST DEVELOPMENT

The breast changes of puberty occur in five defined stages. Firstly, during thelarche, the nipple appears higher. Following this, the breast bud develops behind the areola, causing the nipple and tissue around it to project from the chest wall. Next, the areola enlarges, accompanied by further breast tissue growth. Then, changes in the nipple and areola cause them to protrude forward from the rest of the breast. In the final phase of development, the smooth contour of the breast is established.

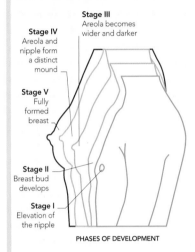

Stage III
Areola becomes wider and darker

Stage IV
Areola and nipple form a distinct mound

Stage V
Fully formed breast

Stage II
Breast bud develops

Stage I
Elevation of the nipple

PHASES OF DEVELOPMENT

HORMONAL CONTROL

The onset of puberty is triggered by the release of gonadotropin-releasing hormone (GnRH) by the brain's hypothalamus. This hormone stimulates the release of two hormones by the pituitary gland—follicle stimulating hormone (FSH) and luteinizing hormone (LH). FSH and LH cause the ovaries to produce two hormones, estrogen and progesterone, which are responsible for the major changes occuring at puberty and for the monthly menstrual cycles in the years that follow (see p.44–45). Their release is controlled by a negative-feedback system: as levels of the ovarian hormones rise, so the levels of the hormones that stimulate their release are reduced.

OVULATION UP CLOSE
The tiny pituitary gland at the base of the brain releases the hormone LH, which stimulates the rupture of a follicle within the ovary to release a mature egg on a monthly basis.

SELF-REGULATION
The hypothalamus and pituitary gland release stimulatory hormones that prompt the ovaries to produce estrogen and progesterone. These feed back to the brain to regulate the release of further hormones.

KEY
⫸⫸ INSTRUCTIONS FROM THE BRAIN
⫸⫸ INHIBITION VIA NEGATIVE FEEDBACK

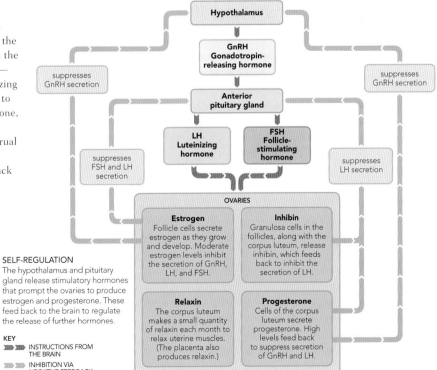

Hypothalamus

GnRH
Gonadotropin-releasing hormone

suppresses GnRH secretion

suppresses GnRH secretion

Anterior pituitary gland

LH
Luteinizing hormone

FSH
Follicle-stimulating hormone

suppresses FSH and LH secretion

suppresses LH secretion

OVARIES

Estrogen
Follicle cells secrete estrogen as they grow and develop. Moderate estrogen levels inhibit the secretion of GnRH, LH, and FSH.

Inhibin
Granulosa cells in the follicles, along with the corpus luteum, release inhibin, which feeds back to inhibit the secretion of LH.

Relaxin
The corpus luteum makes a small quantity of relaxin each month to relax uterine muscles. (The placenta also produces relaxin.)

Progesterone
Cells of the corpus luteum secrete progesterone. High levels feed back to suppress secretion of GnRH and LH.

THE FEMALE REPRODUCTIVE CYCLE

Eggs are constantly developing but only one is released every month. To prepare the uterus for the potential implantation of a fertilized egg, a cycle of hormonal fluctuations and changes to the uterine lining occur each month.

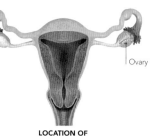

LOCATION OF THE OVARIES

Ovary

HOW A FOLLICLE MATURES AND RELEASES AN EGG

Development of a mature follicle (folliculogenesis), which then releases its egg, from an ovary takes about 28 weeks. Immature eggs remain in the ovaries in an unchanged state from birth until puberty. But once sexual maturity is reached, the egg-containing follicles start to mature in clearly defined stages—from a primordial to primary, secondary, and then tertiary follicle. Finally, a mature egg is released (ovulation), leaving a corpus hemorrhagicum behind, which develops into a corpus luteum. During a woman's reproductive life, only 400 or so mature eggs will be released; many eggs simply die.

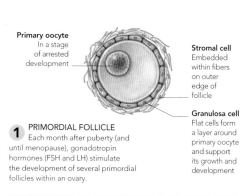

Thecal layer
An organized layer formed by stromal cells

Zona pellucida
A clear layer between primary oocyte and granulosa cells

Theca interna
Layer within which blood vessels develop and whose cells secrete estrogen

Theca externa
An outer layer of stromal cells and fibers

Fully grown primary oocyte

Granulosa cells
Several layers of these cells now surround primary oocyte

Primary oocyte

Primary oocyte
In a stage of arrested development

Stromal cell
Embedded within fibers on outer edge of follicle

Granulosa cell
Flat cells form a layer around primary oocyte and support its growth and development

Antrum
Fluid-filled cavity whose size increases as follicle develops

1 PRIMORDIAL FOLLICLE
Each month after puberty (and until menopause), gonadotropin hormones (FSH and LH) stimulate the development of several primordial follicles within an ovary.

2 PRIMARY FOLLICLE
Granulosa cells multiply greatly and become cube-shaped rather than flat. Receptors that respond to levels of FSH develop, and there is dramatic growth of both oocyte and follicle.

3 SECONDARY FOLLICLE
The thecal layer differentiates further into two layers. Granulosa cells start to secrete follicular fluid, which collects in the antrum. Many follicles develop at the same time but not all mature successfully.

1	1	2	3	4	5	6	7	8	9	10	11	12	13

WEEKS

Days in cycle

THE MENSTRUAL CYCLE

The start of this 28-day cycle sees the shedding of the lining of the uterus. This causes blood to exit via the vagina, known as menstruation, which lasts for a few days. After this, the lining of the uterus begins to thicken again in preparation for the potential implantation of a fertilized egg. The period when the uterine lining is most hospitable for implantation is called the fertile window; it begins five days before ovulation and is about a week long. If an egg is not fertilized, the uterine lining will break down and the cycle begins again. The fluctuations of four interacting hormones—FSH, LH, estrogen, and progesterone—trigger and control this monthly cycle. The first half of the cycle is called the follicular phase; the second half, following ovulation, is known as the luteal phase.

HORMONES
Every month, a rise in FSH causes egg maturation, and then LH surges, causing egg release. Estrogen levels peak just prior to egg release and then progesterone levels rise, causing endometrial thickening.

KEY
— FOLLICLE-STIMULATING HORMONE (FSH)
— LUTEINIZING HORMONE (LH)
— ESTROGEN
— PROGESTERONE

UTERINE LINING
Estrogen and progesterone cause the endometrium to thicken to about ¼ in (6 mm) ready for embryo implantation. If fertilization fails to occur, the functional layer is shed and then rebuilt for the next cycle.

Functional layer of endometrium is shed during menstruation

Functional layer regenerates to provide the perfect environment for implantation

DAYS IN CYCLE	1	2	3	4	5	6	7	8
PHASES OF CYCLE						FOLLICULAR		

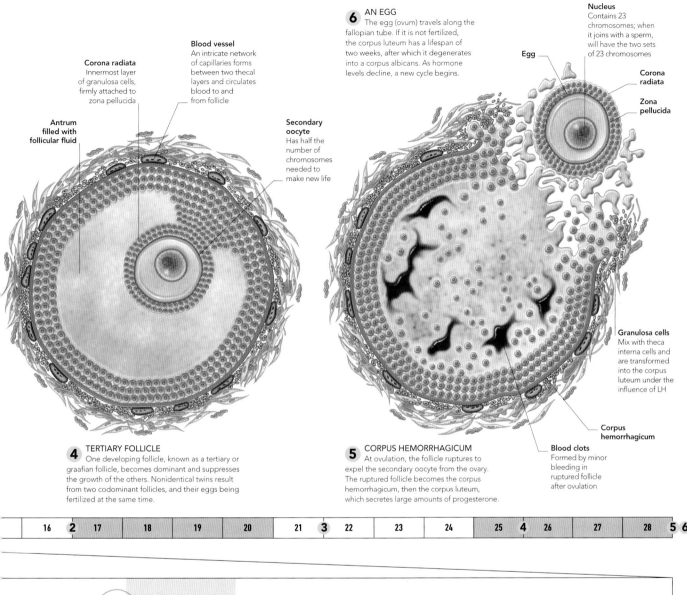

Corona radiata
Innermost layer of granulosa cells, firmly attached to zona pellucida

Blood vessel
An intricate network of capillaries forms between two thecal layers and circulates blood to and from follicle

Antrum filled with follicular fluid

Secondary oocyte
Has half the number of chromosomes needed to make new life

6 AN EGG
The egg (ovum) travels along the fallopian tube. If it is not fertilized, the corpus luteum has a lifespan of two weeks, after which it degenerates into a corpus albicans. As hormone levels decline, a new cycle begins.

Nucleus
Contains 23 chromosomes; when it joins with a sperm, will have the two sets of 23 chromosomes

Egg

Corona radiata

Zona pellucida

Granulosa cells
Mix with theca interna cells and are transformed into the corpus luteum under the influence of LH

Corpus hemorrhagicum

Blood clots
Formed by minor bleeding in ruptured follicle after ovulation

4 TERTIARY FOLLICLE
One developing follicle, known as a tertiary or graafian follicle, becomes dominant and suppresses the growth of the others. Nonidentical twins result from two codominant follicles, and their eggs being fertilized at the same time.

5 CORPUS HEMORRHAGICUM
At ovulation, the follicle ruptures to expel the secondary oocyte from the ovary. The ruptured follicle becomes the corpus hemorrhagicum, then the corpus luteum, which secretes large amounts of progesterone.

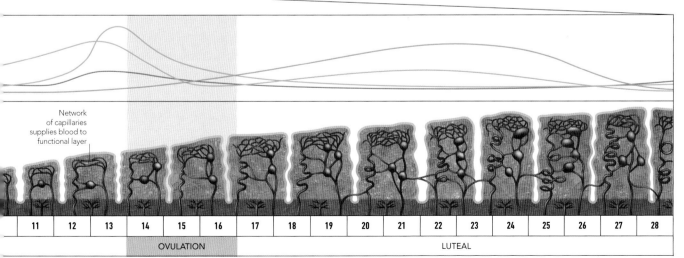

Network of capillaries supplies blood to functional layer

THE BLUEPRINT FOR HOW THE HUMAN BODY GROWS, DEVELOPS, AND FUNCTIONS LIES IN THE MASS OF DNA CURLED UP IN THE NUCLEUS OF EVERY CELL. WHEN A NEW LIFE IS CONCEIVED, HALF OF THE GENETIC INSTRUCTIONS CONTAINED IN ITS DNA IS INHERITED FROM EACH PARENT. ALTHOUGH THE BUILDING BLOCKS OF DNA ARE SIMPLE, THE WAY IN WHICH THIS SET OF INSTRUCTIONS IS READ IS A COMPLEX AND REMARKABLE PROCESS. HOWEVER, THINGS CAN GO WRONG. LEARNING HOW THE DNA CODE WORKS, AND BEING ABLE TO DECIPHER IT, CAN HELP US UNDERSTAND HOW CHILDREN INHERIT THEIR PARENTS' CHARACTERISTICS AND THE CAUSES OF CERTAIN DISEASES.

GENETICS

THE MOLECULES OF LIFE

All living things, including humans, owe their existence to the intricate architecture of a few chemical building blocks that contain the coded instructions needed to build our bodies, keep us alive, and create new life.

DNA, GENES, AND CHROMOSOMES

The human body owes its structure and ability to function to one fundamental chemical unit: deoxyribonucleic acid or DNA. Encoded in the structure of a DNA molecule are our genes, which in turn are knitted into chromosomes. DNA is made of units called nucleotides that come in only four different types—adenine (A), guanine (G), cytosine (C), and thymine (T); these form the letters of the genetic code. At a basic level, a gene is a DNA sequence that codes for a protein. If genes are a cell's instruction code that needs to be "read," proteins are the cell's workers, performing a vital job in keeping our cells functioning. Proteins are building blocks to make enzymes, which oversee every single chemical reaction in the human body.

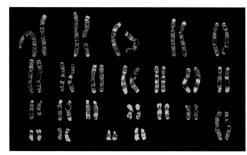

KARYOTYPE
DNA in a higher organism is organized into chromosomes; a full set is called a karyotype. This light micrograph shows a human set of 46 chromosomes arranged in 23 pairs from a female (the XX chromosomes are at bottom right).

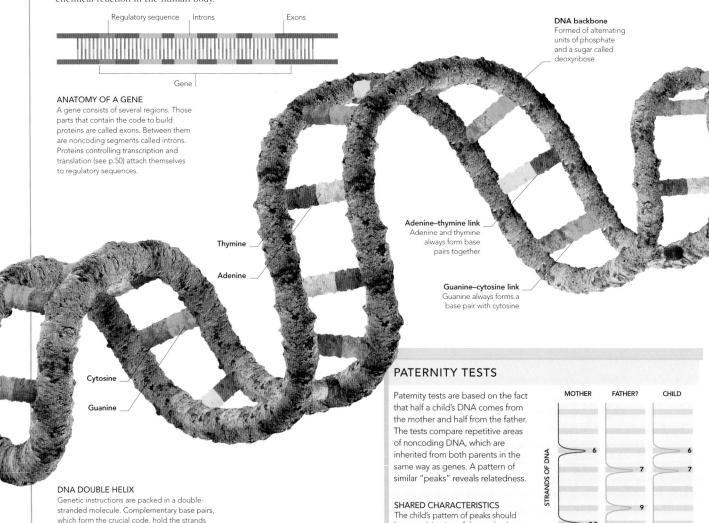

Regulatory sequence | Introns | Exons

Gene

ANATOMY OF A GENE
A gene consists of several regions. Those parts that contain the code to build proteins are called exons. Between them are noncoding segments called introns. Proteins controlling transcription and translation (see p.50) attach themselves to regulatory sequences.

DNA backbone
Formed of alternating units of phosphate and a sugar called deoxyribose

Thymine

Adenine

Adenine–thymine link
Adenine and thymine always form base pairs together

Guanine–cytosine link
Guanine always forms a base pair with cytosine

Cytosine

Guanine

DNA DOUBLE HELIX
Genetic instructions are packed in a double-stranded molecule. Complementary base pairs, which form the crucial code, hold the strands of DNA together by weak bonds that are easily broken when the sequence of bases is read. Until required, the DNA is coiled tightly in the cell's nucleus in a mesh called chromatin.

PATERNITY TESTS

Paternity tests are based on the fact that half a child's DNA comes from the mother and half from the father. The tests compare repetitive areas of noncoding DNA, which are inherited from both parents in the same way as genes. A pattern of similar "peaks" reveals relatedness.

SHARED CHARACTERISTICS
The child's pattern of peaks should be a combination of the peaks shown by both mother and father. Unknown peaks might suggest a different father.

MOTHER | FATHER? | CHILD

STRANDS OF DNA

6

7

9

9.3

6

7

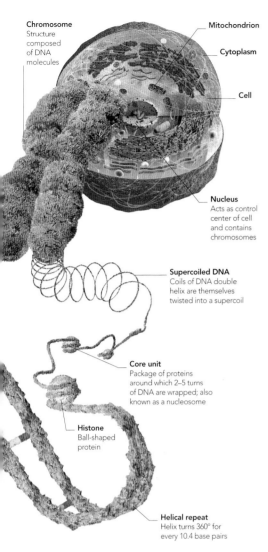

Chromosome
Structure composed of DNA molecules

Mitochondrion

Cytoplasm

Cell

Nucleus
Acts as control center of cell and contains chromosomes

Supercoiled DNA
Coils of DNA double helix are themselves twisted into a supercoil

Core unit
Package of proteins around which 2–5 turns of DNA are wrapped; also known as a nucleosome

Histone
Ball-shaped protein

Helical repeat
Helix turns 360° for every 10.4 base pairs

THE HUMAN GENOME

A genome is an organism's entire genetic code. Starting in 1990, international teams of scientists raced to decode the entire three billion bases or letters of the human genome—an undertaking known as the Human Genome Project—in the hope that once scientists could read an individual's DNA, a better understanding of human health and disease might follow. Conditions such as Alzheimer's, cancer, and heart disease could be tackled differently, and individually tailored drugs could become a reality. The specific sequence of bases differs between individuals, so the Human Genome Project used several anonymous donors to obtain an average sequence. The project was officially completed in 2003. It is thought that there are 20,000–23,000 genes, although there are still small parts of the genome where the sequence of bases is unknown.

CHROMOSOME SEVEN
Chemical staining gives chromosomes a banded appearance, which can be used to map gene locations. Chromosome seven, shown here, contains about 5 percent of the total DNA in human cells.

OPNISW gene
Active in retinal cells, it is needed for color vision, enabling sight at the blue-violet end of the spectrum

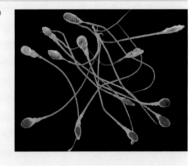

DFNA5 gene
Codes for the DFNA5 protein, which is considered important in the functioning of the cochlea in the inner ear, a structure required for normal hearing

DDC gene
Produces an enzyme in the brain and nervous system that is critical for making two of the brain's neurotransmitters —dopamine and serotonin

KRIT1 gene
Unclear role but plays a part in the development and formation of blood vessels and related structures, including the blood–brain barrier

SHH gene
Produces a protein called sonic hedgehog in the embryo; has a role in the formation of the brain, spinal system, limbs, and eyes

SELECTING GENDER

Men have the final say on gender because each sperm carries only an X or a Y chromosome. It is unclear if gender can be influenced naturally but conditions at conception may play a part. Gender can be manipulated by using sperm sorting to enrich semen with the desired sperm, or selecting embryos during in-vitro fertilization for implantation. Choosing gender for nonmedical reasons is illegal in some countries.

X AND Y SPERM
This electron micrograph has been color coded to show that semen contains virtually equal numbers of X and Y sperm.

SEX DETERMINATION

What is that makes a boy a boy and a girl a girl? Sex is determined by specialized chromosomes known as X and Y. The X chromosome is much longer than the Y chromosome and carries many more genes. These two chromosomes form a pair, sometimes called chromosome 23. In females, both chromosomes in the pair are X, resulting in XX. Males have one X and one Y, which gives them an XY identity. Genes on these chromosomes switch on and off vital processes that make a person male or female. For example, a crucial gene on the Y chromosome known as SRY is responsible for a fetus developing into a male. There are other genes on the Y chromosome thought to be involved in male fertility. Since females have two Xs, one is usually randomly deactivated in the early embryo.

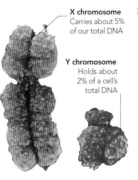

X chromosome
Carries about 5% of our total DNA

Y chromosome
Holds about 2% of a cell's total DNA

SEX CHROMOSOME
Chromosome 23 is made up of either two XXs (female) or an X and a Y (male). The X chromosome has up to 1,400 genes; the Y has only 70 to 200.

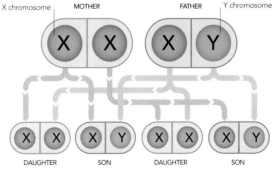

X chromosome MOTHER FATHER Y chromosome

DAUGHTER SON DAUGHTER SON

BOY OR GIRL?
The gender of a baby is determined by the father's sperm. If the sperm that fertilizes the egg has a Y chromosome, the offspring is a boy; an X chromosome produces a girl. A mother donates either one of her X chromosomes to the child.

HOW DNA WORKS

DNA is the master molecule that orchestrates everything that happens in the body's cells. One of its important functions is to replicate itself in order to create new body cells and sex cells, which allow DNA to perpetuate.

TRANSCRIPTION AND TRANSLATION

Before the DNA blueprint can be read, its instructions must first be transcribed into a form that can be decoded. Information from the DNA is copied to form an intermediate type of molecule called messenger RNA (mRNA). The mRNA then moves from the nucleus to protein assembly units called ribosomes. The mRNA acts as a template for the formation of subunits of proteins called amino acids, a process called translation. Their order is specified by lengths of mRNA three bases long called codons.

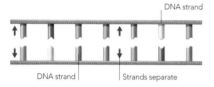

DNA strand

DNA strand   Strands separate

1 SEPARATION
Enzymes separate the two DNA strands. One of these unzipped strands acts as the template for a molecule known as messenger RNA (mRNA), which performs a temporary function in this process.

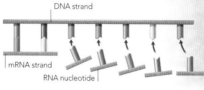

DNA strand

mRNA strand

RNA nucleotide

2 TRANSCRIPTION
Units or nucleotides of mRNA then attach to match the complementary molecules of the DNA code, forming a chain copy of the code, then the mRNA and DNA chains break apart.

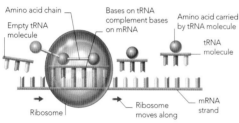

Amino acid chain

Empty tRNA molecule

Bases on tRNA complement bases on mRNA

Amino acid carried by tRNA molecule

tRNA molecule

Ribosome

Ribosome moves along

mRNA strand

3 TRANSLATION
The message is translated into a protein chain outside the nucleus at a ribosome, where small transfer RNAs (tRNAs) collect amino acids coded for by each codon.

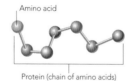

Amino acid

Protein (chain of amino acids)

4 FORMING A PROTEIN
Amino acids link to form a protein chain. Their sequence dictates the unique 3D shape of the protein, which is crucial to the function of the protein.

THE FORMATION OF NEW CELLS
Cells are dividing all the time, so it is crucial that genomes are copied and split correctly. The average cell divides over 50 times before it reaches the end of its life.

Cell membrane
Splits as the cell begins the process of division

Threads of spindle
Connect the center of each chromosome

Centriole
Made of hollow tubules; duplicates prior to cell division

Organelles
Specialized structures within the cell's cytoplasm, which pull apart during cell division

MITOSIS

The human body constantly produces new cells for various purposes: to replace old cells lost through wear and tear or that have reached the end of a finite life span; to increase numbers of cells for particular jobs such as making more immune cells to fight an infection; or simply to grow as muscle tissue is boosted or as children grow in stature. To produce these new cells, cells must replicate themselves exactly—and that means copying DNA instructions with extreme precision. This is done by a process known as mitosis, by which cells create a second set of identical chromosomes, temporarily doubling their DNA as they grow. Just before they divide, the two sets pull apart neatly and precisely so that each new cell has its own perfectly copied blueprint.

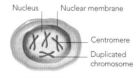

Nucleus   Nuclear membrane

Centromere

Duplicated chromosome

1 PREPARATION
Before mitosis, the parent cell grows and duplicates its genetic material by forming paired chromosomes.

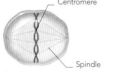

Centromere

Spindle

2 ALIGNMENT
The cell's nucleus vanishes. Paired chromosomes (chromatids) align on a scaffoldlike structure (the spindle).

Single chromosome

3 SEPARATION
The spindle's opposing poles pull apart the chromatids, doubling the parent cell's chromosome number.

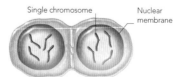

Single chromosome

Nuclear membrane

4 SPLITTING
The new cells pinch apart. Each takes an equal share of the chromosomes, which become enclosed in separate nuclei.

Nucleus

Chromosome

5 NEW CELLS
Two identical cells are formed with a full complement of 46 chromosomes. The cell's chromosomes then "rest" in a coiled form (chromatin) in its nucleus until it divides again.

Cleavage
Point at which the cell begins to divide

Chromosome
Contains most of the cell's genetic material

Centromere
Point at which the chromosome pair splits to form single chromosome

MEIOSIS

A special type of cell division is used to produce sex cells (eggs and sperm). A person inherits half their DNA from each parent, so sex cells are exceptional in that they contain only half the DNA of other cells. Both an egg and sperm cell each contains 23 chromosomes, which come together as a full 46-chromosome set when they fuse to form an embryo. Sex cells are also exceptional in that the chromosomes inherited from each parent are far from identical copies. Instead, the genes on the chromosomes are shuffled like a deck of cards by a process called genetic recombination.

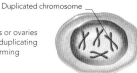
Duplicated chromosome

1 PREPARATION
Parent cells in the testes or ovaries grow, doubling in size and duplicating their genetic material by forming double chromosomes.

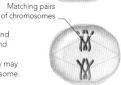
Matching pairs of chromosomes

2 PAIRING
Identical copies of maternal and paternal chromosomes pair up and are interwoven in the process of recombination, during which they may swap genes or pieces of chromosome.

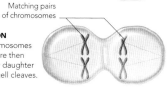
Matching pairs of chromosomes

3 FIRST SEPARATION
The paired chromosomes (sister chromatids) are then pulled into two new daughter cells as the parent cell cleaves.

Duplicated chromosome

4 TWO OFFSPRING
The daughter cells are genetically nonidentical to the parent cell, but each has a set of 46 chromosomes that must be halved to make a sex cell.

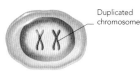
Single chromosome

5 SECOND SEPARATION
The nucleus disappears and the spindle reappears to pull the sister chromatids apart into four new cells. There is no doubling of genetic material at this stage.

Spindle

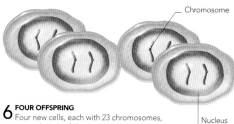
Chromosome

6 FOUR OFFSPRING
Four new cells, each with 23 chromosomes, are created. Each of these cells is genetically unique, containing a random mix of the genes from the original chromosomes (see panel, left).

Nucleus

GENETIC RECOMBINATION

Genes are shuffled randomly during the "pairing" stage of meiosis by a process known as recombination. Each cell has two copies of every chromosome—one from each parent. During recombination, pairs of these chromosomes come together in a process called "crossing over." The pairs intertwine, exchanging pieces of DNA.

CROSSING OVER
Chromosome pairs exchange as little as a few genes, or as much as a whole arm in this genetic lottery, ensuring that gene combinations are mixed up in sex cells.

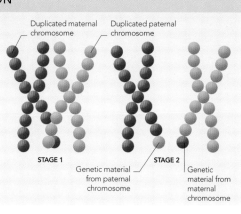

Duplicated maternal chromosome

Duplicated paternal chromosome

STAGE 1

STAGE 2

Genetic material from paternal chromosome

Genetic material from maternal chromosome

PATTERNS OF INHERITANCE

How can people share a great-uncle's nose or a quirky sense of humor? Patterns of gene inheritance help us understand this—although nurture also determines how such traits are expressed.

THE FAMILY TREE

DNA is shuffled randomly from one generation to the next, but there are rules and basic mathematics that reveal a lot about genetic relationships. A person shares half of his or her DNA with each parent, and each of the parents has half of each of their parents' genes. This means that an individual shares a quarter of the genes with each grandparent. Although siblings are different from each other, they share about half their genes. The closest genetic relationship is between identical twins, in whom 100 percent of genes are the same. In contrast, only 12.5 percent of genes are shared with a first cousin.

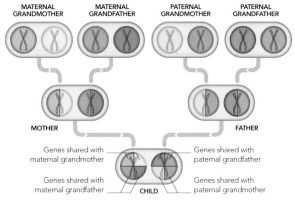

SHARED GENES
The share of common genes is halved in each new generation. Every person inherits half of the genes from each parent, and passes on half of these to the children.

HUMAN DIVERSITY
Humans are amazingly diverse, a fact that is due to both genetic inheritance and variation in environmental factors.

ATTACHED OR UNATTACHED EARLOBES?
Whether earlobes hang free or attach to the sides of the face is thought to hinge upon a single gene, although some scientists have recently suggested that the issue is more complex than that.

SINGLE- AND MULTIPLE-GENE INHERITANCE

A gene comes in different forms (alleles) and, for any particular gene, one allele is inherited from each parent. The gene's expression in the offspring depends on the combination of alleles, and whether or not the gene governs one trait by itself or works in combination with others. The simplest type of inheritance is where one gene is responsible for one trait—for example, some diseases, such as Huntingdon's, are carried on single genes. Typically, an allele can be dominant or recessive. When there is a dominant copy from one parent paired with a recessive copy from the other parent, the dominant one will manifest—only one copy is needed. The recessive trait is expressed only when a person inherits two copies of a recessive allele. But many traits, such as eye color, are governed by many genes and so, although their inheritance works on the same principles as single inheritance, it is harder to predict the outcome.

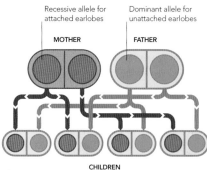

DOMINANT AND RECESSIVE GENES
This chart shows a possible combination of genes in the inheritance of earlobe shape. To manifest the recessive trait, a person needs two recessive alleles for this gene. Here, the children have unattached lobes but carry the recessive trait, so some of their children could have attached earlobes.

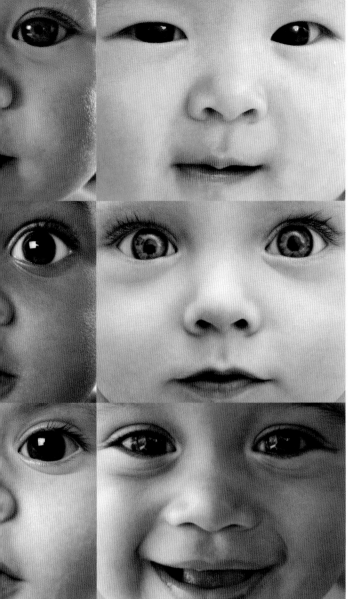

SEX-LINKED INHERITANCE

Some genes for non-sex functions are on the sex chromosomes X and Y. How they are passed on depends on which of these chromosomes they are located, and also whether the alleles are dominant or recessive on that chromosome. For example, because males are XY and carry only one X, they will inherit any X-linked genes and pass these on to all of their daughters, but not their sons. The daughters will all be "carriers" if the allele is recessive, and affected if it is dominant. Females have two Xs, and in any given cell one is randomly deactivated, so they rarely manifest X-linked recessive disorders because they usually have a normal backup copy of the allele active in other cells.

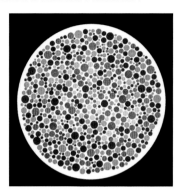

COLOR BLINDNESS
This largely red image with the number 74 appearing in green is a classic test for color blindness. Color blindness is an X-linked recessive disorder, so it is much more common in men than women.

GENES AND THE ENVIRONMENT

Many of the traits people have are honed by a complex, often shifting interchange between genes and environment—the "nature versus nurture" debate. Traits such as personality, intelligence, and height are on a continuum. How people turn out depends on the genetic hand dealt to them by their parents as well as their external environment, such as family upbringing, socioeconomic status, nutrition, physical circumstances, and emotional environment. Many diseases, such as depression, heart disease, schizophrenia, and cancer, may have both genetic and environmental components. So genetics might make someone susceptible to a condition, then negative or positive environmental factors can tip the balance either way. Studies of identical twins have probed the question of what proportions of particular traits are heritable.

INHERITING INTELLECT?
About half the variation in human IQ may be down to genetics. But whether or not a child's genetic promise is fulfilled depends on nurture. A child with a less fortunate hand may make up the difference in a good environment.

TRANSGENERATIONAL INHERITANCE

In recent years, scientists have found that the pattern of how genes are switched on and off in the body in response to the environment—the science of "epigenetics"—is also inherited. This means that changes in gene expression from our grandparents in response to their environment can be inherited. For example, studies have suggested that famine affects the expression of certain genes, which is thought to cause obesity in subsequent generations.

Methyl group
A hydrocarbon, which attaches itself to DNA; huge numbers of these can silence a gene

SWITCHING OFF GENES
Genes can turn off when methyl groups attach to bases. Heavily methylated areas of DNA have been shown to be inactive. This pattern of methyl groups can be inherited.

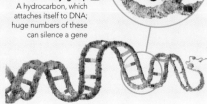

ORIGINS OF CHARACTER TRAITS		
ENVIRONMENTAL	**INTERACTIONAL**	**GENETIC**
• Specific language (entirely)	• Height	• Blood group (entirely)
• Specific religion (entirely)	• Weight	• Eye color (entirely)
• Specific culture (entirely)	• Intelligence	• Hair color (entirely)
• Sensitivity to environment-based stresses like radiation (mostly environmental)	• Personality	• Certain genetic diseases, such as Huntingdon's (entirely)
	• Certain multifactorial diseases like heart disease	• Baldness (mostly genetic)

GENETIC PROBLEMS AND INVESTIGATIONS

DNA replicates itself millions of times and with amazing accuracy over a lifetime but, sometimes, things can go wrong.

WHERE GENETIC PROBLEMS ARISE

When there is a change in DNA (due to an internal error in the normal functioning of a cell or an attack from an external environmental cause, known as a mutagen), problems can arise on three main levels. In the first level, a change in a gene affects the protein for which it codes. The next level sees a change in the number of chromosomes. Thirdly, problems can occur when there are alterations in several genes plus environmental triggers. There is a fourth level, affecting mitochondrial DNA, but this is unusual compared with the other three levels.

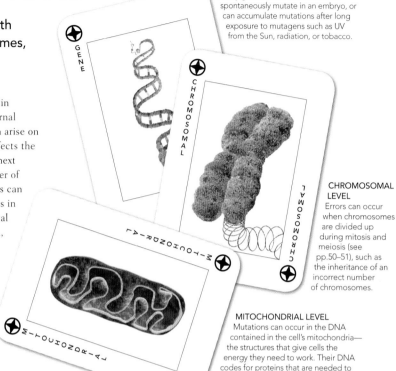

GENE LEVEL
Faulty genes can be inherited, can spontaneously mutate in an embryo, or can accumulate mutations after long exposure to mutagens such as UV from the Sun, radiation, or tobacco.

CHROMOSOMAL LEVEL
Errors can occur when chromosomes are divided up during mitosis and meiosis (see pp.50–51), such as the inheritance of an incorrect number of chromosomes.

MULTIFACTORIAL LEVEL
Some diseases are influenced by mutations in a number of genes plus environmental factors that affect susceptibility. For example, Alzheimer's disease and breast cancer are multifactorial in their origins.

MITOCHONDRIAL LEVEL
Mutations can occur in the DNA contained in the cell's mitochondria—the structures that give cells the energy they need to work. Their DNA codes for proteins that are needed to keep mitochondria working properly.

MUTATIONS

Any permanent alteration in the code of a DNA sequence is known as a mutation. It can be as small as a one-"letter" change in a gene or as large as a chunk of chromosome. The effect of a chromosomal mutation depends on the size and location of the structural change and whether any DNA is lost. It usually occurs either in the sperm or eggs, or early in embryonic development. Gene mutations can be inherited or they can occur spontaneously in an embryo. But, often, they occur in body cells when the intricate system for replicating DNA slips up somewhere. A gene mutation can have a negative effect if it impairs the normal functioning of a gene.

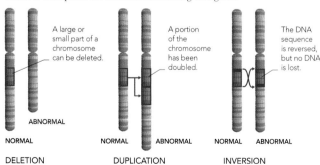

A large or small part of a chromosome can be deleted.

A portion of the chromosome has been doubled.

The DNA sequence is reversed, but no DNA is lost.

DELETION
A chromosome segment can break off. The effect on function depends on the amount of genetic material deleted and its function.

DUPLICATION
A segment of chromosome can be copied more than once erroneously. That segment may then be duplicated several times.

INVERSION
A chromosome can break in two places, and the "lost" portion can be reinserted, but the wrong way around. Usually, no DNA is lost.

DIFFERENT TYPES OF GENE MUTATION

Gene mutations can be caused by particular types of errors. The way in which they affect the gene's function depends on if and how they alter the "reading" of the DNA code, and if any subsequent changes affect the protein made by the gene.

TYPE OF MUTATION	CORRECT CODE	FAULTY CODE
Frameshift mutation DNA is read in "frames" of three letters that translate into amino acids. A mutation shifts this frame and will change the amino acids.	CAT CAT CAT CAT ↑ Three-letter frame	AT**C** **AT**C ATC ATC ↑ Sequence shifted one frame to the right, so CAT becomes ATC
Deletion mutation Any small or large loss of the DNA bases or letters of a gene is a deletion.	CAT CAT CAT	CAT CTC ATC ↑ "A" removed
Insertion mutation Any insertion of extra DNA, from a single unit (nucleotide) to larger pieces, could potentially disrupt the function of a gene.	CAT CAT CAT CAT	CAT CAT **A**CA TCA ↑ "A" inserted
Increased repeat mutation This is a type of insertion mutation that adds in a short repeating DNA sequence that can impair gene function.	TAG GCC CAG GTA	TAG GCC CAG **CAG** ↑ CAG frame repeated
Missense mutation One letter of the code is swapped for another, introducing the sequence codes for a different amino acid than intended.	CAT CAT CAT	CAT CAT C**C**T ↑ "C" incorrectly added instead of "A"

GENETIC COUNSELING

A person with a genetic disease in the family, such as cystic fibrosis or some cancers, might visit a genetic counselor to seek advice on the risk of developing the disease or passing it onto their children. A genetic counselor offers guidance on ways to prevent the disease if there are environmental components, testing family members if appropriate, or treatment options, if available. Pregnant women may visit genetic counselors if they have had an abnormal prenatal-test result. Parents of children with medical or learning difficulties where a genetic condition may be involved may also attend for an assessment. A genetic counselor can also give information on the chances of an unborn child carrying a potentially problematic gene. The genetic counselor informs the mother-to-be of what a genetic test result during pregnancy means and outlines the options for treatment and management of the condition.

MEDICAL FAMILY TREE
To assess the risk of developing an inherited disorder, a genetic counselor takes a detailed medical history of a patient and his or her family's health to produce a family tree like this one.

KEY
- ■ AFFECTED BY CANCER
- □ NOT AFFECTED BY CANCER

DIED OF UNKNOWN CANCER

DIED FROM BOWEL CANCER

DIED FROM BOWEL CANCER

DIAGNOSED WITH ENDOMETRIAL (UTERINE) CANCER

THIS WOMAN IS VISITING A GENETIC COUNSELOR

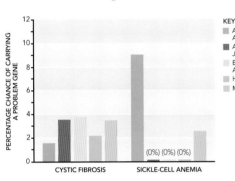

KEY
- ■ AFRICAN-AMERICAN
- ■ ASHKENAZI JEW
- ▦ EUROPEAN-AMERICAN
- ■ HISPANIC
- ▦ MEDITERRANEAN

(0%) (0%) (0%)

CYSTIC FIBROSIS   SICKLE-CELL ANEMIA

PERCENTAGE CHANCE OF CARRYING A PROBLEM GENE

GENETIC DISORDERS AND ETHNICITY
Some ethnic groups are much more likely to carry problem genes than others. This graph shows that Americans of African descent have a much higher chance (9 percent) of carrying sickle-cell anemia than any of the other ethnic groups tested.

GENETIC SCREENING AND TESTING

Genetic tests for disorders are performed early in pregnancy, or in newborns, to catch disorders that can be treated early in life (such as phenylketonuria), or later in life to screen for disease-susceptibility genes (such as BRCA1 for breast cancer) before any symptoms have occurred. Prenatal tests include amniocentesis, which involves taking fluid from the amniotic sac, which contains free-floating cells from the fetus. These cells are examined for chromosomal abnormalities, and can pick up certain conditions, for example Down syndrome.

PREIMPLANTATION TESTS
Where the risk of a serious genetic condition is high, tests may be carried out in some countries on embryos fertilized in a laboratory to select a healthy embryo for implantation.

"SAVIOR" SIBLING

Occasionally, embryos are selected for implantation to create "savior siblings" for existing children with severe, life-threatening disorders, such as Diamond blackfan anemia. Using preimplantation genetic tests, they are chosen on the basis of being disease-free themselves and being able to provide a tissue match for their sibling. When these children are born, stem cells from their umbilical cords or bone marrows may be used to treat their older siblings.

BORN TO CURE
In 2003, the parents of Zain Hashmi (pictured) won a legal battle in the UK to attempt to conceive a tissue-matched healthy sibling for Zain to help cure his debilitating beta-thalassemia.

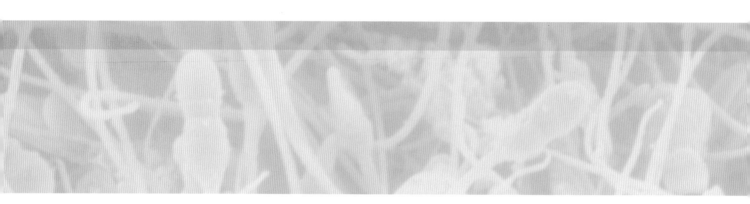

THE EVENTS THAT LEAD UP TO THE BEGINNING OF
A PREGNANCY ARE MORE COMPLEX THAN THEY MAY
APPEAR. IN HUMANS, SEX STARTS WITH AN INTERPLAY
BETWEEN SENSORY STIMULI AND HORMONES, WHICH
RESULTS IN ATTRACTION. DESIRE, AROUSAL, AND
ORGASM FOLLOW, AS THE GENITALS AND THE BRAIN
CONTINUOUSLY COMMUNICATE WITH EACH OTHER VIA
THE NERVOUS SYSTEM. HUMANS DIFFER FROM MOST
OTHER ANIMALS IN HAVING SEX FOR PLEASURE AND
NOT JUST FOR PROCREATION. THE NEED TO AVOID THE
UNWANTED PREGNANCIES THAT OFTEN ARISE FROM
THIS PLEASURABLE ACT HAS LED TO THE DEVELOPMENT
OF VARIOUS FORMS OF CONTRACEPTION.

THE SCIENCE
OF SEX

THE EVOLUTION OF SEX

The word "sex" is used to distinguish males from females, and can also mean the act of reproduction. Evolution is involved in both of these definitions of sex, allowing species to adapt to their environment and maximize the spread, and therefore survival, of their genes.

WHAT IS SEX?

The sex of a human is obvious from the external genitalia, but for many animals gender can only be determined by either the sex chromosomes or the size of the sex cells (gametes). Females usually have the larger sex cells (eggs) and males the smaller (sperm), yet early in the evolution of organisms sex cells of the same size combined to produce offspring. Differing sizes are thought to have evolved because some gametes found it advantageous to become smaller and quicker, while others had to enlarge to produce offspring of the same fitness.

Egg
Large and relatively immobile cell

Sperm
Small cell trades size for swimming ability

RELATIVE SIZE OF SEX CELLS
Some species, such as yeast, still reproduce by the fusion of equal-sized gametes, but many organisms that evolved more recently have much larger female sex cells than male sex cells.

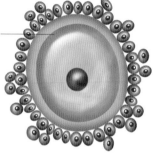

WHY HAVE SEX?

The primary reason for sex is to make new genetic copies of ourselves—offspring are the only way for our genes to survive. Many animals will only have sex during a female's fertile period, yet humans, and some other species such as dolphins, also have sex for pleasure. This human instinct may have evolved to help bond men and women as a couple ("pair bonding"), which would have been vital in the past when it was harder for one person to look after a baby. Sex triggers the release of oxytocin from the pituitary gland; this hormone is thought to play a key role in pair bonding.

GENETIC COPIES
Offspring are a way of a parent's genes surviving, yet two parents can each give only 50 percent of their genes to a child.

SEX FOR PLEASURE
Humans have engaged in sex for pleasure for much of their evolutionary history. Arstistic depictions are common, such as this ancient Greek erotic scene.

SPERM COMPETITION

Only the female can guarantee that a baby is hers; males have no such security. To ensure that they have a better chance of fertilizing the female's egg, males must ensure that their sperm are fitter than a potential rival's. Certain animals, including some butterflies, produce two types of sperm; one that fertilizes and another that helps it along (accessory sperm). Producing more sperm can also help to ensure successful fertilization. Promiscuous species produce more sperm and therefore have larger testes. Humans are more promiscuous than some other apes, such as gorillas, so male humans have proportionally larger testes than male gorillas.

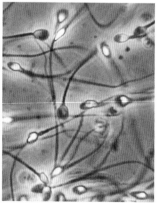

SURVIVAL OF THE FITTEST
There are normally over 15 million sperm per milliliter of human semen. Each sperm races the others to be the one to fertilize the egg, and only the fittest sperm will win.

HERMAPHRODITES

Humans born with dual genitalia, due to rare hormonal disturbances, can appear to be both male and female. Yet they cannot use both sets of genitals to reproduce. True hermaphrodites have both male and female reproductive organs and are able to fertilize one another. This is an evolutionary advantage for animals such as slugs or snails in which individuals are solitary and rarely meet, so a single encounter carries double the chance of successful reproduction.

ASEXUAL REPRODUCTION

Some organisms reproduce asexually by making copies of themselves. There are a variety of different methods of asexual reproduction (see right), but each process bypasses the need for fertilization, making it far quicker to produce offspring than by sexual reproduction. Offspring are therefore genetically identical to their parents. Reproducing asexually does not create the genetic variety ideal for overcoming environmental change, but for many organisms it is a successful strategy. It is best suited to organisms that face little competition from more adaptable species or that live in environments that see little change.

ADVANTAGES AND DISADVANTAGES

Asexual reproduction is most common in single-celled organisms, such as bacteria, but many plants and fungi, as well as larger animals such as the whiptail lizard, use it too.

Advantages	• No need to look for a mate • Energy can be devoted to making new copies • Fast reproductive method • Parent genes are not diluted by those of a partner
Disadvantages	• No genetic variation (bad genes persist) • No adaption to environmental change

CLONING
Some animals, such as corals, can reproduce by producing exact genetic copies (clones) of themselves. Corals can also use sexual reproduction.

REGENERATION
This is the result when an animal forms from a broken-off fragment of a parent. Starfish can develop in this way, but only if part of the center is included in the fragment.

PARTHENOGENESIS
Parthenogenesis is the development of an offspring from a female egg that has not been fertilized by a male. Whiptail lizards reproduce in this way.

SEXUAL REPRODUCTION

Sexual reproduction occurs when a male and female combine the genes contained within their sex cells through fertilization. This does not need to involve penetrative sex: some fish sex cells combine in the water, outside the female's body. All sex cells are haploid, meaning they have half the correct number of chromosomes; they then combine to form a diploid cell containing the full complement. Reproducing sexually creates offspring with huge genetic variety, enabling natural selection to occur. As environments change, individuals with genes that help in their new environment adapt and survive; those without them die off. This makes organisms that reproduce sexually more likely to be able to evolve over time.

COMBINING SEX CELLS
Parent cells with 46 chromosomes divide by meiosis to produce haploid sex cells with only 23. One sex cell from each parent combines to produce a diploid offspring cell; this divides by mitosis to form an organism (see pp.50–51).

Haploid sex cells
sperm and egg cells; each have one set of 23 chromosomes

Meiosis → Fertilization

Multicellular organism
capable of producing sex cells

Diploid zygote
contains 2 sets of 23 chromosomes

Mitosis

ADVANTAGES AND DISADVANTAGES

Sexual reproduction is currently a major form of reproduction among organisms. It is primarily, but not exclusively, seen in members of the animal kingdom.

Advantages	• Two parents create genetic variation • Species adapt easily to changes in their environment • Less chance of genetic disease
Disadvantages	• Time must be invested in finding a mate • Fertilization is not always successful • Parents can only pass on 50 percent of their genes

LACTOSE TOLERANCE
Humans only started consuming dairy produce in recent evolutionary history. In early societies, some people had genes that allowed them to digest lactose—the sugar found in milk. When they began to farm dairy animals, these people thrived and their lactose-tolerant genes became prevalent. In societies that have not traditionally farmed dairy animals, lactose intolerance is common.

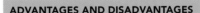

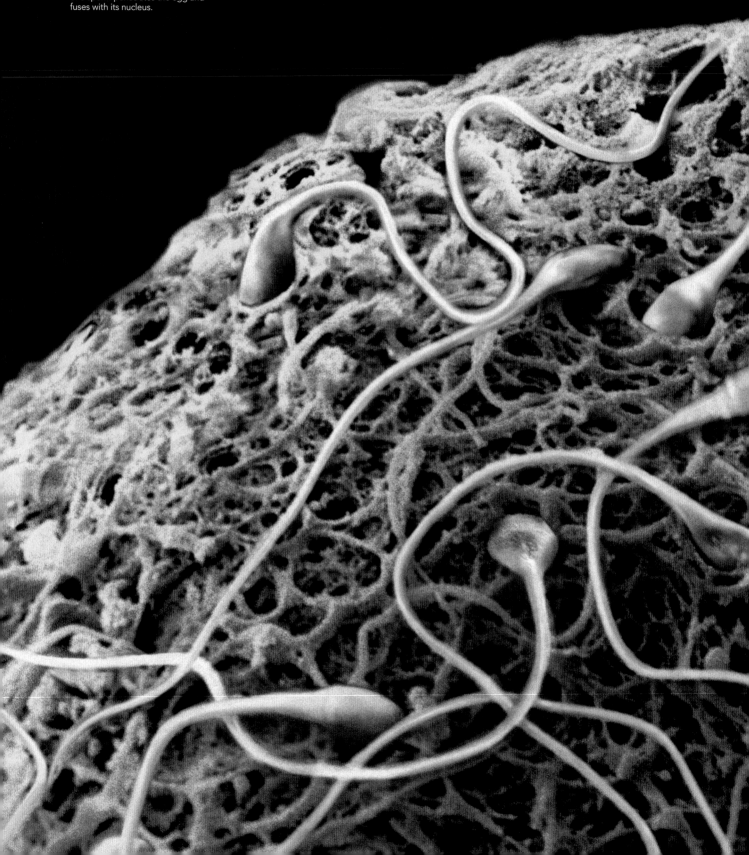

POINT OF FERTILIZATION
This electron micrograph shows
tadpole-like sperm surrounding
the much larger egg. Fertilization,
which occurs in one of the fallopian
tubes, takes place when the head
of a sperm penetrates the egg and
fuses with its nucleus.

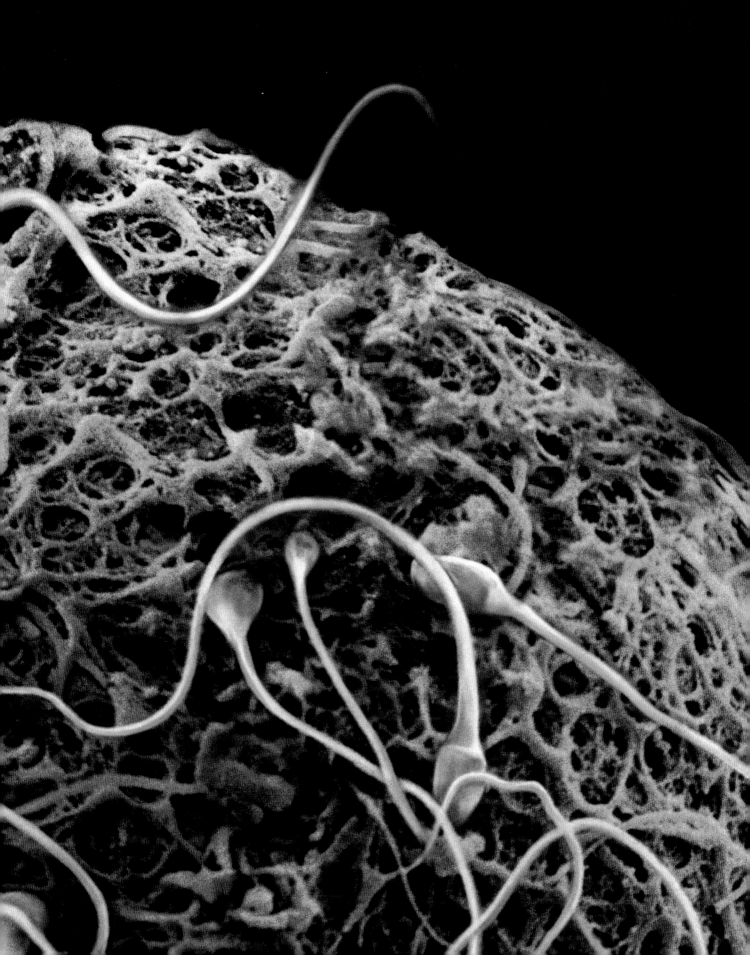

ATTRACTIVENESS

Sexual attraction is often assumed to be an inexplicable instinct, yet the interaction of many factors lies behind this seemingly mysterious chemistry. Chemical cues, thought to be pheromones, add to hormonal effects, visual cues, and other as yet unidentified factors, which lead us to be attracted to others.

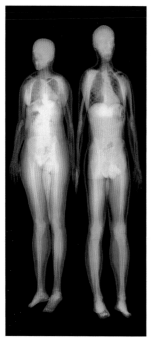

PROPORTIONAL BODY SIZES
In promiscuous mating systems, or those where long-term pair bonds form, such as in humans, males and females are similar in appearance.

HOW MATING SYSTEMS AFFECT APPEARANCE

The environment in which animals live has had a large impact on the development of their mating systems, which, in turn, have strongly shaped their appearance. In environments that support many animals, a large group of females may be guarded by a single male. These males are often larger than the females and have developed weapons, such as large antlers, with which to fight other males for possession of the females. When the environment does not support a large group, and there is no advantage in fighting, some males attract females using showy physical attributes, such as colored feathers, to signal that they are fit to mate with.

SHOWING OFF
If a peacock's tail has more eye spots than a rival's, it signals to the female that the male is genetically fit, and will pass on good genes.

WEAPONS
Male red deer compete to mate with females. If a competing male is not scared off by a rival's appearance, fierce fighting ensues.

POSITIVE ASSORTATIVE MATING

Positive assortative mating is the inclination of organisms to select a mate that displays similar attributes to themselves. Humans, subconsciously, choose their partners in this way—people who are similar in appearance and intellect often tend to form couples. This instinct may have evolved in humans because it promotes long-term, stable relationships. These were necessary in early human evolutionary history because offspring had a better chance of survival if both parents were able to look after them.

PHYSICAL SIMILARITIES
It is easiest to observe positive assortative mating when looking at the physical similarities, such as race or body height, between partners in a couple.

THE MENSTRUAL CYCLE AND MATE CHOICE

Hormone fluctuations during the menstrual cycle affect how women rate the attractiveness of men. In their most fertile period (around ovulation), women tend to be attracted to men with highly masculine attributes who are most genetically different from them. This attraction is subconscious and thought to be because these men will produce offspring who are the most genetically fit. Yet at other phases of the menstrual cycle, women tend to be attracted to men who are genetically similar and less masculine, but more likely to enter into a partnership and look after any offspring. It seems that women have therefore evolved to try to mate with genetically fit men, and form long-term partnerships with those perceived to have a more nurturing nature.

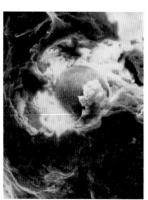

OVULATION AND ATTRACTION
This colored scanning electron micrograph shows the moment of ovulation when the egg (pink) is released. Around this time, women are subconsciously attracted to men who are most genetically fit and suitable to father offspring.

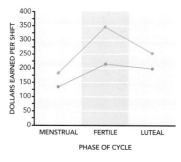

Y-axis: DOLLARS EARNED PER SHIFT (0, 50, 100, 150, 200, 250, 300, 350, 400)
X-axis: MENSTRUAL, FERTILE, LUTEAL
PHASE OF CYCLE

KEY
WOMEN NOT ON ORAL CONTRACEPTIVES
WOMEN ON ORAL CONTRACEPTIVES

"CONCEALED" OVULATION
A study shows how lap dancers get better tips when they ovulate. This suggests that subtle changes in behavior around ovulation enable men to determine whether women are in their fertile phase.

THE EFFECTS OF ORAL CONTRACEPTIVES

The oral contraceptive pill usually suppresses ovulation, meaning that the subtle cues that attract women to masculine, genetically dissimilar men during ovulation are disturbed. The long-term effects of this are not yet known. However, it might lead women to be more likely to produce offspring with men that are genetically similar to them, which could lead to less fit offspring. It could also have implications for relationship stability, because when a woman comes off oral contraceptives, she may start to view her partner in a different way.

PHEROMONES

Pheromones are chemicals that animals of the same species emit to communicate with one another. Some animals use them to mark an area of territory. Ants use them to set trails to guide other ants to food or alert them of danger. Pheromones play a role in mating. In many species, possibly including humans, they signal that a female is ready to mate; one study showed that men were more attracted to the clothing of ovulating women. Pheromones may also account for people being drawn to potential partners who are genetically different from themselves, allowing for maximum genetic diversity of any potential offspring.

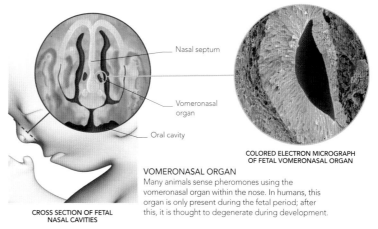

Nasal septum

Vomeronasal organ

Oral cavity

CROSS SECTION OF FETAL NASAL CAVITIES

COLORED ELECTRON MICROGRAPH OF FETAL VOMERONASAL ORGAN

VOMERONASAL ORGAN
Many animals sense pheromones using the vomeronasal organ within the nose. In humans, this organ is only present during the fetal period; after this, it is thought to degenerate during development.

FACIAL SYMMETRY

Facial features are rated as attractive based on whether they are seen as more masculine, for a man, or feminine, for a woman. Subconsciously, facial symmetry has an effect on the percieved masculinity or femininity of a face. People with more symmetrical faces, and those with faces characteristic of their sex, report fewer health problems, so facial characteristics may be a way of signaling this fitness to others. Only high-quality males or females are symmetrical and are seen to have more masculine or feminine facial features.

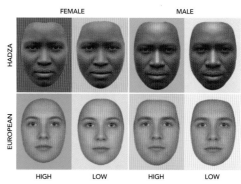

FEMALE   MALE

HADZA

EUROPEAN

HIGH   LOW   HIGH   LOW

HIGH AND LOW FACIAL SYMMETRY
These composite faces, made from photos of people from two ethnic groups, represent high and low symmetry for each group.

LINES OF SYMMETRY
To judge whether a face is symmetrical, people assess the distance from the middle of the face to points such as the eyes, margins of the face, and edges of the nose.

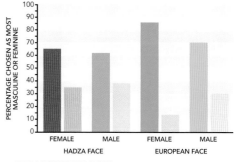

PERCENTAGE CHOSEN AS MOST MASCULINE OR FEMININE

FEMALE   MALE   FEMALE   MALE

HADZA FACE   EUROPEAN FACE

ATTRACTIVENESS RATING
In this study, those with high-symmetry faces are judged to be more masculine or feminine than those with low-symmetry faces.

KEY
HIGH SYMMETRY
LOW SYMMETRY

DESIRE AND AROUSAL

Desire and arousal are the conscious preludes to sex. To experience these basic human instincts requires complex interactions between the brain, nerve networks, and hormones, which coordinate the body's response to sensory and physical stimulation.

WHAT TRIGGERS DESIRE?

Sexual desire is usually instigated by the combined effect of numerous sensory desire cues. Sight, smell, sound, touch, and even taste all help trigger desire. Stimuli are detected by the peripheral nervous system, which transmits nerve impulses to the brain's somatosensory cortex where we "feel" these senses. Imagination and thoughts of reward, involving several areas of the brain collectively known as the limbic system, also play key roles in desire. Once the senses and imagination are stimulated, impulses from the relevant areas of the brain travel to the hypothalamus where they are processed, resulting in feelings of desire and arousal.

KISSING
A kiss is a highly effective trigger for desire. Involving lips and tongue (key erogenous zones), it requires close physical proximity and activates sensations of touch, taste, and smell.

Somatosensory cortex
The body's sensory system, located along the parietal lobe of the brain

Hypothalamus
Coordinates sensual stimuli to trigger desire and arousal

Genital area
Breast area
Lips and tongue area

BRAIN

Lips and tongue
Breasts
Genitals

Lips and tongue
Genitals

MALE
FEMALE

KEY EROGENOUS ZONES
These are densely packed with nerves that detect touch. The areas of the brain that process signals from these nerves are proportionate in size to the number of nerve endings in each erogenous zone.

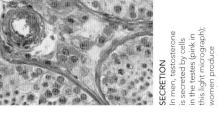

FLUCTUATIONS IN DESIRE

Levels of desire fluctuate throughout life. These fluctuations have many causes, including hormonal and psychological factors. For women, levels of desire regularly fluctuate with the short-term hormone variations of the menstrual cycle. The hormone testosterone is also linked to long-term desire in both men and women. Feelings of desire increase rapidly after puberty, when testosterone levels first rise, although both will decline with age. Levels of testosterone in men peak in the mid-thirties and slowly decline; in women, levels of all sex hormones fall sharply after menopause.

SECRETION
In men, testosterone is secreted by cells in the testes (pink in this light micrograph); women produce it in their ovaries.

Bleeding begins
During a period, sexual desire is often at its lowest point

28 0 DAYS
6

Premenstrual stage

Fertile time
Around ovulation (day 14), women experience a sharp rise in desire

12

15

MENSTRUAL CYCLE
Feelings of desire and arousal generally increase around the time of ovulation, which is when women are most likely to become pregnant.

AROUSAL PATHWAYS

Signals are carried between the brain and genitals by sensory nerves and nerves of the parasympathetic and sympathetic nervous system (part of the autonomic nervous system, which regulates internal processes). The signaling is coordinated by the hypothalmus, which sends signals down the spinal cord to interact with the parasympathetic nerves that carry signals to the genitals to trigger arousal. Sensory nerves travel back from the genitals to the spinal cord and relay messages of sexual pleasure. These act directly on the parasympathetic nerves to heighten genital arousal, including engorgement of erectile tissue, and also signal to the brain to boost arousal. This builds until a tipping point is reached, when the sympathetic nerves take over and cause orgasm.

SEXUAL RESPONSE
In both men and women, arousal is controlled by impulses traveling between the spinal cord and the brain. A complex interaction of nerve signals leads to arousal that can culminate in orgasm. To prevent arousal at inappropriate times, the pons (located in the brainstem) sends inhibitory signals via the sympathetic nerves.

KEY
SYMPATHETIC NERVE FIBERS
PARASYMPATHETIC NERVES FIBERS
PUDENDAL NERVE

1 BRAIN SIGNALS
Impulses from the hypothalamus pass down the spinal cord to instigate arousal in the genitals; pleasurable sensations are later relayed back to the brain. Inhibitory signals are sent by the pons.

Hypothalamus
Pons
Spinal cord

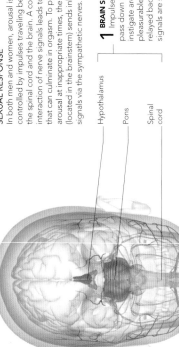

ENGORGEMENT

When arousal begins, the erectile tissue, which is found in the man's penis and the clitoris and labia in the woman, starts to fill with blood in response to signals sent along the parasympathetic nerve fibers. As the penis becomes engorged, it becomes erect and hard, which is necessary for penetration. Engorgement of the clitoris and labia heightens the pleasure a woman experiences from sex.

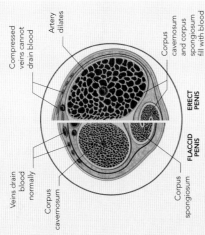

Compressed veins cannot drain blood

Artery dilates

Corpus cavernosum and corpus spongiosum fill with blood

ERECT PENIS

FLACCID PENIS

Corpus spongiosum

Veins drain blood normally

Corpus cavernosum

MALE ERECTILE TISSUE
During arousal, arteries that supply the penis dilate, allowing excess blood to engorge the spongy erectile tissue. Veins become compressed, preventing blood from leaving the penis; this maintains the erection.

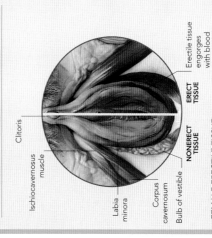

Erectile tissue engorges with blood

ERECT TISSUE

NONERECT TISSUE

Clitoris

Ischiocavernosus muscle

Labia minora

Corpus cavernosum

Bulb of vestibule

FEMALE ERECTILE TISSUE
Erectile tissue in women is similar to that in men, but the volume is far smaller. The clitoris becomes erect when blood engorges the corpus cavernosum. The external genitalia (vulva) also become engorged with blood during arousal.

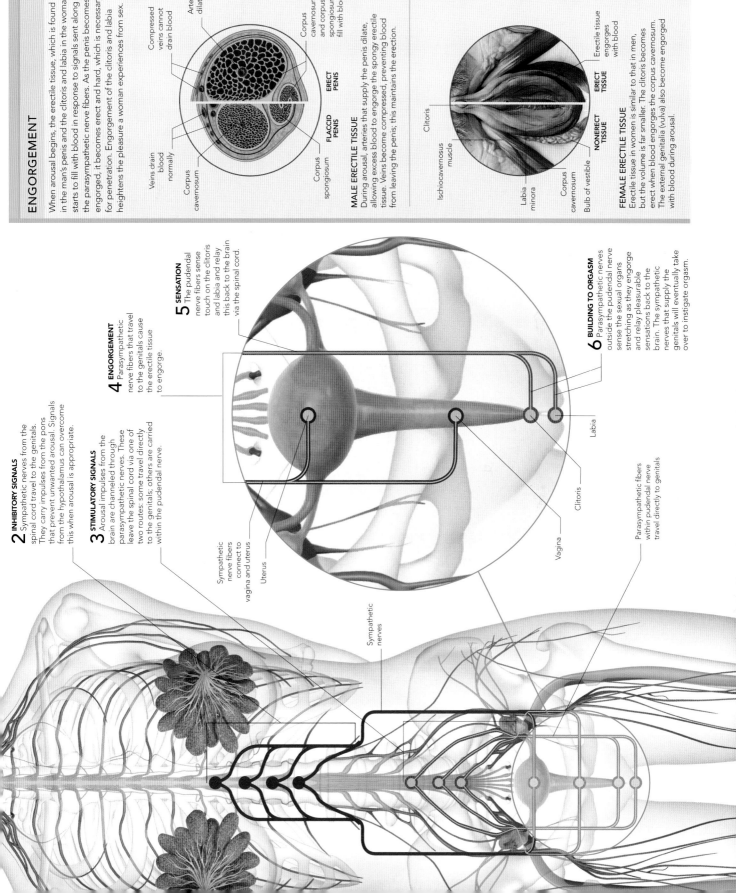

2 INHIBITORY SIGNALS
Sympathetic nerves from the spinal cord travel to the genitals. They carry impulses from the pons that prevent unwanted arousal. Signals from the hypothalamus can overcome this when arousal is appropriate.

3 STIMULATORY SIGNALS
Arousal impulses from the brain are channeled through parasympathetic nerves. These leave the spinal cord via one of two routes: some travel directly to the genitals; others are carried within the pudendal nerve.

4 ENGORGEMENT
Parasympathetic nerve fibers that travel to the genitals cause the erectile tissue to engorge.

5 SENSATION
The pudendal nerve fibers touch on the clitoris and labia and relay this back to the brain via the spinal cord.

6 BUILDING TO ORGASM
Parasympathetic nerves outside the pudendal nerve sense the sexual organs stretching as they engorge and relay pleasurable sensations back to the brain. The sympathetic nerves that supply the genitals will eventually take over to instigate orgasm.

Sympathetic nerve fibers connect to vagina and uterus

Uterus

Sympathetic nerves

Parasympathetic fibers within pudendal nerve travel directly to genitals

Vagina

Clitoris

Labia

THE ACT OF SEX

Humans engage in sexual intercourse to enable conception as well as for physical pleasure and emotional bonding. Most other animals, by contrast, use sex purely as a means of reproduction.

SEXUAL INTERCOURSE

Sexual intercourse usually involves penetration of the vagina by the penis. This requires the penis to be erect and to have enough lubrication to pass into the vagina easily and painlessly. Secretions from glands in the vagina lubricate the passage; accessory sex glands in the male genital areas, such as the bulbourethral glands, help lubricate the male urethra. The head of the penis (glans penis) contains hundreds of sensory nerve endings, which are stimulated as the penis moves into and out of the vagina.

This movement also stimulates nerve endings in the clitoris and vagina. Sexual pleasure builds and usually culminates in orgasm—a state achieved more easily by men than women.

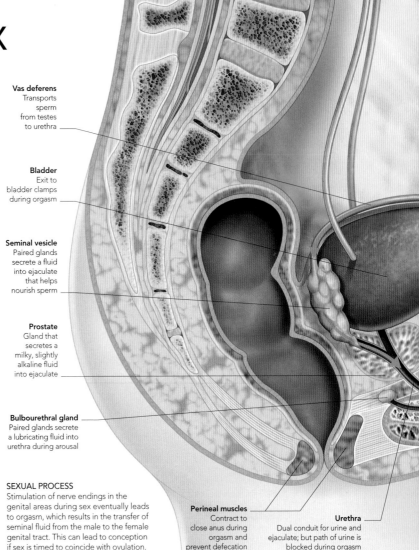

Vas deferens
Transports sperm from testes to urethra

Bladder
Exit to bladder clamps during orgasm

Seminal vesicle
Paired glands secrete a fluid into ejaculate that helps nourish sperm

Prostate
Gland that secretes a milky, slightly alkaline fluid into ejaculate

Bulbourethral gland
Paired glands secrete a lubricating fluid into urethra during arousal

Perineal muscles
Contract to close anus during orgasm and prevent defecation

Urethra
Dual conduit for urine and ejaculate; but path of urine is blocked during orgasm

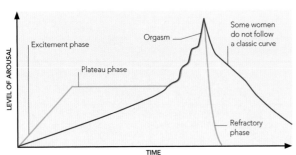

PENETRATION
This MRI scan taken during sex shows the penis mostly lies outside of the vagina. It can assume a boomerang shape during penetration.

SEXUAL PROCESS
Stimulation of nerve endings in the genital areas during sex eventually leads to orgasm, which results in the transfer of seminal fluid from the male to the female genital tract. This can lead to conception if sex is timed to coincide with ovulation.

THE PHASES OF SEX

There are four classic phases of sex in both men and women. The first is excitement, where erotic physical or mental stimulation causes arousal, resulting in lubrication and swelling of erectile tissue. Second is the plateau phase, when the erectile tissue has swollen to its maximal size and arousal is at its greatest. These two phases last for varying lengths of time. The third phase is short, when orgasm occurs. After orgasm is the refractory phase, where the erectile tissue relaxes and erection in the man cannot be resumed for a time.

LEVEL OF AROUSAL

Excitement phase

Plateau phase

Orgasm

Some women do not follow a classic curve

Refractory phase

TIME

KEY
— CLASSIC CURVE
— A FEMALE VARYING FROM CLASSIC CURVE

SEXUAL AROUSAL
This graph shows the classic curve of the four phases of sex (green). Most people pass through each of these phases in a similar way, but some women's sexual response curve (purple) can vary from the classic curve.

THE LOVE HORMONE

Oxytocin is a hormone that is released from the pituitary gland into the bloodstream, where it is transported to organs such as the breasts and uterus. Among many other actions, oxytocin has effects on sexual behavior, orgasm, pregnancy, labor, breastfeeding, and also relationships. It is thought to be oxytocin that helps couples form a stable pair bond after sex (see p.58).

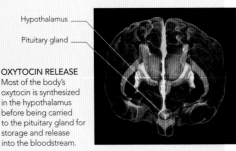

Hypothalamus

Pituitary gland

OXYTOCIN RELEASE
Most of the body's oxytocin is synthesized in the hypothalamus before being carried to the pituitary gland for storage and release into the bloodstream.

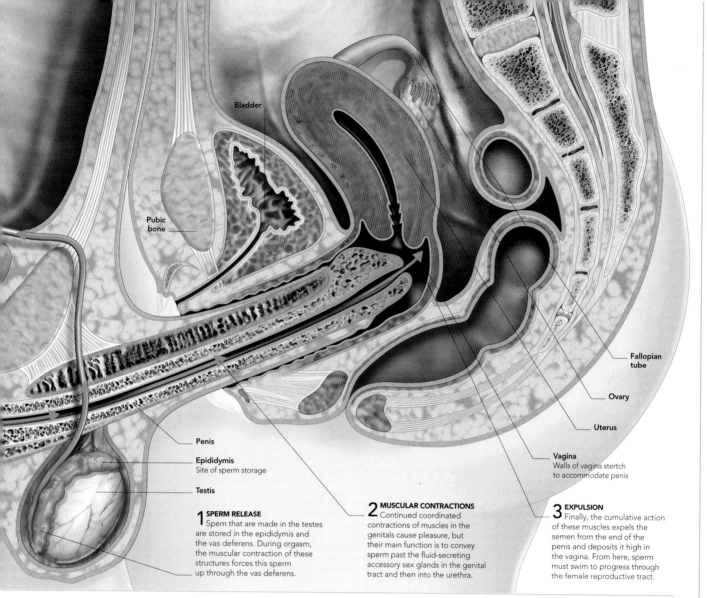

Bladder

Pubic
bone

Fallopian
tube

Ovary

Uterus

Penis

Epididymis
Site of sperm storage

Testis

Vagina
Walls of vagina stertch
to accommodate penis

1 SPERM RELEASE
Spem that are made in the testes
are stored in the epididymis and
the vas deferens. During orgasm,
the muscular contraction of these
structures forces this sperm
up through the vas deferens.

2 MUSCULAR CONTRACTIONS
Continued coordinated
contractions of muscles in the
genitals cause pleasure, but
their main function is to convey
sperm past the fluid-secreting
accessory sex glands in the genital
tract and then into the urethra.

3 EXPULSION
Finally, the cumulative action
of these muscles expels the
semen from the end of the
penis and deposits it high in
the vagina. From here, sperm
must swim to progress through
the female reproductive tract.

ORGASM

Orgasm is the intense climax of sexual
pleasure that is caused by activation of
the sympathetic nerves (see pp.64–65)
that leave the sacral area of the spinal
cord, in the lower back. These nerves
travel to muscles within the lower pelvis
and cause them to rhythmically contract.
Sympathetic nerves also cause the
muscles at the exit of the bladder
to close, so that urination does not
simultaneously
occur during
orgasm. The
number of muscle
contractions can
vary, but they
usually total
between 10 and
15 per orgasm.

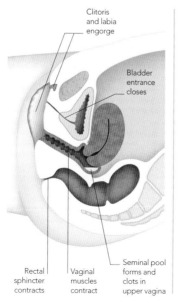

Clitoris
and labia
engorge

Bladder
entrance
closes

Rectal
sphincter
contracts

Vaginal
muscles
contract

Seminal pool
forms and
clots in
upper vagina

**SPERM DURING
FEMALE ORGASM**
Semen clots in the
upper vagina, and
sperm must swim
to continue past
the cervix. Orgasmic
contractions may help
open the cervix and
move semen toward
the fallopian tubes.

EJACULATION

In males, the rhythmic contraction
of muscles in the lower pelvis, such
as the bulbospongiosus muscle at
the base of the penis, propels semen
through the genital tract. Semen
comprises sperm and fluid from
the vas deferens, as well as fluid
secreted by the accessory sex glands,
which include the seminal vesicles,
prostate, and bulbourethral glands.
Semen is alkaline to counteract
the acidity of the vagina and enable
sperm to swim. It is ejected between
the first and seventh contraction
of orgasm into the top of the vagina.
Sperm will only be able to fertilize
an egg once they have become
activated in a process called
capacitation (see p.80).

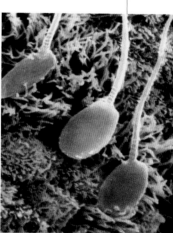

JOURNEY OF SPERM
This colored micrograph
shows sperm in the female
reproductive tract. Mucosa
cells (purple) secrete a fluid
to coat and protect sperm.

BIRTH CONTROL

Birth control has been used for generations as a means of avoiding unwanted pregnancies. Today there is a range of methods available, and most people will find something to suit them.

THE IMPORTANCE OF BIRTH CONTROL

For many people, birth control simply allows them to have sex without the fear of pregnancy. However, birth control has been a factor in empowering women around the world, and has greatly contributed to the improvement in sexual health. In developing countries, avoiding unwanted pregnancies has given women the chance to educate themselves and to find work outside the home.

PREGNANCY THROUGH CHOICE
Oral contraceptives and other methods of birth control allow people to enjoy sex and to time pregnancy for when it is convenient.

METHODS OF BIRTH CONTROL

Natural methods, such as *coitus interruptus*, and some barrier methods have been used for hundreds of years. Modern methods started to become widely available in the 1960s. The main types currently used are barrier methods, hormonal methods, and intrauterine devices (IUDs). These all operate either as contraceptives, which prevent fertilization of the egg, or as contragestives, which prevent a fertilized egg from implanting in the uterus.

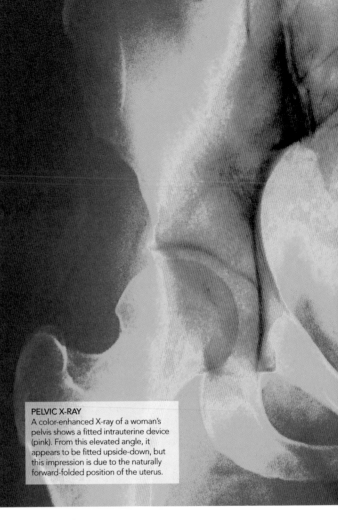

PELVIC X-RAY
A color-enhanced X-ray of a woman's pelvis shows a fitted intrauterine device (pink). From this elevated angle, it appears to be fitted upside-down, but this impression is due to the naturally forward-folded position of the uterus.

BARRIER METHODS

Devices that form a physical barrier between the sperm and egg are known as barrier methods. The four main types are male and female condoms, the cervical cap, and the diaphragm. Condoms are usually disposable, while caps and diaphragms can be used many times. They all prevent pregnancy by stopping the sperm from entering the uterus via the cervix. Condoms also prevent sexually transmitted diseases. Barrier methods are popular because they are cheap and easy to use, but are less reliable than other methods. Over a year, a woman has a 2 in 100 chance of getting pregnant if a condom is used each time she has sex. Caps and diaphragms are even less reliable, but their effectiveness is improved by using a spermicide (a gel that kills sperm).

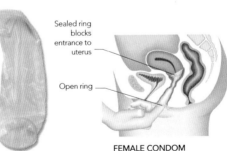

Sealed ring blocks entrance to uterus

Open ring

MALE CONDOM
Usually made of latex, a male condom is worn over the penis during sex and then discarded.

FEMALE CONDOM
A thin plastic or rubber pouch connects two flexible rings; one is placed deep inside the vagina, and the other stays outside.

THE INTRAUTERINE DEVICE (IUD)

IUDs have to be fitted by a doctor or nurse and can stay in place for several years to give long-term contraception. There are two main forms: those made of copper, and those that contain progesterone. Both stimulate the release of prostaglandins from the uterus, making it inhospitable to eggs and sperm. Progesterone-secreting IUDs (also called the intrauterine system, or IUS) also thin the uterine lining, increase cervical mucus, and inhibit ovulation. IUDs work as contraceptives, but may also prevent implantation (contragestive).

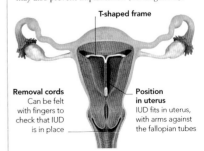

T-shaped frame

Removal cords
Can be felt with fingers to check that IUD is in place

Position in uterus
IUD fits in uterus, with arms against the fallopian tubes

FITTING AN IUD
Uterus size is measured with a small device before an IUD can be fitted. Progesterone IUDs tend to be large and can be difficult to insert into women who have not given birth.

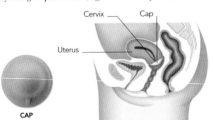

Cervix   Cap

Uterus

CAP

CERVICAL CAP
The small, flexible cap made of rubber is placed high up in the vagina. It fits securely over the cervix, blocking the entrance to the uterus.

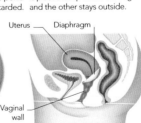

Uterus   Diaphragm

Vaginal wall

DIAPHRAGM

DIAPHRAGM
The diaphragm is larger than the cap. Its dome-shaped body is bounded by a flexible ring that sits against the vaginal walls, blocking the entrance to the uterus.

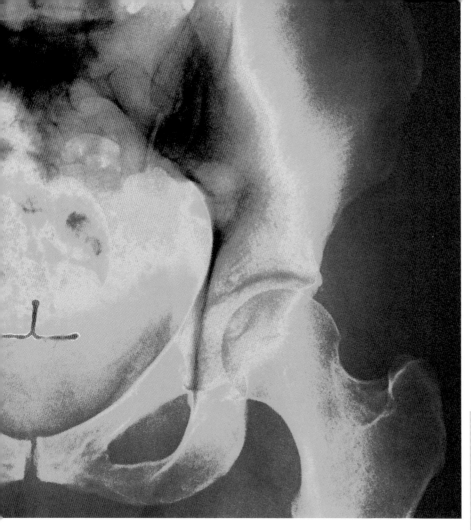

HORMONAL METHODS

The most well-known hormonal method is the combined oral contraceptive pill (the pill). It contains higher levels of estrogen and progesterone than are normally found in the body. Every month, as natural levels of progesterone and estrogen fall, the pituitary gland produces follicle-stimulating hormone (FSH) and luteinizing hormone (LH) to trigger ovulation. High levels of estrogen and progesterone from the pill will prevent this sequence of events. Contraceptive implants, patches, and the vaginal ring also block ovulation, releasing a steady stream of hormones. The progesterone-only pill (mini pill) can prevent ovulation although less successfully than the combined pill; its main function is to thicken cervical mucus and prevent sperm from reaching the fallopian tubes.

USING HORMONES TO PREVENT PREGNANCY
Contraceptive hormones can disrupt the menstrual cycle in a range of different ways, enabling them to be tailored to individual preferences.

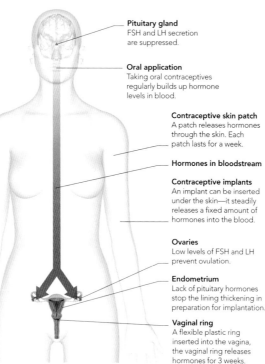

Pituitary gland
FSH and LH secretion are suppressed.

Oral application
Taking oral contraceptives regularly builds up hormone levels in blood.

Contraceptive skin patch
A patch releases hormones through the skin. Each patch lasts for a week.

Hormones in bloodstream

Contraceptive implants
An implant can be inserted under the skin—it steadily releases a fixed amount of hormones into the blood.

Ovaries
Low levels of FSH and LH prevent ovulation.

Endometrium
Lack of pituitary hormones stop the lining thickening in preparation for implantation.

Vaginal ring
A flexible plastic ring inserted into the vagina, the vaginal ring releases hormones for 3 weeks.

THE ROLE OF ESTROGEN

There are several kinds of estrogen, and all are produced in the ovaries in response to stimulation by the hormones FSH and LH. They are involved in the fertility cycle of all vertebrates. Estrogen is also an important component of the combined oral contraceptive pill, as well as the morning-after pill. Estrogen found in contraceptives is usually synthetic, but some estrogen prescribed to humans is extracted from the urine of pregnant horses.

ESTRADIOL
This light micrograph shows estradiol crystals. Estradiol is one of the estrogen hormones that controls the menstrual cycle.

EMERGENCY CONTRACEPTION

The morning-after pill is a term for a variety of different drugs that are used to prevent pregnancy after unprotected sex. Some pills contain a progesterone-like hormone, others combine estrogen with progesterone, and pills such as mifepristone (used for emergency contraception only in certain countries) block the effects of progesterone. Although different in composition, these drugs all prevent fertilization by two methods: either by delaying ovulation; or by making it difficult for sperm to reach the egg. However, their main method of action is to delay ovulation, so the morning-after pill is much less effective if ovulation has already occurred. Slightly more effective than pills, IUDs can also be used for emergency contraception because they stop a fertilized egg from implanting.

Effects of emergency contraceptive

Ovulation
Estrogen, progesterone, or low-dose mifepristone pills stop the rise in LH—this prevents the egg from developing and delays ovulation.

Fertilization
Progesterone pills make the inside of the uterus too alkaline for sperm to swim and also thicken cervical mucus. This prevents the sperm from reaching and fertilizing the egg.

Implantation
IUDs can prevent the fertilized egg from implanting in the uterus lining. High-dose mifepristone prevents implantation, but in low doses has no effect.

OCCASIONAL USE
Emergency contraception is designed for use when other contraception has failed. A range of drugs or IUDs can be used after sex to prevent pregnancy.

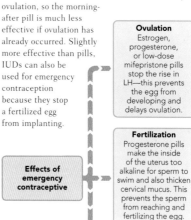

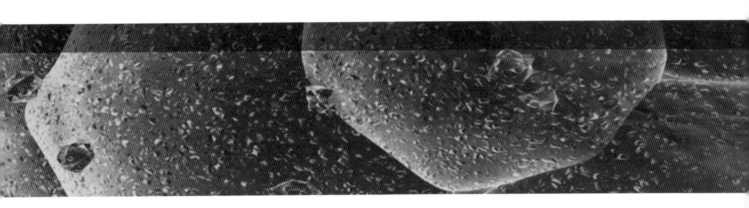

AT OVULATION, A MATURE FOLLICLE IN THE WOMAN'S OVARY RUPTURES TO RELEASE AN EGG. IF IT MEETS SPERM ON ITS JOURNEY DOWN ONE OF THE FALLOPIAN TUBES TO THE UTERUS, FERTILIZATION MAY OCCUR. THROUGH A MULTITUDE OF COMPLEX PROCESSES, THE FERTILIZED EGG FIRST BECOMES A BALL OF CELLS. OVER TIME IT DEVELOPS INTO AN EMBRYO WITH A BASIC HUMAN SHAPE, THEN A FETUS THAT CAN MOVE AND RESPOND, AND FINALLY INTO A FULLY DEVELOPED BABY, READY FOR LIFE OUTSIDE ITS MOTHER. THROUGHOUT PREGNANCY, THE MOTHER'S BODY UNDERGOES A RANGE OF CHANGES IN ORDER TO SUPPORT AND NOURISH THE GROWING FETUS.

CONCEPTION TO BIRTH

Eye bud and liver A primitive eye can be seen on this eight-week old fetus. The dark area in the abdomen is the developing liver.

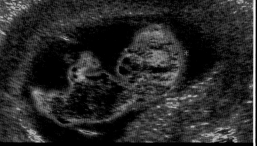

12-week ultrasound An ultrasound scan enables the fetus to be measured, which helps estimate pregnancy dates and keep track of growth.

External ear and digits By the 12th week, the tiny outer ear is recognizable at the side of the head, and separate fingers and toes have formed.

TRIMESTER 1
MONTHS 1–3 | WEEKS 1–12

During the first trimester the single-celled, fertilized egg embeds in the uterus and is transformed into a tiny, yet recognizably human, embryo with all its major organ systems in place.

The first trimester is a time of remarkable growth and development. The single-celled, fertilized egg divides rapidly into an embryo, then a fetus. Although there is much growth and maturation to come, by the end of this trimester the fetus has a recognizably human form, with facial features, sense organs, fingers and toes on the end of tiny limbs, and even tooth buds, fingerprints, and toenails. The brain, nervous system, and muscles are all functioning and the fetus can perform involuntary reflexes, such as moving about vigorously, swallowing, hiccupping, yawning, and urinating.

This period of initial human development can be fraught with danger. As its organs are forming, the embryo is especially sensitive to harmful influences, including drugs, pollutants, and infections. The first trimester is the time when congenital damage is most likely and pregnancy loss is most common, but by the end of the trimester the threat is far less. Although the woman may not appear noticeably pregnant until the third month, she may notice her waistline expanding and other early symptoms, such as nausea. It is at the end of this trimester that many women announce their pregnancy.

TIMELINE

MOTHER

MONTH 1

WEEK 1
Menstruation marks the start of pregnancy if conception occurs in the month following it. Ovarian follicles start to ripen in readiness for ovulation.

WEEKS 2–3
Follicle stimulating hormone (FSH) causes an egg to ripen inside a follicle. The follicle moves to the surface of the ovary and ruptures, releasing the ripe egg.

The uterine lining thickens in preparation for possible pregnancy.

On ovulation, basal body temperature rises and cervical mucus becomes stringy.

WEEK 4
The thickened uterine lining is ready to receive and nourish the blastocyst.

A mucus plug forms in the cervix to protect the uterus from infection.

MONTH 2

WEEKS 5–6
A pregnancy test may show a positive result even before a period is missed.

Early pregnancy symptoms may include nausea, increased urinary frequency, fatigue, and sensitive breasts.

WEEK 1	WEEK 2	WEEK 3	WEEK 4	WEEK 5	WEEK 6

FETUS

WEEKS 1–2
When a mature egg is released from an ovary, it travels down the fallopian tube toward the uterus. If the woman has sex during this fertile time, the sperm will swim up the fallopian tube to meet the egg, and fertilization may occur.

WEEK 3
If conception occurs, the fertilized egg starts to divide as it travels down the fallopian tube.

The hormone hCG is produced to "switch off" the menstrual cycle.

WEEK 4
The blastocyst implants in the uterine lining. It develops a fluid-filled core that will become the yolk sac and will separate the embryonic cells from placental cells.

WEEKS 5–6
The embryo divides into three layers, and cells start to specialize. The outer layer forms the neural tube that develops into the brain and the spinal cord.

A bulge in the middle layer forms the heart, which starts to divide into four chambers and to circulate blood around the body.

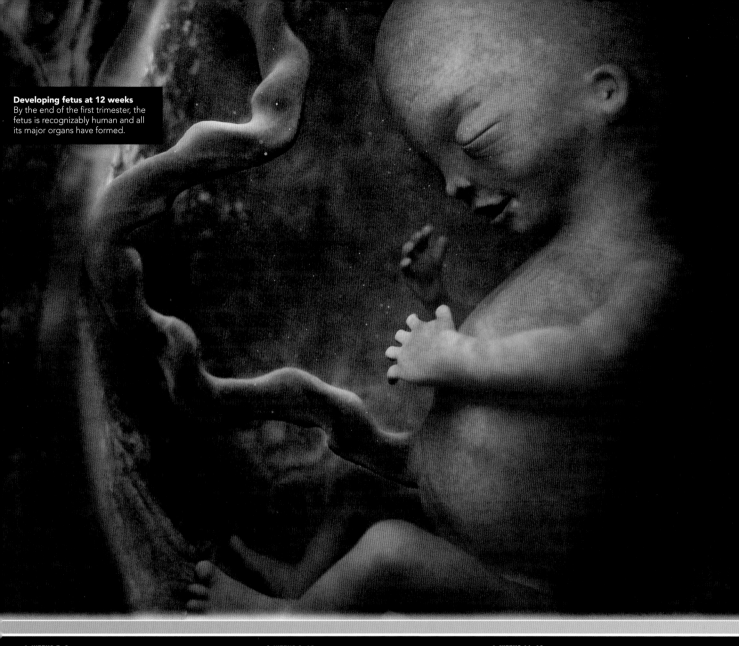

Developing fetus at 12 weeks
By the end of the first trimester, the fetus is recognizably human and all its major organs have formed.

WEEKS 7–8

Metabolism speeds up, the heart and lungs become more efficient, and blood volume expands to meet the increased demands of pregnancy.

Weight gain may become noticeable.

Some women develop nausea, a heightened sense of taste and smell, and food cravings.

WEEKS 9–10

Breasts and waistline expand, and clothing may feel tight.

The growing uterus may press on the lower spine, causing backache.

Increased blood circulation may make some women feel uncomfortably hot.

Hormonal changes increase vaginal discharge.

WEEKS 11–12

The uterus moves upward out of the pelvis; it can now be felt above the pubic bone. The "bump" may now show.

Energy increases and urinary symptoms lessen.

Varicose veins or hemorrhoids may be troublesome.

The nipples, areolae, and freckles darken. Brown patches may also appear on the face.

WEEK 7	WEEK 8	MONTH 3	WEEK 9	WEEK 10	WEEK 11	WEEK 12

WEEK 7

The intestines bulge to form a stomach.

Limb buds develop paddle-shaped ends.

WEEK 8

The yolk sac starts to disappear, as the placenta develops.

Limbs lengthen and develop elbows and webbed digits. The primitive tail shrinks.

WEEKS 9–10

The nose, mouth, and lips are almost fully formed, and the eyes have moved to the front of the face. Eyelids fuse over the eyes.

Buds from the bladder grow upward to join with developing kidneys.

The gonads develop into either testes or ovaries, and the ovaries begin to produce eggs.

WEEKS 11–12

The mouth can open and close, enabling swallowing, and has tiny tooth buds.

A heartbeat may be detected.

Brain cells multiply rapidly as the brain develops two halves, or hemispheres.

The fetus has reflexes and may move if the abdomen is pressed.

MONTH 1 | WEEKS 1–4

Pregnancy is dated from the start of the woman's last menstrual period. For the first two weeks from this time, the body prepares for conception. A fertilized egg then undergoes rapid cell division as it travels to implant in the uterus, where the development of an embryo begins.

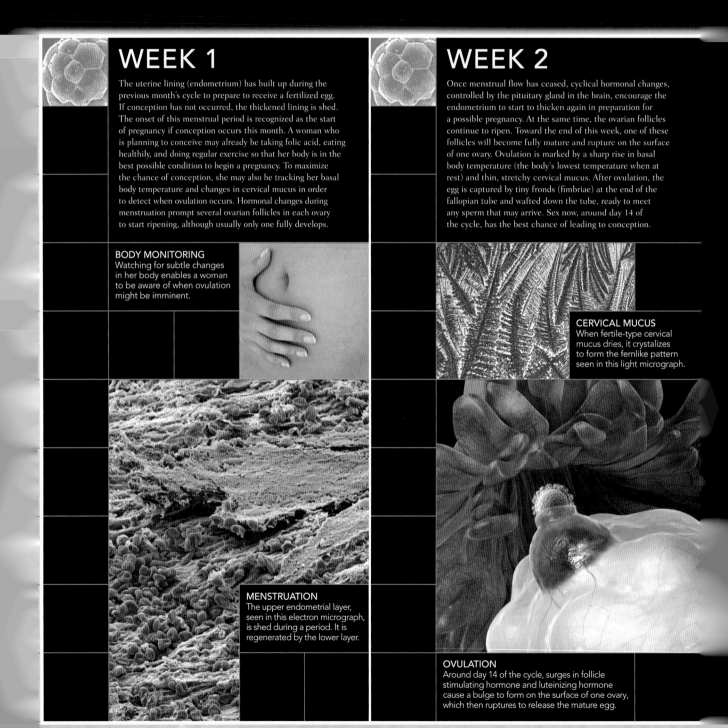

WEEK 1

The uterine lining (endometrium) has built up during the previous month's cycle to prepare to receive a fertilized egg. If conception has not occurred, the thickened lining is shed. The onset of this menstrual period is recognized as the start of pregnancy if conception occurs this month. A woman who is planning to conceive may already be taking folic acid, eating healthily, and doing regular exercise so that her body is in the best possible condition to begin a pregnancy. To maximize the chance of conception, she may also be tracking her basal body temperature and changes in cervical mucus in order to detect when ovulation occurs. Hormonal changes during menstruation prompt several ovarian follicles in each ovary to start ripening, although usually only one fully develops.

BODY MONITORING
Watching for subtle changes in her body enables a woman to be aware of when ovulation might be imminent.

MENSTRUATION
The upper endometrial layer, seen in this electron micrograph, is shed during a period. It is regenerated by the lower layer.

WEEK 2

Once menstrual flow has ceased, cyclical hormonal changes, controlled by the pituitary gland in the brain, encourage the endometrium to start to thicken again in preparation for a possible pregnancy. At the same time, the ovarian follicles continue to ripen. Toward the end of this week, one of these follicles will become fully mature and rupture on the surface of one ovary. Ovulation is marked by a sharp rise in basal body temperature (the body's lowest temperature when at rest) and thin, stretchy cervical mucus. After ovulation, the egg is captured by tiny fronds (fimbriae) at the end of the fallopian tube and wafted down the tube, ready to meet any sperm that may arrive. Sex now, around day 14 of the cycle, has the best chance of leading to conception.

CERVICAL MUCUS
When fertile-type cervical mucus dries, it crystalizes to form the fernlike pattern seen in this light micrograph.

OVULATION
Around day 14 of the cycle, surges in follicle stimulating hormone and luteinizing hormone cause a bulge to form on the surface of one ovary, which then ruptures to release the mature egg.

WEEK 3

Up to 350 million sperm are released in a single ejaculation, but fewer than 1 in 1,000 manage to pass through the cervix into the uterus, and only around 200 reach the correct fallopian tube to meet the egg. At the moment of conception, a single sperm is drawn into the egg, which then blocks entry to others. The fertilized egg produces a hormone called human chorionic gonadotropin (hCG) that "switches off" the menstrual cycle by stimulating continued production of progesterone, the hormone needed to maintain the endometrium. The egg moves down the fallopian tube and divides, forming a two-celled zygote and then a cluster of smaller cells called blastomeres. By the time it reaches the uterus, it is a ball of around 100 cells called a blastocyst.

WEEK 4

The blastocyst arrives in the uterus on average six days after conception—the endometrium is now thickened, ready to receive and nourish it. Hormones also thicken cervical mucus so that it forms a plug in the cervix, which protects the uterus during pregnancy from infections that might otherwise travel up from the vagina. The blastocyst now develops a fluid-filled cavity, creating two layers of cells. The outer layer (trophoblast) burrows into the endometrium and will become the placenta. The inner cell mass forms the early embryo (embryoblast)—these cells then differentiate into a two-layered embryonic disk. The fluid-filled cavity develops into a yolk sac that will provide nourishment for the embryo during the early weeks, until the placenta has developed.

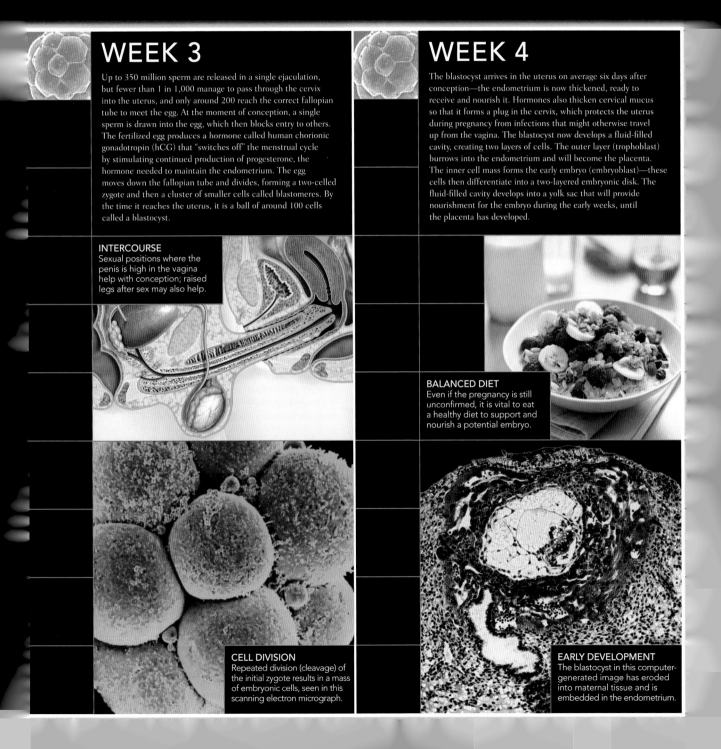

INTERCOURSE
Sexual positions where the penis is high in the vagina help with conception; raised legs after sex may also help.

BALANCED DIET
Even if the pregnancy is still unconfirmed, it is vital to eat a healthy diet to support and nourish a potential embryo.

CELL DIVISION
Repeated division (cleavage) of the initial zygote results in a mass of embryonic cells, seen in this scanning electron micrograph.

EARLY DEVELOPMENT
The blastocyst in this computer-generated image has eroded into maternal tissue and is embedded in the endometrium.

MONTH 1 | WEEKS 1–4
MOTHER AND EMBRYO

At the start of each menstrual cycle, the mother's body prepares for a potential pregnancy. During the first two weeks, there will be no outward signs of ovulation or the changes in the uterine lining in preparation for pregnancy. After the uterine lining has been shed, a rejuvenated lining emerges that thickens over the following one or two weeks. Under the influence of progesterone and estrogen, the lining becomes sticky and nutrient-rich in order to encourage and support successful implantation of the fertilized blastocyst. In each cycle, the chance of conception is around 40 percent. The first clue that conception has occurred may be a slight implantation bleed, which can be confused with a very light period, although a missed period is usually the first definite sign that a woman is pregnant. A pregnancy test taken around week four will confirm the pregnancy.

MOTHER AT FOUR WEEKS
Normal female anatomy is shown here because it is still too early to see any visible changes in the internal arrangement and size of the mother's major organs.

Lungs
The lungs here are in their normal position. During pregnancy, the diaphragm pushes up, and the lungs adapt to their new position.

Bowel
The transverse colon, below the stomach and above the small bowel, is in its normal position. As the pregnancy advances, the bowel is displaced up as the uterus expands out of the pelvis.

Uterus
The uterus is approximately the size of a pear and is protected within the bony pelvis.

MOTHER

🔲 65 beats per minute
🔲 107/70
🔲 7½ pints (4.2 l)

The hormone hCG is released when the embryo implants. It is detected in the mother's urine by pregnancy tests.

20%

Around 20 percent of women become more sensitive to odors in the first few weeks of pregnancy.

Once it has been released, an egg survives for 24 hours if it is not fertilized.

EMBRYO

The sex of the embryo is determined by the sperm at the point of conception. If the sperm is carrying a Y chromosome, the embryo is male; an X chromosome will lead to a female embryo.

The heart starts beating at three weeks at a relatively slow 20–25 beats per minute. By the third month, this has increased to an incredible rate of 157 beats per minute.

1/16 in

By day 28 the embryo grows 1/16 in (1 mm) a day, but is smaller than a match head.

As a result of cell division, the blastocyst contains 100–150 cells by the time it enters the uterine cavity to implant. The cells are arranged in a three-layered sphere.

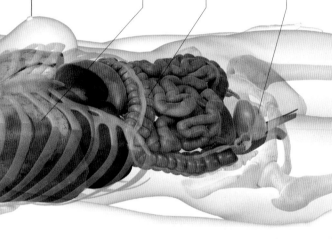

| 1 | 2 | 3 | 4 | 5 | 6 | 7 | 8 | 9 | 10 | 11 | 12 | 13 | 14 | 15 | 16 | 17 | 18 | 19 | 20 | 21 | 22 | 23 | 24 | 25 | 26 | 27 | 28 | 29 | 30 | 31 | 32 | 33 | 34 | 35 | 36 | 37 | 38 | 39 | 40 |

2 WEEKS

Numerous sperm reach the egg and will attempt to enter it as soon as they arrive. When a single sperm has successfully burrowed through the outer wall of the egg, the wall undergoes a change known as depolarization. This prevents any further sperm from penetrating and fertilizing the egg.

Corona radiata
The large egg cell is surrounded by smaller corona radiata cells.

Fallopian tube
Fertilization occurs in the widest portion of the fallopian tube, known as the ampulla.

Fertilization
Sperm is about to break through the outer wall and fertilize the egg.

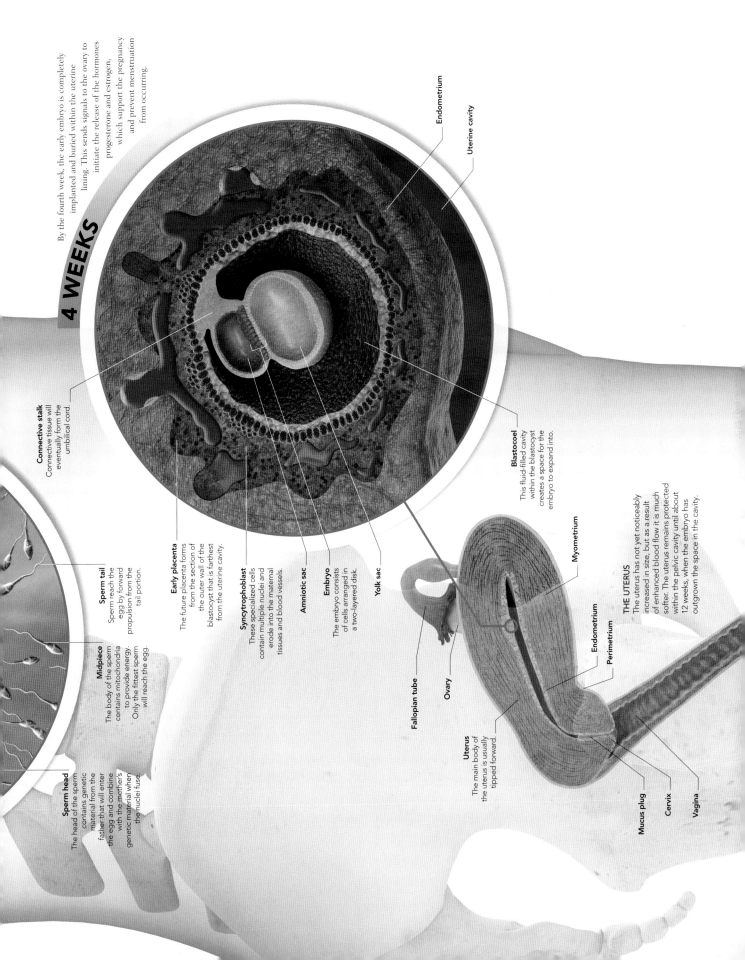

4 WEEKS

By the fourth week, the early embryo is completely implanted and buried within the uterine lining. This sends signals to the ovary to initiate the release of the hormones progesterone and estrogen, which support the pregnancy and prevent menstruation from occurring.

Endometrium

Uterine cavity

Connective stalk
Connective tissue will eventually form the umbilical cord.

Sperm head
The head of the sperm contains genetic material from the father that will enter the egg and combine with the mother's genetic material when the nuclei fuse.

Midpiece
The body of the sperm contains mitochondria to provide energy. Only the fittest sperm will reach the egg.

Sperm tail
Sperm reach the egg by forward propulsion from the tail portion.

Early placenta
The future placenta forms from the section of the outer wall of the blastocyst that is farthest from the uterine cavity.

Syncytrophoblast
These specialized cells contain multiple nuclei and erode into the maternal tissues and blood vessels.

Amniotic sac

Embryo
The embryo consists of cells arranged in a two-layered disk.

Yolk sac

Blastocoel
This fluid-filled cavity within the blastocyst creates a space for the embryo to expand into.

Fallopian tube

Ovary

Uterus
The main body of the uterus is usually tipped forward.

Mucus plug

Cervix

Vagina

Myometrium

Endometrium

Perimetrium

THE UTERUS
The uterus has not yet noticeably increased in size, but as a result of enhanced blood flow it is much softer. The uterus remains protected within the pelvic cavity until about 12 weeks, when the embryo has outgrown the space in the cavity.

MOTHER

WHEN PREGNANCY STARTS

From the moment of conception, hormonal changes in the mother's body prepare the uterus to accept the pregnancy (implantation) and adapt to the future needs of the developing embryo. To accommodate the pregnancy, the volume of the uterus will expand by over 500 times and numerous hormonal and metabolic changes take place in order to balance the mother's needs with those of the fetus. On average, a pregnancy lasts 266 days (38 weeks) from the time of ovulation. For simplicity, the weeks of pregnancy are counted from the first day of the last menstrual period, which is usually two weeks earlier, so that the average length of a pregnancy becomes 280 days (40 weeks).

Ovulation
Conception is possible after this date, usually day 14 of cycle

Last menstrual period
Calculated as the first day of pregnancy

Post term
Period that exceeds the usual 40 weeks of pregnancy

Term
Period at which the fetus is considered full-term

MENSTRUATION
This electron micrograph shows the uterus lining (red layers) being shed. The red dots are blood cells from underlying vessels.

Due date
280 days after the first day of last menstrual period

Preterm
Period from 24–37 weeks, when the fetus is viable but not fully developed

OBSTETRIC CALCULATOR
This simple calendar wheel is used to estimate the expected date of delivery from the date of the last menstrual period.

CHANGES IN THE UTERUS

From conception, the uterus has only six days to prepare to receive the blastocyst. The menstrual cycle stops as the ovary secretes estrogen and progesterone from the empty follicle (corpus luteum) at the site of ovulation. The uterine lining (endometrium) becomes thicker, more receptive, and "sticky" to encourage implantation. Glandular activity increases, levels of estrogen and progesterone rise, and blood supply increases. Not all fertilized eggs implant, and implantation occasionally occurs outside the uterus as an ectopic pregnancy. The endometrium is only actually receptive to implantation for a mere one to two days.

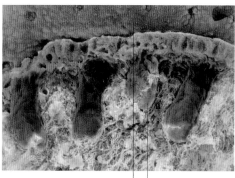

Endometrium
Regenerates following menstruation

Endometrial gland
Produces secretions that prepare the endometrium for implantation

ENDOMETRIUM
The outermost layer of the endometrium (uterine lining) is shed at the end of each menstrual cycle. The deeper glandular layer is retained for the next menstrual cycle.

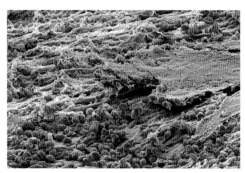

MONITORING FERTILITY

The timing of ovulation can vary with the length of the menstrual cycle but in the absence of pregnancy it is followed by menstruation 14 days later. For women with very irregular cycles, where the timing of ovulation is difficult to predict, measuring basal body temperature and assessing the quality of the cervical mucus can provide clues to the fertility window. Once released, the egg will survive for only 24 hours if unfertilized. The fertile window is open slightly longer than this because sperm remain active in the fallopian tubes for 48 hours, with some remaining active for up to 80 hours.

MALE FERTILITY

Males remain fertile throughout life from early puberty. Fertility is not strongly linked with the volume of ejaculate, but is based on the overall sperm count, sperm shape, and motility. Laboratory semen analysis is vital for the investigation of a couple's fertility problems. Although sperm counts decline with age, this does not usually significantly impair fertility. Abstinence for several days prior to intercourse around the time of ovulation will improve conception chances. Several conditions can reduce fertility (see pp.222–23), and fertility can be improved with lifestyle changes such as reducing smoking and alcohol intake.

FERTILE WINDOW

CHANGES WITHIN THE BODY

Basal body temperature
An accurate thermometer can measure the tiny 0.4–0.9° F (0.2–0.5°C) rise in basal body temperature that signifies that ovulation has taken place. The temperature should be recorded daily, because the sudden increase is crucial.

Menstral cycle
Carefully recording the details of each menstrual cycle will reveal if they are regular. Average cycle length is 28 days, but the normal range is 21–35 days. Irregular cycles are different lengths most months, making it difficult to calculate the time of ovulation.

Cervical mucus
Under the influence of estrogen, cervical mucus changes at ovulation to facilitate the passage of sperm through the cervical canal. The mucus becomes stretchy, thinner, and less acidic in order to promote sperm motility. Later, under the influence of progesterone, these changes are reversed and the passage of sperm is restricted by thicker mucus.

WET DAY TEST
Stretching cervical mucus between the thumb and forefinger will test its quality. If it is thin, watery, slightly stretchy, and forms a string, it is likely that ovulation is taking place.

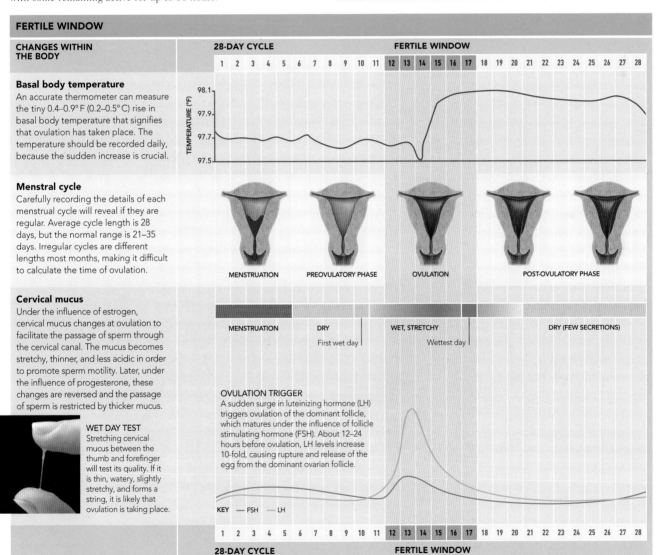

28-DAY CYCLE — **FERTILE WINDOW**

MENSTRUATION | PREOVULATORY PHASE | OVULATION | POST-OVULATORY PHASE

MENSTRUATION | DRY — First wet day | WET, STRETCHY — Wettest day | DRY (FEW SECRETIONS)

OVULATION TRIGGER
A sudden surge in luteinizing hormone (LH) triggers ovulation of the dominant follicle, which matures under the influence of follicle stimulating hormone (FSH). About 12–24 hours before ovulation, LH levels increase 10-fold, causing rupture and release of the egg from the dominant ovarian follicle.

KEY — FSH — LH

28-DAY CYCLE — **FERTILE WINDOW**

CONCEPTION

For conception to take place just one of the millions of sperm released must penetrate the egg. However, the sperm must first swim through the cervix and uterus into the fallopian tube, and only a few will successfully reach their goal.

After release from the ovary, the egg is swept into the fallopian tube by the frondlike fimbriae. Fertilization usually takes place in the wider mid-portion of the fallopian tube, called the ampulla. Most of the millions of sperm released, however, do not make the journey this far. This is an important factor because it improves the chance of only the fittest sperm reaching and fertilizing the egg.

DAY 15 **PASSAGE OF THE EGG** Swept up by the fimbriae at the end of the fallopian tube, the egg passes along the tube to rest within the wider ampullary portion. Successful fertilization usually occurs at this position and can occur up to one or two days after ovulation.

Path of egg

Ampulla
Usual site of fertilization

200–300 sperm
enter each tube

Ovary

Fimbriae

100,000 sperm
enter uterine cavity

60–80 million sperm
pass cervix

200–300 million sperm
enter vagina

DAYS 12–14

SPERM RACE Millions of sperm are released in the ⁷⁄₁₀₀–¹⁄₅ fl oz (2–6 ml) of ejaculate. Movement is limited, but the cervical mucus and receptive uterine environment lead to forward progression of up to ¹⁄₁₆–¹⁄₈ in (2–3 mm) per minute.

CAPACITATION OF SPERM

Sperm can move once in the vagina, but their movement is restricted until they reach the more favorable, less acidic environment of the uterus. They are unable to fertilize an egg until they have undergone the process of capacitation. This involves removal of the protein coat over the head of the sperm (acrosome), allowing it to fuse with the egg. Capacitation does not last long and only occurs once to each sperm. Usually only the strongest, most mature sperm will complete capacitation as they journey toward the egg.

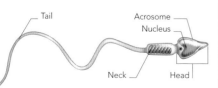

Tail

Acrosome

Nucleus

Neck   Head

DAY 14 **OVULATION** Usually a single dominant follicle is matured to the point of ovulation. In a 28-day cycle, ovulation typically occurs on day 14. Ovulation will occur on an earlier day in shorter cycles and later in longer cycles. In any given cycle, the chances of successful fertilization are approximately 40 percent.

DAY 16 FERTILIZATION

Numerous sperm are required to stimulate the corona radiata layer surrounding the egg in order to start the acrosomal reaction. This allows penetration by a single sperm through the inner zona pellucida into the egg. Unless sperm counts are very low, several hundred sperm will reach the egg approximately 5–20 minutes after ejaculation.

Double-layered egg membrane

Burrowing sperm

DAYS 16–17 FUSION OF GAMETES

When the sperm enters the egg, it precipitates a reaction in the zona pellucida that prevents other sperm from entering. The female pronucleus completes its final meiotic division, and as the pronuclei approach each other their membranes die and they fuse.

1. CORONA RADIATA
Enzymes in the acrosome combined with fast tail movements enable the sperm to pass through this outer layer and reach the zona pellucida.

2. ACROSOMAL REACTION
On contact, glycoprotein in the zona pellucida binds to proteins in the sperm's head to trigger the release of acrosomal contents.

3. DIGESTING A PATHWAY
Acrosomal enzymes digest a pathway that allows the sperm passage through the zona pellucida. The tail propels it forward.

4. PENETRATION OF EGG
As the sperm's head pierces the egg membrane, the zona pellucida alters its stucture to block entry to any other sperm. Only the sperm's head and tail enter the egg, leaving the cytoplasm behind.

5. PRONUCLEI FORM
The sperm's head becomes the male pronucleus; the egg nucleus becomes the female pronucleus.

6. FUSION
The pronuclei meet and fuse, resulting in a single nucleus containing the full number of 46 chromosomes (23 from each pronucleus).

Polar bodies

Sperm
Swim along fallopian tube to reach egg

Fimbriae
Guide released egg into fallopian tubes

Ovary
Ruptures to release a ripe egg

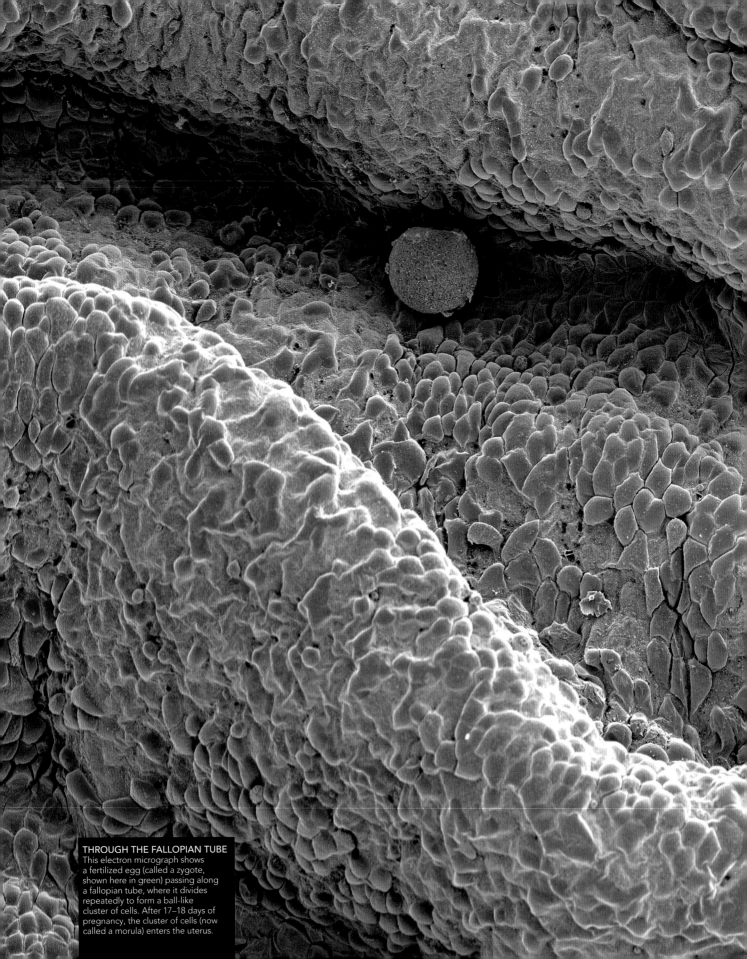

THROUGH THE FALLOPIAN TUBE
This electron micrograph shows a fertilized egg (called a zygote, shown here in green) passing along a fallopian tube, where it divides repeatedly to form a ball-like cluster of cells. After 17–18 days of pregnancy, the cluster of cells (now called a morula) enters the uterus.

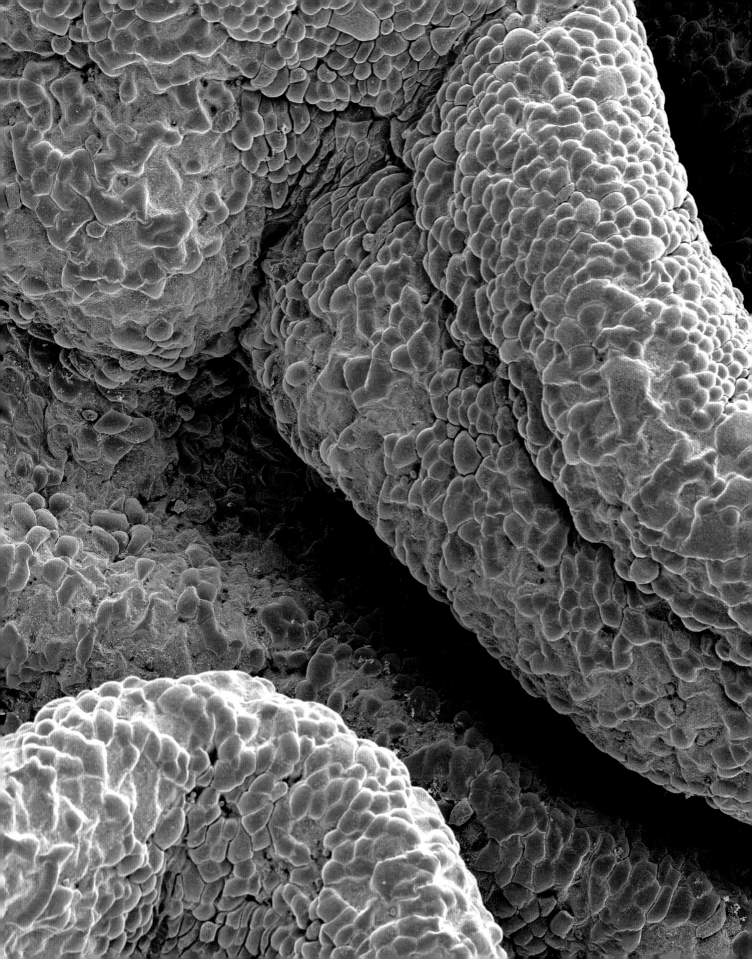

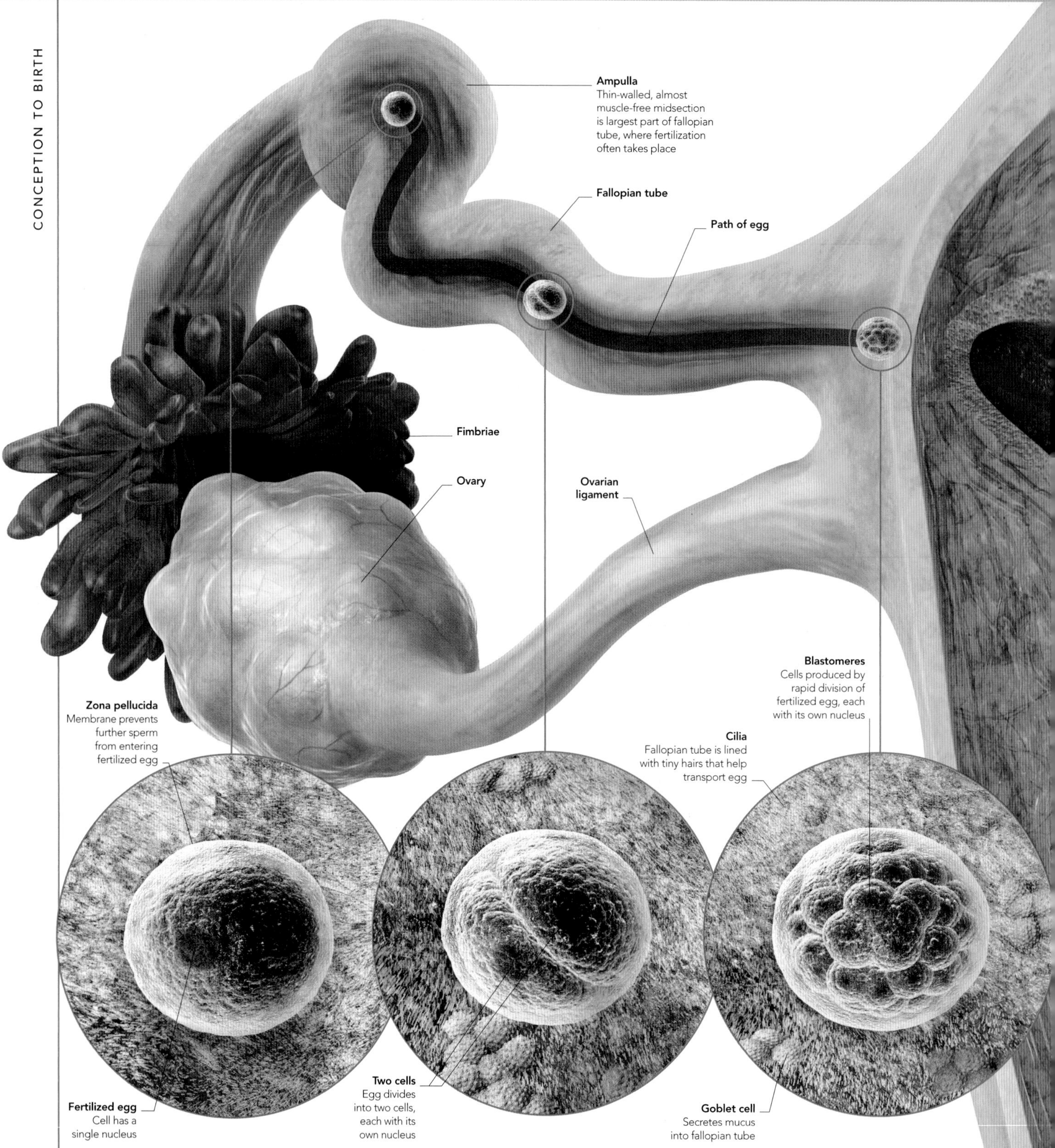

Ampulla
Thin-walled, almost muscle-free midsection is largest part of fallopian tube, where fertilization often takes place

Fallopian tube

Path of egg

Fimbriae

Ovary

Ovarian ligament

Blastomeres
Cells produced by rapid division of fertilized egg, each with its own nucleus

Cilia
Fallopian tube is lined with tiny hairs that help transport egg

Zona pellucida
Membrane prevents further sperm from entering fertilized egg

Fertilized egg
Cell has a single nucleus

Two cells
Egg divides into two cells, each with its own nucleus

Goblet cell
Secretes mucus into fallopian tube

DAY 17 **FERTILIZED EGG**
The zona pellucida now depolarizes, preventing further sperm from entering the egg. The male and female pronuclei combine to produce the "zygote," which prepares for the first cell division. In rare instances, two sperm simultaneously fertilize the egg, resulting in a molar pregnancy (see p.227).

DAY 18 **ZYGOTE**
Within 24 hours of fertilization the zygote duplicates the nuclear genetic material then divides into two cells by mitosis (see p.50). Through a sequence of rapid cell divisions, 16–32 cells, called blastomeres, are produced. These form the morula, which is Latin for "mulberry."

DAY 20 **MORULA**
The morula is still contained within the zona pellucida at this stage. This is possible because cell division has occurred without cell growth. The morula travels the length of the fallopian tube to emerge into the uterine cavity for implantation.

FERTILIZATION TO IMPLANTATION

Before implantation, the fertilized egg divides rapidly, but stays the same size and encased in the protective zona pellucida. In order to implant and grow further, the blastocyst erodes a hole in the zona pellucida so that it can squeeze through and bury itself in the uterine lining.

Not all fertilized eggs will implant successfully. The uterine lining is stimulated for implantation by progesterone, which is produced by the ovary responsible for ovulation. This reaction makes the lining sticky and full of nutrients to support the blastocyst. If the passage of the egg is blocked, it may implant in the fallopian tube, resulting in an ectopic pregnancy (see p.227). The hormone hCG is triggered by implantation. This leads the corpus luteum to produce hormones that support the pregnancy for the first 11–12 weeks.

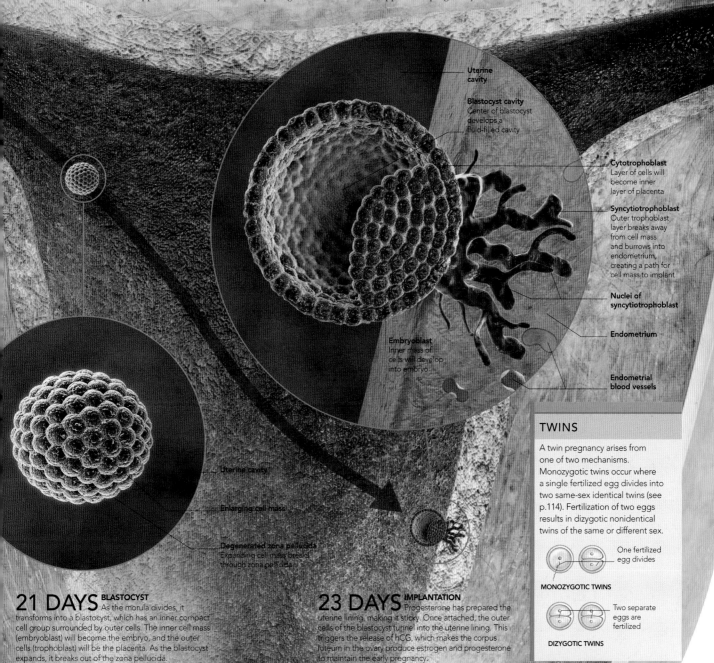

Uterine cavity

Blastocyst cavity
Center of blastocyst develops a fluid-filled cavity

Cytotrophoblast
Layer of cells will become inner layer of placenta

Syncytiotrophoblast
Outer trophoblast layer breaks away from cell mass and burrows into endometrium, creating a path for cell mass to implant

Nuclei of syncytiotrophoblast

Embryoblast
Inner mass of cells will develop into embryo

Endometrium

Endometrial blood vessels

Uterine cavity

Enlarging cell mass

Degenerated zona pellucida
Expanding cell mass breaks through zona pellucida

21 DAYS BLASTOCYST
As the morula divides, it transforms into a blastocyst, which has an inner compact cell group surrounded by outer cells. The inner cell mass (embryoblast) will become the embryo, and the outer cells (trophoblast) will be the placenta. As the blastocyst expands, it breaks out of the zona pellucida.

23 DAYS IMPLANTATION
Progesterone has prepared the uterine lining, making it sticky. Once attached, the outer cells of the blastocyst tunnel into the uterine lining. This triggers the release of hCG, which makes the corpus luteum in the ovary produce estrogen and progesterone to maintain the early pregnancy.

TWINS

A twin pregnancy arises from one of two mechanisms. Monozygotic twins occur where a single fertilized egg divides into two same-sex identical twins (see p.114). Fertilization of two eggs results in dizygotic nonidentical twins of the same or different sex.

One fertilized egg divides

MONOZYGOTIC TWINS

Two separate eggs are fertilized

DIZYGOTIC TWINS

Blastocyst cavity
Cells from embryoblast spread out to line cavity, which now becomes yolk sac

Amniotic cavity

Amnion
Layer of embryonic cells that line amniotic cavity

Syncytiotrophoblast

Endometrial vein

Yolk sac
Lined with cells derived from embryoblast, this sac will provide early sustenance for embryo

Connective tissue
Loose tissue forms from cells of yolk sac

Embryonic disk
Original cell mass (embryoblast) has developed into sa defined, two-layered disk

Endometrial capilliary

Embryoblast
Cells of embryoblast differentiate into two distinct types

Endometrium

Cytotrophoblast

Lacunae
Isolated cavities form in syncytiotrophoblast and fill with maternal blood and fluid from endometrial glands

Amniotic cavity

25 DAYS UTERINE INVASION

The blastocyct continues to invade the uterine wall, facilitated by the outer trophoblast (syncytiotrophoblast), which will become the future placenta. The inner cell mass (embryoblast), which will become the future embryo, differentiates into two distinct layers. At implantation the woman may experience a slight bleed that occasionally can be confused with a light menstrual period.

26 DAYS IMPLANTATION

The blastocyst is completely buried in the uterine wall at this stage, and the implantation point is closed over by a blood clot. By this time, the trophoblast has differentiated into an inner cytotrophoblast layer and a more invasive syncytiotrophoblast layer. The syncytiotrophoblast starts to invade maternal blood vessels. As fluid collects, the amniotic cavity expands.

25 days
26 days
29 days
30 days

THE JOURNEY OF A FERTILIZED EGG
The fertilized egg takes around seven days, from the time of conception, to travel down the fallopian tube to the uterus. Along the way, it grows from a single cell to a cluster of cells called a blastocyst. On reaching the uterus, the blastocyst becomes attached to the sticky uterine wall; it then buries itself in the endometrium. This not only offers protection but also allows the blastocyst to access nutrients for further cell growth. Once the blastocyst is fully embedded, the only sign of the entry point is a small protective blood clot.

EMBRYONIC DEVELOPMENT

Successful implantation is vital for the growth of the blastocyst into the early embryo. Once the blastocyst has successfully implanted in the uterus, it undergoes internal reorganization and burrows deep into the uterine lining.

The blastocyst differentiates into two internal cell types: the embryoblast, which will form the fetus; and two trophoblast layers, which form the placenta. The two-layered trophoblast has an inner cell layer (the cytotrophoblast) that has defined cell walls and will form the final barrier between maternal and fetal blood. The cells of the outer cell layer (the syncytiotrophoblast) do not have cell walls, which allows the interconnected cells to spread out and aggressively invade and destroy the maternal tissue. This enables the blastocyst to embed deeply in the lining of the uterus.

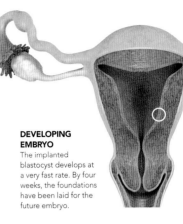

DEVELOPING EMBRYO
The implanted blastocyst develops at a very fast rate. By four weeks, the foundations have been laid for the future embryo.

Syncytiotrophoblast
Made up of numerous interconnected cells

Cavity
Spaces form within connective tissue; these gradually enlarge and fuse, displacing connective tissue

Cytotrophoblast
Each cell in this layer is encased within an intact cell membrane

Chorionic cavity
Fused cavities eventually form chorionic cavity (large fluid-filled space that surrounds amniotic and yolk sacs)

Connecting stalk
Area of connective tissue remains after chorionic cavity has formed; will form future umbilical cord

29 DAYS CAVITY FORMATION
Further separation of the yolk sac from the outer cell wall occurs. The syncytiotrophoblast layer continues to invade maternal blood vessels, creating networks of nutrient-rich blood. Cavities start to form and fuse within the connective tissue

Blood networks
Networks form as blood capillaries continue to be eroded and fuse with each other

Chorion
Comprises both layers of trophoblast, plus remaining connective tissue; will form principal part of placenta

30 DAYS CHORIONIC CAVITY
The future embryo is now attached by a connecting stalk. Although smaller than the yolk sac, the amniotic cavity continues to expand—by the eighth week, it encircles the embryo. The yolk sac will nourish the fetus and become the first site for production of red blood cells.

Amniotic sac

Yolk sac
Gradually decreases in size as chorionic cavity enlarges

SAFETY IN PREGNANCY

During pregnancy the world can seem like a dangerous place, filled with potential hazards for the growing fetus. Everything, from infections and medications to animals, domestic chemicals, and even some food, can cause concern. Fortunately, a few sensible precautions can minimize the hazards and help ensure a healthy pregnancy.

INFECTIOUS HAZARDS

During pregnancy a woman's immune system is suppressed to ensure that her body does not reject the fetus. Unfortunately, this means she is more susceptible to certain infections, and to developing complications from any infections she contracts. In addition to affecting the woman's health, some infectious agents can pass across the placenta and may harm the developing fetus. Particular risks are contaminated food, infectious diseases, and diseases carried by animals, especially cats.

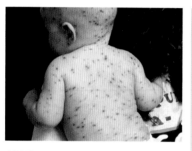

CONGENITAL INFECTIONS

Infectious diseases, including rubella (German measles), chickenpox, measles, and cytomegalovirus (CMV), can cross the placenta and cause congenital infections in the fetus, which may result in various birth defects. Although it is relatively rare, the risk is highest if infection occurs during the first trimester. Vaccinations should be kept up-to date, and people with infections be avoided.

CONTACT WITH ANIMALS

Some animals, and their feces, carry diseases that can harm a developing fetus. Pregnant women should stay away from cat litter, birdcages, reptiles, rodents, and sheep at lambing time. Cats must be kept away from food preparation and eating areas, and hands should be washed after touching them. Gardening with bare hands should also be avoided in case cats have soiled the area.

COLDS, FLU, AND VACCINATIONS

Because the immune system is suppressed during pregnancy, women are more likely to catch colds and flu, and to succumb to further complications. The risk of infection can be reduced by staying away from people with cold or flu symptoms, avoiding crowds when possible, and washing hands after touching communal surfaces, such as faucets, telephones, and door handles. Having the annual flu shot protects against complications and also reduces infection of the newborn during the first six months of life.

TOXOPLASMOSIS

This rare infection is caused by a parasite that is found in animal feces, bird droppings, poorly cooked meat or fish, soil, and contaminated fruit and vegetables. Infection, especially in the second trimester, can cause eye and brain damage, congenital abnormalities, miscarriage, stillbirth, premature birth, and low birth weight. The most common sources of infection are domestic cats and undercooked meat, so precautions should be taken with food hygiene.

CHEMICALS

It is almost impossible to completely avoid chemical exposure, but it is sensible to take simple precautions. Keep the use of chemicals to a minimum, use them in well-ventilated areas, wear protective clothing, and follow the safety precautions on the packaging.

IN THE HOME

Although many pregnant women worry about risks from cleaning products, they are in fact relatively risk free. Bleach should not be mixed with other cleaners, however, and cleaning the oven should be avoided if possible. Pesticides and insecticides—even organic ones—have been liked to birth defects, pregnancy complications, and miscarriage. If possible they should be avoided altogether, especially during the first trimester. Prolonged exposure to paint chemicals may also increase the risk of miscarriage and possibly birth defects. While there is currently no firm evidence that hair dyes harm the fetus, it also makes sense to minimize exposure to these kinds of chemicals. Highlights or streaks are a better alternative to overall hair treatments, and vegetable-based dyes, such as henna, are also a good option.

DRUGS

Any prescribed medicine, over-the-counter treatments, herbal remedies, or recreational drugs taken during pregnancy can reach the fetus through the placenta. It is not always possible to avoid medication, but a doctor can advise which ones are safe during pregnancy. Care should be taken with over-the-counter remedies, which can contain multiple ingredients.

SMOKING

Smoking during pregnancy is bad for both the mother and the fetus. It has been linked with numerous problems, including an increased risk of miscarriage, premature birth, low birth weight, crib death, and breathing problems in the newborn.

PHYSICAL HAZARDS

During pregnancy, the woman's body generally provides a safe cocoon for the fetus. Special care should still be taken to avoid physical hazards. Some women may find that a combination of an altered center of gravity and loose ligaments can make injuries, such as sprains and strains, more likely to occur. It is advisable to be extra safety-conscious at this time, and to take sensible precautions such as wearing supportive, flat shoes, avoiding contact sports and other dangerous activities, and wearing a seat belt while driving. Medical advice should be sought immediately after a bad fall, accident, or other injury.

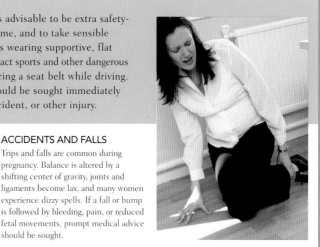

TRAVEL

Travel poses two main risks—infectious diseases and accidents. To reduce the risks, the destination should be researched and a doctor consulted about immunizations and other protective measures against diseases. The safety of the water supply should be checked and care taken with food hygiene. Pregnant women are at risk of clots in the legs (DVT), so they should avoid sitting for too long on flights.

Flying
Most airlines allow pregnant women to fly until the end of the 35th week. Women with medical conditions should check with a doctor before flying.

Seat belts
A seat belt should be worn with the lap strap under the bump and resting on the hip bones. The diagonal strap should be to the side of the bump.

INCORRECT   CORRECT

ACCIDENTS AND FALLS

Trips and falls are common during pregnancy. Balance is altered by a shifting center of gravity, joints and ligaments become lax, and many women experience dizzy spells. If a fall or bump is followed by bleeding, pain, or reduced fetal movements, prompt medical advice should be sought.

WORKING ENVIRONMENT

Most women continue working during pregnancy with few adjustments, but it is an employer's duty to ensure that there is no exposure to harmful substances or excessive physical demands. Some employers may allow pregnant staff to cut back on their hours, take more breaks, reduce the time spent standing, and provide supportive seating for them.

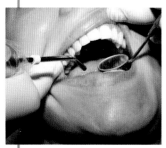

DENTAL CARE

Good oral hygiene is especially important during pregnancy. Hormonal changes increase the risk of gum disease, which has in turn been linked with an increased chance of premature birth. Most dental treatments can be given safely during pregnancy, but it is important that the dentist knows if a woman is pregnant, because some procedures and treatments, such as X-rays and certain antibiotics, are best avoided during pregnancy.

STRESS

Stress can cause increased heart rate, blood pressure, and stress hormones. There is limited evidence that severe stress, especially early in pregnancy, is linked to premature birth, low birth weight, and even miscarriage or stillbirth. Relaxation, regular exercise, a healthy diet, and sufficient sleep should be part of a daily routine.

RADIATION

X-rays can damage a developing fetus, so it is important for a woman to tell her doctor or dentist if she thinks she may be pregnant. If a chest or abdominal X-ray, CAT scan, or radiation test or treatment is needed, the benefits must be weighed up against the hazards. Most scientists believe there is minimal risk from ultrasound or from electromagnetic fields emitted by computers, mobile phones or masts, power lines, and airport screening devices.

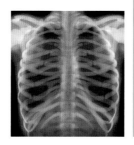

OVERHEATING

Raised body temperature during the first trimester has been linked with an increased risk of spinal deformity in the fetus. Avoid saunas and hot tubs—just 10–20 minutes' exposure can increase body temperature to dangerous levels. A hot bath does not pose the same danger because the top half of the body is exposed to cooler air, and the water gradually cools.

BEDTIME

Pregnant women can find it hard to find a comfortable sleeping position, particularly later in pregnancy when lying on the back should be avoided because pressure from the uterus can squash blood vessels. It can be even more difficult getting out of bed. This should be done slowly to avoid giddiness, straining abdominal muscles, or aggravating back pain.

DIET AND EXERCISE

Diet and exercise play an important part in overall health during pregnancy. Eating well and exercising regularly will help the fetus grow and develop healthily, and ensure that the mother's body is in peak condition and ready for the birth.

WEIGHT GAIN

Most pregnant women gain 22–29 lb (10–13 kg) during pregnancy. Gaining more increases the risk of complications, such as preeclampsia and diabetes, while insufficient weight gain is linked to premature birth and low birth weight. Weight is also an important consideration before becoming pregnant. If there are any concerns, a midwife or physician can advise on a sensible target weight.

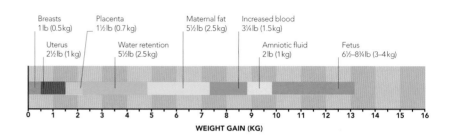

Breasts
1 lb (0.5 kg)

Placenta
1½ lb (0.7 kg)

Maternal fat
5½ lb (2.5 kg)

Increased blood
3¼ lb (1.5 kg)

Uterus
2½ lb (1 kg)

Water retention
5½ lb (2.5 kg)

Amniotic fluid
2 lb (1 kg)

Fetus
6½–8¾ lb (3–4 kg)

WEIGHT GAIN (KG)

LIMIT OR AVOID

Some foods that can normally be eaten as part of a healthy diet pose a risk during pregnancy, either because they carry a higher than average risk of food poisoning or because they may contain specific organisms or toxins that could harm the fetus. Ideally, guidance on healthy eating during pregnancy should be followed from the moment a women starts trying to get pregnant. If the pregnancy is unplanned, however, then a healthy eating regime should begin as soon as it is confirmed.

SOFT CHEESE AND DAIRY

Pregnant women are at risk of contracting listeriosis from unpasteurized dairy products, notably soft and blue-veined cheeses such as Brie, Stilton, and Camembert. It can cause miscarriage, stillbirth, or neonatal death. Hard cheeses and cottage cheese are safe and good sources of calcium.

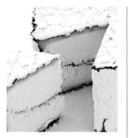

PATÉ AND LIVER

All meat and vegetable paté may contain listeria and should not be eaten. Liver, sausages, and paté contain high levels of vitamin A (retinol), which can cause birth defects. (High-strength multivitamins or cod liver oil containing vitamin A should also be avoided.)

UNDERCOOKED EGGS

Raw or partly cooked eggs may harbor salmonella, one cause of food poisoning. Eggs should be cooked until the yolks are solid, not runny, and foods containing raw eggs, such as homemade mayonnaise, or partly cooked eggs, should be avoided.

CAFFEINE AND ALCOHOL

High doses of caffeine have been linked with low birth weight and miscarriage, so caffeine consumption should be limited. It remains uncertain whether there is a safe level of alcohol intake, so it is best avoided completely.

OILY FISH

Sardines, mackerel, and other oily fish should be eaten as part of a healthy diet. But the oils concentrate pollutants, which can harm a fetus; pregnant women should have only two portions weekly, and avoid shark, marlin, and swordfish.

FOOD HYGIENE

Food poisoning can be hazardous, and some forms, such as toxoplasmosis (see p.88), pose special risks. Kitchen surfaces should be kept clean and hands washed after using the toilet, before preparing food, after handling raw meat or poultry, and before eating anything.

SHOPPING, STORAGE, AND PREPARATION

"Use by" dates should never be exceeded. Raw foods should be kept separate, and any kind of raw meat stored at the bottom of the refrigerator so it does not drip on other food. A separate chopping board must be used for raw meat, and salad, fruit, and vegetables should be washed or peeled.

REHEATING FOOD

Food that has been warmed and allowed to cool is far more likely to harbor harmful bugs. Reheated food should be heated for at least two minutes until it is steaming hot. It must be piping hot all the way through before it is served, and eaten immediately. Food should never be reheated more than once. For precooked ready meals, it is important to follow cooking instructions.

COOKING FOOD

Undercooked meat, poultry, and fish may contain bacteria, viruses, or parasites that cause food poisoning and other diseases. Frozen food should be thawed, cooked at the correct temperature for the right time, and heated all the way through before being eaten.

HEALTHY EATING

Eating healthily before conception and during pregnancy helps ensure that the body has the necessary stores of nutrients for a healthy pregnancy. Eating the right balance of the main food groups will also ensure that weight gain in pregnancy remains within healthy limits.

NUTRITION

A healthy, balanced diet includes plenty of unrefined, carbohydrate-rich starches (potatoes, wholegrain bread, and whole grains), at least five portions of fruit and vegetables daily, and sufficient meat, fish, or other high-class protein (eggs, nuts, or legumes). Milk and dairy produce or other sources of calcium are especially important for the growing fetus.

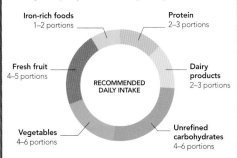

Iron-rich foods
1–2 portions

Protein
2–3 portions

Fresh fruit
4–5 portions

Dairy products
2–3 portions

RECOMMENDED
DAILY INTAKE

Vegetables
4–6 portions

Unrefined carbohydrates
4–6 portions

SUPPLEMENTS

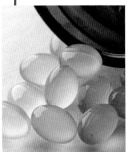

A woman trying to get pregnant is advised to take 400mcg of folic acid daily, from the time she stops contraception until the end of the first trimester. It can help prevent birth defects such as spina bifida. Some women may be advised to take a multivitamin, vitamin D, aspirin, omega-3 oils, or iron in addition.

HERBS

Most herbs are safe in cooking, but avoid basil, sage, oregano, and rosemary in high doses. Never use pennyroyal, because it can cause miscarriage, or feverfew and aloe. Raspberry leaf tea taken in late pregnancy can ease and speed labor.

ACTIVITY AND EXERCISE

Unless there are medical- or pregnancy-related problems, it is generally safe for women to continue with most of the physical activities they did before becoming pregnant. The exception is any activity that could involve an injury or jolting forces. A fall, jolt, or blow to the abdomen can lead to premature labor, and women who have had a previous miscarriage may be advised to avoid energetic sports and activities. If there is any doubt, a physician or midwife should be consulted.

EXERCISE FOR HEALTHY PREGNANCY

Exercise has many benefits during pregnancy and in preparing the body for the birth. It maintains fitness, strengthens muscles, boosts circulation, and helps prevent varicose veins, constipation, and backache. However, strenuous exercise may be more difficult. Levels of fatigue and breathlessness are a good guide, as are previous levels of fitness—pregnancy is not the time to start a demanding regime. Exercise should be stopped at once if pain or dizziness are experienced.

EXERCISING APPROPRIATELY

HIGH RISK	PROCEED WITH CAUTION	RECOMMEND
Activities involving high impact or jolting, or reduced oxygen availability, are best avoided, especially after week 12, as are those with a high risk of accidents: • Horseback riding • Skydiving • Skiing or skating • Diving	As pregnancy progresses, some activities may become harder. Be guided by how you feel and stop if you develop any symptoms: • Tennis • Running • Going to the gym • Dancing • Intense aerobics	As weight increases and center of gravity shifts, non-weight-bearing activities and those with gentle, rhythmic movements are best: • Swimming • Bicycling • Walking • Yoga (not supine positions) • T'ai chi

KEGEL EXERCISES

Exercising the pelvic floor helps prevent weakening from the weight of the uterus and strengthens the muscles used for giving birth. It reduces the risk of postpartum incontinence or prolapse. Kegel, or pelvic floor, exercises are simple and can be done in any position. The muscles involved can be identified by squeezing as if to stop urinating midflow, without clenching the abdominal or buttock muscles. They should be squeezed for a count of three, relaxed for three, and repeated 10 times. This should be done three times a day and gradually increased to 10s squeezes and 25 repetitions as often as possible.

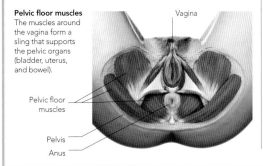

Pelvic floor muscles
The muscles around the vagina form a sling that supports the pelvic organs (bladder, uterus, and bowel).

Vagina

Pelvic floor muscles

Pelvis

Anus

EXERCISES FOR BIRTH

Giving birth demands energy and the fitter you are, the more likely labor will proceed smoothly. Any regular exercise is helpful—at least half an hour three times a week. Squats can help strengthen thigh muscles, and sitting cross legged improves flexibility in your pelvic joints.

SEX

It is usually safe to have sex in pregnancy. Positions may have to be adapted to avoid the bump, but the fetus will be safely cushioned inside the amniotic fluid and protected from infection by the cervical plug. Physicians may advise against sex if there is a history of miscarriage, premature birth, bleeding, or other complications.

MONTH 2 | **WEEKS 5–8**

In this period of remarkable growth, the embryo grows from the size of a grain of rice to that of a raspberry, with correspondingly rapid growth of its vital organs. The woman's uterus reaches the size of a grapefruit, her waist thickens, and her breasts enlarge.

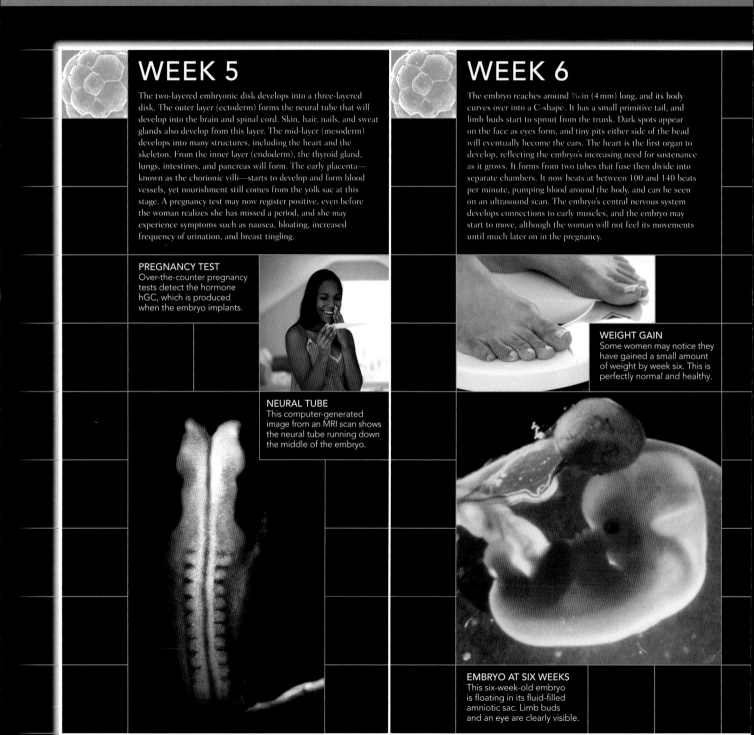

WEEK 5

The two-layered embryonic disk develops into a three-layered disk. The outer layer (ectoderm) forms the neural tube that will develop into the brain and spinal cord. Skin, hair, nails, and sweat glands also develop from this layer. The mid-layer (mesoderm) develops into many structures, including the heart and the skeleton. From the inner layer (endoderm), the thyroid gland, lungs, intestines, and pancreas will form. The early placenta—known as the chorionic villi—starts to develop and form blood vessels, yet nourishment still comes from the yolk sac at this stage. A pregnancy test may now register positive, even before the woman realizes she has missed a period, and she may experience symptoms such as nausea, bloating, increased frequency of urination, and breast tingling.

PREGNANCY TEST
Over-the-counter pregnancy tests detect the hormone hGC, which is produced when the embryo implants.

NEURAL TUBE
This computer-generated image from an MRI scan shows the neural tube running down the middle of the embryo.

WEEK 6

The embryo reaches around ⁵⁄₁₆ in (4 mm) long, and its body curves over into a C-shape. It has a small primitive tail, and limb buds start to sprout from the trunk. Dark spots appear on the face as eyes form, and tiny pits either side of the head will eventually become the ears. The heart is the first organ to develop, reflecting the embryo's increasing need for sustenance as it grows. It forms from two tubes that fuse then divide into separate chambers. It now beats at between 100 and 140 beats per minute, pumping blood around the body, and can be seen on an ultrasound scan. The embryo's central nervous system develops connections to early muscles, and the embryo may start to move, although the woman will not feel its movements until much later on in the pregnancy.

WEIGHT GAIN
Some women may notice they have gained a small amount of weight by week six. This is perfectly normal and healthy.

EMBRYO AT SIX WEEKS
This six-week-old embryo is floating in its fluid-filled amniotic sac. Limb buds and an eye are clearly visible.

WEEK 7

The embryo continues to grow rapidly to around ³⁄₁₆ in (8 mm) long—about the size of a kidney bean. The limb buds develop paddle-shaped ends from which fingers and toes will form. The lens and retina begin to develop in the rudimentary eyes, and the liver forms and begins producing red blood cells. Veins become apparent under the fetus's skin. The yolk sac starts to shrink as the developing chorionic villi increasingly supply the embryo with oxygen and nutrients from the maternal bloodstream. The woman's clothing may now start to feel uncomfortably tight around the waist. Dietary tastes often change, and some women develop aversions to particular foods. In some women, increased circulating blood volume may give rise to headaches.

WEEK 8

By the end of the second month, the embryo measures about ½ in (1.4 cm) long, the size of a raspberry, and all major organs have started to form. The primitive tail begins to disappear, and limbs lengthen and develop webbed fingers and toes. Unique fingerprints have already formed. As the elbows develop, the arms curve and can move. The brain matures further, and heart valves form so that the primitive circulation flows in the right direction. The lungs continue to grow, and airways develop that connect them to the back of the throat. The mother's uterus is now the size of a small grapefruit and may press on the lower spine, sometimes causing backache. Her waistline is now thicker, and her breasts may appear bigger, although she will still not look noticeably pregnant to others.

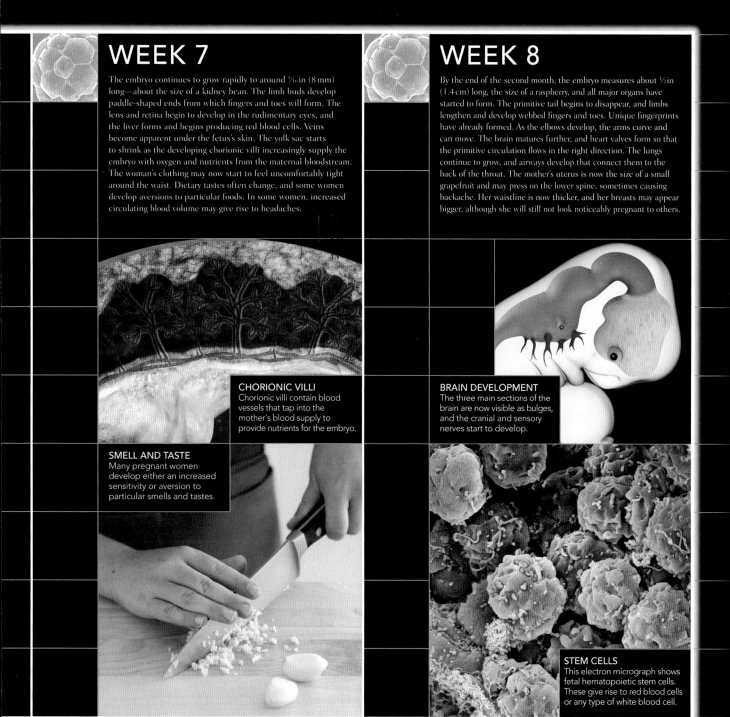

CHORIONIC VILLI
Chorionic villi contain blood vessels that tap into the mother's blood supply to provide nutrients for the embryo.

BRAIN DEVELOPMENT
The three main sections of the brain are now visible as bulges, and the cranial and sensory nerves start to develop.

SMELL AND TASTE
Many pregnant women develop either an increased sensitivity or aversion to particular smells and tastes.

STEM CELLS
This electron micrograph shows fetal hematopoietic stem cells. These give rise to red blood cells or any type of white blood cell.

MONTH 2 | WEEKS 5–8
MOTHER AND EMBRYO

At this stage of pregnancy, a large number of mothers notice a feeling of nausea (often not exclusively in the morning), increased fatigue, and the need to urinate more frequently. These common early symptoms of pregnancy may not pass until after 12 weeks, and some women experience them for longer. Many of these symptoms are side effects caused by the hormones that are produced within the ovary to support the development and growth of the early embryo. Over the next two weeks, the embryo takes on a more recognizably human form. The growth of the brain is especially rapid, leading to a head size that is half the length of the body. The embryo remains curled up and floats weightlessly within the amniotic sac. By the eighth week all the organ systems have been formed—they are entire, but minute, and their function is very limited.

MOTHER

- 66 beats per minute
- 106/69
- 7½ pints (4.33l)

400 mg

Pregnant women should continue to take 400mg of folic acid a day, up to week 12.

MOTHER AT EIGHT WEEKS
Some women may notice no changes during early pregnancy, whereas others may experience quite strong reactions to the huge physiological changes.

Stomach
Nausea is common from six weeks, but it usually passes by about 12 weeks. Progesterone may increase acid reflux or heartburn.

Bowel
Progesterone relaxes the smooth muscle of the intestines, which slows the passage of waste products and can lead to constipation.

Uterus
The uterus has enlarged slightly, but remains well within the pelvis.

| 1 | 2 | 3 | 4 | 5 | 6 | 7 | 8 | 9 | 10 | 11 | 12 | 13 | 14 | 15 | 16 | 17 | 18 | 19 | 20 | 21 | 22 | 23 | 24 | 25 | 26 | 27 | 28 | 29 | 30 | 31 | 32 | 33 | 34 | 35 | 36 | 37 | 38 | 39 | 40 |

EMBRYO

- 144 beats per minute
- ⅝ in (1.6cm)
- 1/32 oz (1g)

3/8 in

The fetus is growing rapidly at this stage. In just two weeks, between six and eight weeks, it increases in length by ⅜ in (1 cm).

By eight weeks, development of the heart is finally complete, with all four chambers beating.

During the second month, the embryo is at its most sensitive to the effects of drugs and other toxins. Certain drugs taken by the mother at this stage can result in birth defects or even fetal death.

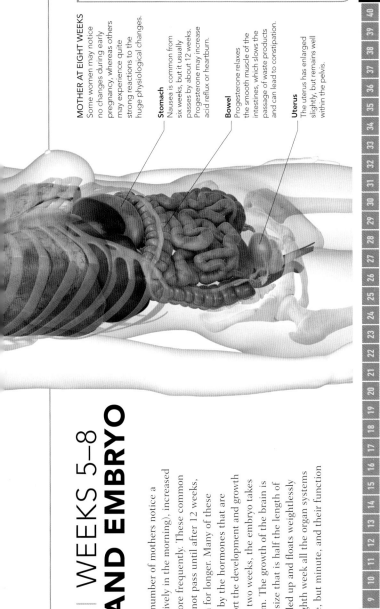

6 WEEKS

The embryo has started to take on a more human appearance. A number of internal organs can be seen and, externally, the ear, eye, and limb buds are now apparent. Growth is very rapid at this stage, with the embryo doubling in size over the following two weeks.

Yolk sac
The earliest blood cells and capillaries form in the wall of the yolk sac.

Villi
Simple villi make up the placenta, which is growing more rapidly than the embryo at this stage.

Umbilical cord
The short umbilical cord is not yet coiled. Blood vessels are clearly visible.

Somite
The somites develop into the spine, vertebral column, trunk muscles, and skin.

Eye

Brachial arches
These are the precursors of the lower jaw and neck structures.

Embryo
The embryo is suspended in the amniotic fluid.

Heart
Development of the heart is almost complete; circulation is established and the heart has started beating.

Upper limb bud
The higher limb buds will eventually develop into arms.

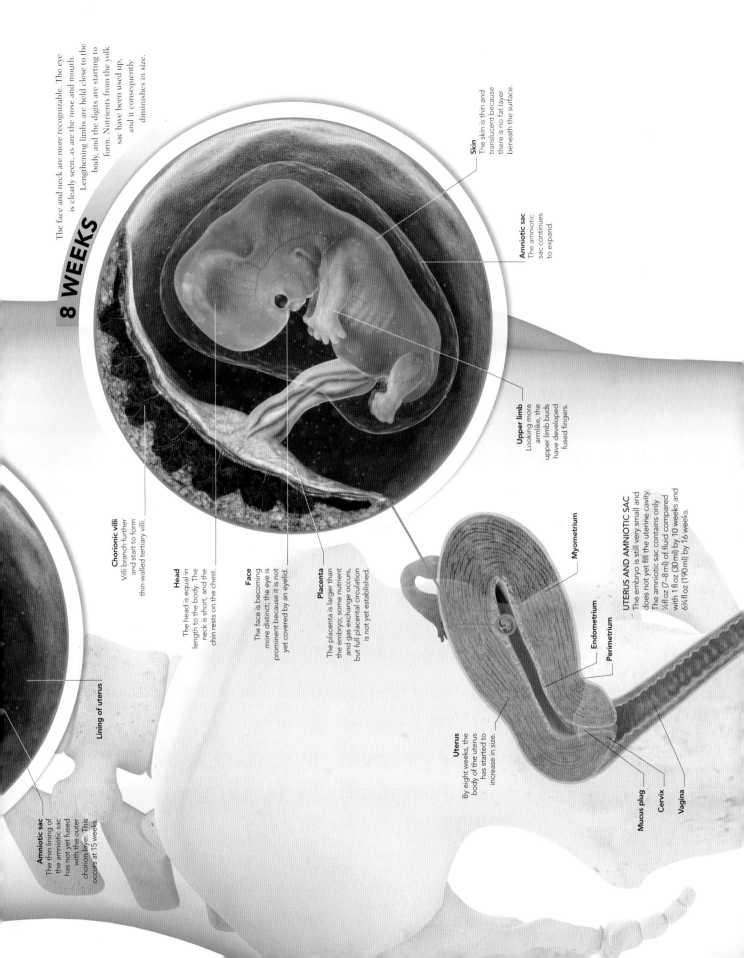

8 WEEKS

The face and neck are more recognizable. The eye is clearly seen, as are the nose and mouth. Lengthening limbs are held close to the body, and the digits are starting to form. Nutrients from the yolk sac have been used up, and it consequently diminishes in size.

Skin
The skin is thin and translucent because there is no fat layer beneath the surface.

Amniotic sac
The amniotic sac continues to expand.

Upper limb
Looking more armlike, the upper limb buds have developed fused fingers.

Chorionic villi
Villi branch further and start to form thin-walled tertiary villi.

Head
The head is equal in length to the body. The neck is short, and the chin rests on the chest.

Face
The face is becoming more distinct; the eye is prominent because it is not yet covered by an eyelid.

Placenta
The placenta is larger than the embryo; some nutrient and gas exchange occurs, but full placental circulation is not yet established.

Myometrium

Endometrium

Perimetrium

UTERUS AND AMNIOTIC SAC
The embryo is still very small and does not yet fill the uterine cavity. The amniotic sac contains only ¼ floz (7–8ml) of fluid compared with 1 floz (30ml) by 10 weeks and 6¾ floz (190ml) by 16 weeks.

Uterus
By eight weeks, the body of the uterus has started to increase in size.

Mucus plug

Cervix

Vagina

Lining of uterus

Amniotic sac
The thin lining of the amniotic sac has not yet fused with the outer chorion layer. This occurs at 15 weeks.

MONTH 2 | **KEY DEVELOPMENTS**

MOTHER

PREGNANCY TESTING

A pregnancy test responds to human chorionic gonadotropin (hCG), a hormone that is produced after conception and is detectable in the urine within two weeks. It contains alpha and beta protein molecules (subunits)—the beta subunit is unique to hCG, and it is this element that is measured in the pregnancy test. Tests are now so sensitive, they can recognize a pregnancy even a few days before menstruation is due.

Positive result

Control panel

Negative result

READING A RESULT
In this test, a positive result requires a blue plus sign in one window and a blue line in the control panel. Other tests may display the results in a different way.

THE CERVICAL MUCUS PLUG

Under the influence of the hormones stimulated by fertilization, cervical mucus alters consistency. At around four weeks it changes from thin mucus into a thick, firm plug that sits in the cervical canal, sealing the entrance to the uterus. This forms a barrier against any infections ascending from the vagina to the uterus.

BARRIER
The mucus plug sits securely in the cervix throughout pregnancy. One of the early signs of labor occurs when it is released as the cervix begins to shorten and open.

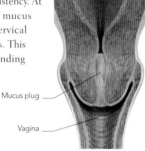

Mucus plug

Vagina

TOLERATING THE FETUS

Pregnancy is a delicate balancing act, and most miscarriages occur in the first 12 weeks. The mother's immune system needs to accept the developing embryo, which would otherwise be detected as foreign and attacked, while maintaining defenses against potential infectants. The mechanism that protects the embryo from maternal immunity is not fully understood, but the role of progesterone is critical. It forms a blocking antibody that mops up any antigen—substances that provoke an immune response—released by the embryo; it also renders white blood cells less able to attack foreign tissue.

FOREIGN TISSUE
Some endometrial white blood cells are naturally more tolerant than those in the general circulation, and this helps protect the developing embryo.

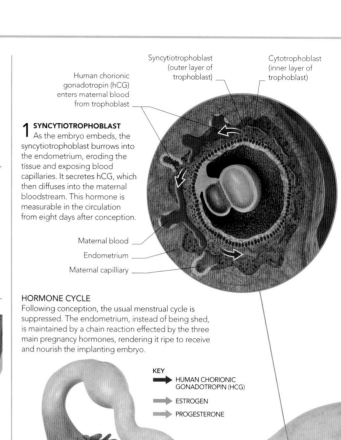

Syncytiotrophoblast (outer layer of trophoblast)

Cytotrophoblast (inner layer of trophoblast)

Human chorionic gonadotropin (hCG) enters maternal blood from trophoblast

1 SYNCYTIOTROPHOBLAST
As the embryo embeds, the syncytiotrophoblast burrows into the endometrium, eroding the tissue and exposing blood capillaries. It secretes hCG, which then diffuses into the maternal bloodstream. This hormone is measurable in the circulation from eight days after conception.

Maternal blood

Endometrium

Maternal capilliary

HORMONE CYCLE
Following conception, the usual menstrual cycle is suppressed. The endometrium, instead of being shed, is maintained by a chain reaction effected by the three main pregnancy hormones, rendering it ripe to receive and nourish the implanting embryo.

KEY

➤ HUMAN CHORIONIC GONADOTROPIN (HCG)

➤ ESTROGEN

➤ PROGESTERONE

hCG in maternal blood stops corpus luteum from disintegrating

Maternal blood vessels

2 CORPUS LUTEUM
High levels of hCG in the blood stimulate continued growth of the corpus luteum in the ovary— it would otherwise disintegrate. It secretes progesterone and estrogen into the maternal bloodstream.

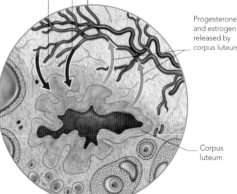

Progesterone and estrogen released by corpus luteum

Corpus luteum

HORMONAL CHANGES

One of the key hormones at the start of pregnancy is hCG, which is released as the embryo embeds in the endometrium. This hormone is responsible for maintaining the corpus luteum in the ovary, which in turn produces relatively small but crucial quantities of estrogen and progesterone. Although hCG declines after 12 weeks, the graph below shows that low levels persist, meaning that a pregnancy test remains positive throughout pregnancy. After 12 weeks, the placenta takes over production of estrogen and progesterone, secreting both hormones in massive quantities. Progesterone levels are higher until around 28 weeks, after which estrogen levels dominate.

KEY
— HUMAN CHORIONIC GONADOTROPIN (HCG)
— ESTROGEN
— PROGESTERONE
● OVULATION

PREGNANCY HORMONES
The above graph shows fluctuations in the three main hormones that act throughout a 40-week pregnancy.

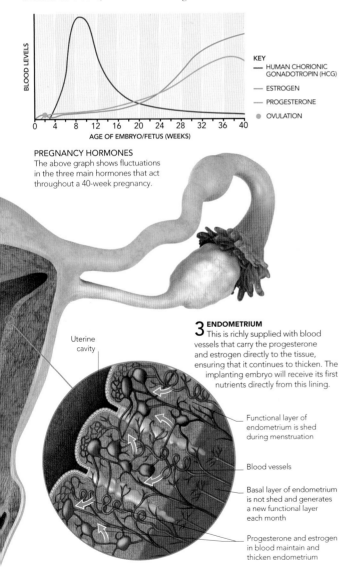

Uterine cavity

3 ENDOMETRIUM
This is richly supplied with blood vessels that carry the progesterone and estrogen directly to the tissue, ensuring that it continues to thicken. The implanting embryo will receive its first nutrients directly from this lining.

Functional layer of endometrium is shed during menstruation

Blood vessels

Basal layer of endometrium is not shed and generates a new functional layer each month

Progesterone and estrogen in blood maintain and thicken endometrium

EARLY SYMPTOMS OF PREGNANCY

Many of the early symptoms of pregnancy are actually side effects of the surging hormones that are necessary for a successful pregnancy. Symptoms vary between individuals in both timing and intensity. Furthermore, no two pregnancies seem to be alike, and a woman may find that certain symptoms may be severe in one pregnancy but not the next. Many of the symptoms will improve over time and seem to be related to hCG levels, which naturally decline after 12 weeks. The most common early pregnancy symptoms are described in the table below.

ALLEVIATING NAUSEA
Morning sickness is extremely common and can be highly disruptive. Eating regularly can help alleviate nausea, as can soothing herbal teas, in particular mint or ginger teas.

EARLY SYMPTOMS

Missed period	A woman's period should follow around two weeks after ovulation unless fertilization has occurred; this is most likely if sex took place near to ovulation. A pregnancy test at the time of a missed period is sensitive enough to detect the presence of an early pregnancy.
Tender and enlarged breasts	Changes to the breasts begin soon after conception and include an increase in breast size, sensitivity, and vascular patterns. Under the influence of the early pregnancy hormones, the ductal system is the first area to proliferate, with the glandular tissue increasing much later in pregnancy. Breast soreness experienced within the first trimester tends to ease as the pregnancy progresses.
Fatigue	The exact cause of fatigue during the early weeks is unknown. It does not affect all women and usually improves by 12 weeks. Fatigue may be related to early hormonal changes and the body's gradual acclimatization to the pregnancy.
Urinary frequency	An increase in blood flow to the kidneys and improvements in their filtering capacity occur early in pregnancy. Urination may occur more often as a result, although excessively high frequency or pain on urinating may indicate an infection requiring help.
Nausea and vomiting	Commonly known as "morning sickness," nausea and vomiting are the classic early symptoms of pregnancy. They can be present at any time of the day or night, and may be exacerbated by certain foods or smells. Usually it takes a mild form, but in rare cases the more severe hyperemesis gravidarum can occur.
Metallic taste in mouth	Changes to taste sensation, such as a metallic taste in the mouth or the preference for certain foods, may be experienced. These usually settle during the pregnancy or, if not, very soon afterward.
Spotting and bleeding	Spotting may occur at the time of implantation—this coincides with the time that menstruation is due and can be confused with a light period. The cervix also softens during pregnancy, and this may lead to some spotting following intercourse.
Constipation	Progesterone prevents the uterus from contracting before term, but it also slows down the contraction of all smooth muscle. This causes digestion to be sluggish, leading to constipation.

EMBRYO

THE DEVELOPMENT OF PRIMARY GERM LAYERS

After implantation, the two-layered embryonic disk undergoes a rapid transformation into a three-layered disk, following the formation of a band of cells called the primitive streak, from which the third cell layer derives. These three primary germ layers are the building blocks from which every body system is derived. The ectoderm forms the upper layer, the endoderm forms the lower layer, and the mesoderm, which is the final layer to appear, is sandwiched between them. They represent the first simple differentiation of cells as they follow separate developmental pathways. Many structures are composed of a combination of all three layers, although some are entirely formed from a single germ layer.

Head-end of embryo

Embryonic disk

1 PRIMITIVE STREAK FORMATION
During the fifth week, a strip of cells known as the primitive streak forms and lengthens along the surface of the embryonic disk. At its head is the primitive node, which moves toward the future head-end of the embryo.

Line of cross section

Head–tail axis
Progression of primitive streak establishes head–tail axis of embryo

Primitive node

Ectoderm
Upper layer of embryonic disk

Amniotic sac

Future mouth

Tail-end of embryo

Primitive streak

Cells move between layers

Mesoderm
Middle layer of embryonic disk

Endoderm
Lower layer of embryonic disk

2 PRODUCTION OF MESODERM
As the primitive streak extends, it forms a depression (the primitive groove). Cells from this groove, known as mesoderm, move between existing layers of endoderm and ectoderm to become the third layer of the embryonic disk.

CROSS SECTION THROUGH THE PRIMITIVE STREAK

BODY SYSTEMS AND THEIR PRIMARY GERM LAYERS

ENDODERM	MESODERM	ECTODERM
• Digestive tract	• Skin (dermis)	• Skin (epidermis)
• Respiratory tract	• Bone	• Hair
• Urinary tract	• Muscle	• Nails
• Liver	• Cartilage	• Tooth enamel
• Glands, such as thyroid and pancreas	• Connective tissue	• Central nervous system
• Reproductive tract	• Heart	• Mammary glands
	• Blood cells and vessels	• Sense-organ receptor cells
	• Lymph cells and vessels	• Parts of eyes, ears, and nasal cavities
	• Kidneys and ureters	

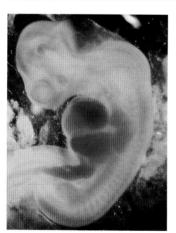

FOLDED SIX-WEEK-OLD EMBRYO
At six weeks, the embryo has a clearly identifiable shape. The heart and liver can be seen through the translucent skin—the heart is in the center and the liver is to its right.

EMBRYONIC FOLDING

By the end of the fifth week of pregnancy, differentiation into a flat, three-layered disk is complete, and the embryo then undergoes a complex three-dimensional folding, from head-end to tail-end and from side to side. This creates the shape of the early human embryo. Embryonic folding results in an enclosed primitive gut tube, which extends from the foregut at the head-end of the embryo, through the midgut, which at this stage is linked to the future yolk sac, and then terminates at the tail-end with the hindgut. The connection between the midgut and the yolk sac gradually narrows until the yolk sac enters into the embryo at the site of the umbilical cord. The connecting stalk to the early placenta is what develops into the umbilical cord. A small tube (the allantois) develops from the hindgut and protudes into the connecting stalk—this will eventually connect to the bladder. At this early stage of development, many species look similar as the most basic body parts are gradually mapped out.

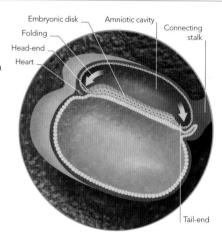

Embryonic disk

Amniotic cavity

Folding

Connecting stalk

Head-end

Heart

Tail-end

1 31 DAYS
Rapid growth at the head- and tail-ends of the disk results in the onset of embryonic folding. The primitive heart, one of the first organs to develop, forms a small bulge that is initially positioned near the head end.

STEM CELLS

Human stem cells have the potential to develop into any cell type in the body. This function is usually lost after the cells commit to developing along a specific pathway, for example, to become a skin cell, nerve cell, or muscle fiber. Umbilical-cord blood is a rich source of fetal stem cells; these provide an exact genetic match for an individual and, because they can be cultured into any cell type, offer great potential for the future treatment of disease.

SPECIALIZING CELLS
This electron micrograph shows embryonic stem cells—their ability to specialize makes them a key focus of scientific research.

NEURAL TUBE FORMATION

The neural tube will form the central nervous system, consisting of the brain and spinal cord. Its development begins with the appearance of the notochord, a column of cells that extends along the back of the embryo and solidify. Ectoderm cells above the notochord sink to form a depression, the edges of which fuse and become a tube. This tube forms centrally then extends outward along the length of the embryo. It finally closes at the top of the embryo on the 38th day of pregnancy, and closes at the base of the spine two days later. As the embryo folds, the neural tube adopts a C-shape— it is not uniform in diameter, but develops dilations at the head-end, identifying the forebrain, midbrain, and hindbrain as distinct divisions from the spinal cord.

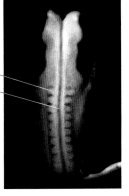

Somite

Neural tube

SOMITES
Mesoderm segments become condensed into pairs called somites. These first form in the fifth week of pregnancy. Three or four pairs then appear each day, starting at the head, until there are 42 pairs by the sixth week.

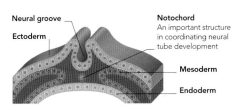

Neural groove

Ectoderm

Notochord
An important structure in coordinating neural tube development

Mesoderm

Endoderm

1 FORMATION OF NEURAL GROOVE
The solid notochord derives from the mesoderm layer. Ectoderm cells directly above sink down to form the neural groove.

Neural folds meet

Early neural tube
Site of future spinal cord

2 FUSION OF NEURAL FOLDS
As the neural groove deepens, its edges (the neural folds) gradually come together, forming an early neural tube.

Neural crest
Specialized cells will migrate to initiate numerous structural developments

Neural tube
Neural folds fuse and neural tube is complete

3 FORMATION OF NEURAL TUBE
The neural folds meet, fuse, and finally break away from the overlying ectoderm. Failure to fuse results in spina bifida.

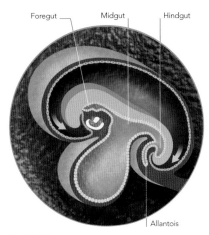

Foregut   Midgut   Hindgut

Allantois

2 38 DAYS
The head rapidly expands as the embryo lengthens, causing it to curl around the cardiac bulge. Within the embryo, neural crest cells are spreading out to form components of the eyes, skin, nerves, and adrenal glands.

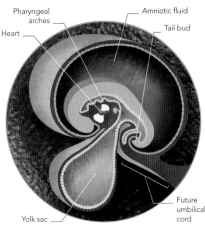

Pharyngeal arches

Heart

Amniotic fluid

Tail bud

Yolk sac

Future umbilical cord

3 42 DAYS
The amniotic cavity now almost fully encircles the embryo. The tail bud will gradually regress as the head continues to expand, and pharyngeal arches of tissue start to form in the future neck and lower jaw region.

HUMAN TAILS

A human tail is an extremely rare finding and its origins are not entirely understood. Unlike a true tail, there are no bones within it, and it simply consists of a length of skin, with a variable amount of nerve tissue, extending from the lowest portion of the spinal column. The condition is often associated with a failure of the lower part of the spine to close around the spinal cord.

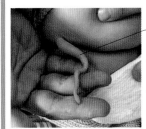

Soft tail

VESTIGIAL REMAINS
Human tails are usually quite short, whereas this image shows a particularly long example of this rare condition.

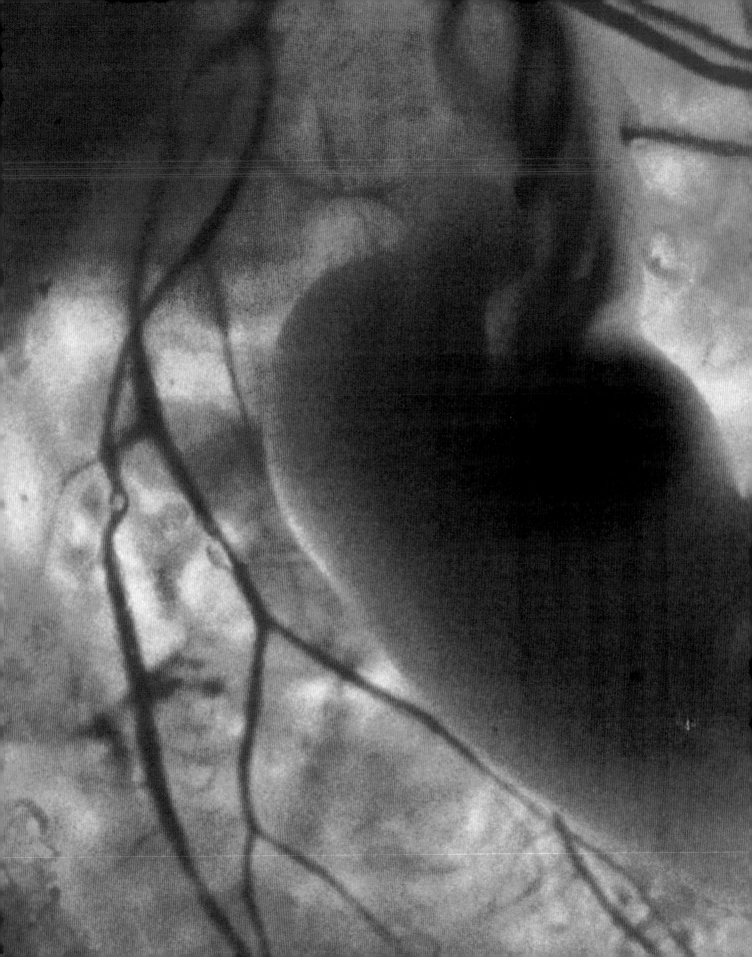

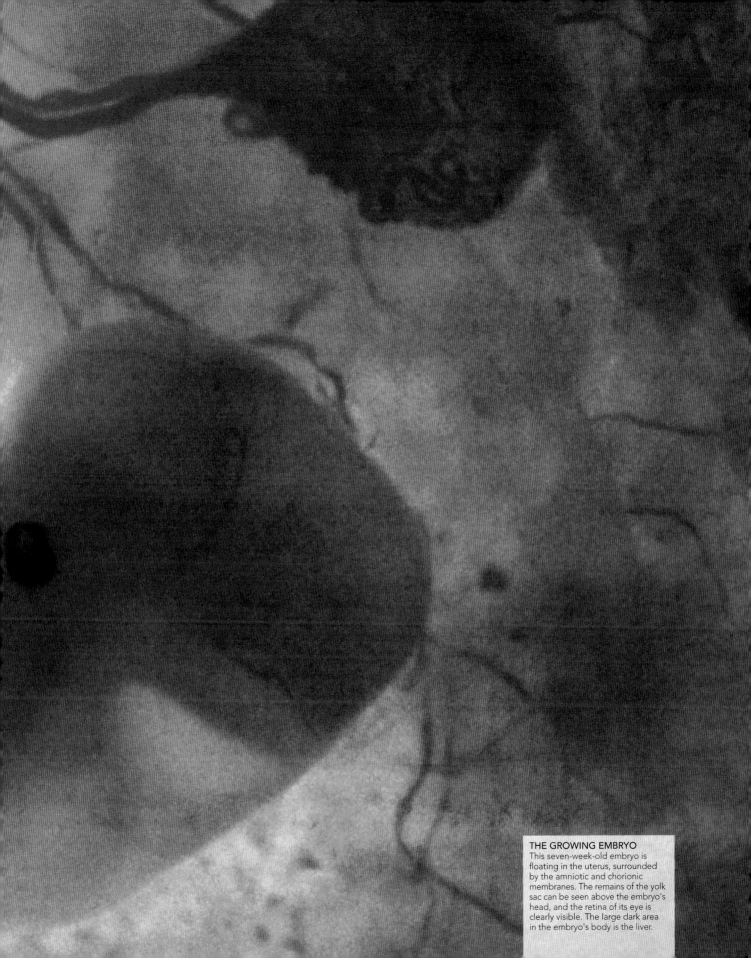

THE GROWING EMBRYO
This seven-week-old embryo is floating in the uterus, surrounded by the amniotic and chorionic membranes. The remains of the yolk sac can be seen above the embryo's head, and the retina of its eye is clearly visible. The large dark area in the embryo's body is the liver.

EMBRYO

NOURISHING THE EMBRYO

Initially the embryo receives nutrients from the yolk sac and eliminates waste products by simple diffusion. This soon becomes inadequate, and a placental interface between the maternal and fetal circulations is established. The outer trophoblast layer invades the uterine lining, eroding maternal capillaries and forming lakes of blood in the rudimentary placenta. The placental tissue sends out fingerlike projections, or villi, to maximize the surface area exposed to blood. These become finer and more numerous, and by the third week they contain simple fetal capillaries. A week later, the early placenta surrounds the entire embryo, but the more distant villi disappear as the mature placenta becomes centered on the umbilical cord. Nutrient exchange is still restricted until a formal circulation develops in the 10th week, with the tertiary chorionic villi becoming filled with circulating fetal blood.

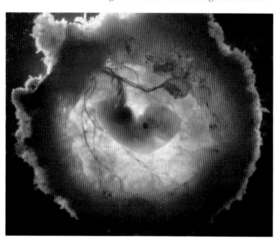

THE CHORION
The outer wall of the blastocyst is called the chorion. The chorion starts to fuse with the amniotic sac by eight weeks (a process that may last up until 15 weeks). This forms a double layer of membranes around the fetus. These membranes rupture during labor when "the waters break."

FUNCTION OF THE YOLK SAC

The yolk sac is a structure outside the embryo that is involved in the care and maintenance of the immature fetus. In the early days of the pregnancy, when the placenta's ability to transfer nutrients is limited, the yolk sac plays a key role in the provision of nutrition by simple diffusion. In this way, and in other ways, it has a similar function to the liver. The first simple capillaries grow within the yolk sac walls and rudimentary, oxygen-carrying blood cells are formed there too. As the placenta begins to function, the yolk sac reduces in size and by the end of the pregnancy, it has disappeared.

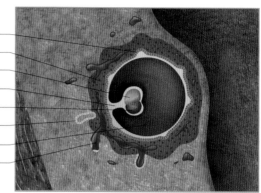

Primary chorionic villus
Protrusions form on inner layer of trophoblast

Outer trophoblast layer

Yolk sac

Connecting stalk

Amniotic sac

Endometrial gland

Chorionic sac

Erosion
Maternal blood from endometrial capillaries fills endometrial glands

1 PRIMARY CHORIONIC VILLUS
By day 26, the invading outer trophoblasts form simple fronds as they invade maternal tissues. Maternal blood leaches into endometrial glands.

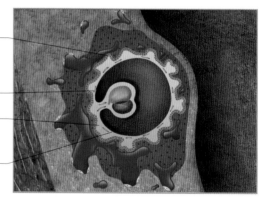

Secondary chorionic villus
Protrusions enlarge to form fingerlike projections

Vessel formation
Early blood vessels begin to form within connective tissue

Connective tissue
Forms a core within secondary chorionic villus

Wall of chorionic sac
Formed by the two layers of trophoblast and the connective tissue

2 SECONDARY CHORIONIC VILLUS
By day 28, small lakes of maternal blood have formed as the capillary walls dissolve. The maternal barrier to nutrient exchange has been broken down.

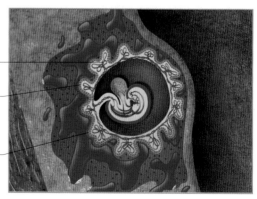

Blood vessels
Form a network within the villi, connecting the stalk and embryo

Barrier
Inner layer of trophoblast prevents maternal and fetal blood from mixing

Diffusion
Development of villi creates a larger surface area for nutrients and oxygen to diffuse across

3 TERTIARY CHORIONIC VILLUS
Further branching refines the structure of the villi to form tertiary villi. These project into the lakes of maternal blood. Fetal capillaries have not grown so nutrient transfer is still inefficient.

AMNIOTIC FLUID

The amniotic fluid protects the fetus from trauma and provides space for it to grow and move. It aids lung development and helps maintain the fetus at a constant temperature. At first it is similar to the plasma in the fetal circulation, but as the fetus's kidneys start to produce urine, this passes into the amniotic fluid. By the end of the pregnancy the fluid is more concentrated and similar to urine. Fetal swallowing and fluid absorption within the gut remove amniotic fluid. As pregnancy advances, amniotic fluid volume steadily increases, reaching 1¾ pints (1 l) by 32 weeks, but can be as much as 3½ pints (2 l). By the end of pregnancy, ⁹⁄₁₀–1¾ pints (0.5–1 l) of amniotic fluid is removed by fetal swallowing and replaced by urine every day.

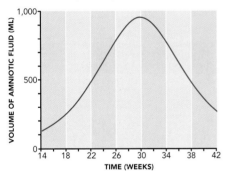

Yolk sac

Umbilical cord

Amniotic sac

CHANGING VOLUME
Toward the end of pregnancy, amniotic fluid volume decreases as the fetal kidneys produce smaller volumes of more concentrated urine.

AMNIOTIC SAC
The amniotic sac completely encloses the embryo. The yolk sac—being a transient structure—remains outside the amniotic sac.

BLOOD DEVELOPMENT

From day 31, primitive red blood cells arise in the wall of the yolk sac—they are formed in blood islands with simple surrounding capillaries. The earliest primitive red blood cells contain embryonic hemoglobin, and they have a central nucleus unlike mature red blood cells. By day 74, the fetal liver will have taken over blood cell production from the yolk sac. Unlike the first primitive red blood cells, the cells produced in the fetal liver can differentiate into any of the components of fetal blood. By the end of pregnancy blood cell production also occurs within the bone marrow.

FETAL BLOOD CELLS
This electron micrograph shows a type of stem cell. In the fetus it gives rise to red blood cells or any type of white blood cell.

BLOOD CELLS

Blood cell production in the liver starts at 37 days. Some blood production occurs in the bone marrow from as early as 10 weeks but the liver remains the dominant site until after birth. Red blood cell production is high. Each fetal red blood cell survives for only 60 days—half that in the adult. The embryo needs iron, folic acid, and vitamin B12 to produce sufficient red blood cells.

White blood cell

Red blood cell

CELL TYPES
Fetal red blood cells resemble the adult type but their hemoglobin binds more avidly to oxygen.

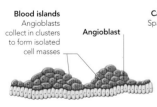

Blood islands
Angioblasts collect in clusters to form isolated cell masses

Angioblast

Cavity formation
Spaces form within blood islands

1 BLOOD ISLANDS
Blood islands or aggregates arise in the yolk sac and connecting stalk. The inner cells form primitive red blood cells while the outer cells make capillary walls.

2 CAVITY DEVELOPMENT
The differentiation between capillary wall and early red blood cells begins with the appearance of spaces inside the blood islands.

Lumen
Cavities grow and fuse to form lumen of blood vessel

Blood cells
Lining of vessel lumen forms cells

3 BLOOD VESSEL FORMATION
The first blood cells produced are almost exclusively primitive red blood cells. A simple network of capillaries is complete by the end of the third week.

EMBRYO

ORGANOGENESIS

Organogenesis is a process of rapid embryonic development, at the end of which all the major organs and external structures have appeared. It lasts from the sixth to the 10th week. Different systems develop concurrently. The respiratory system emerges from an out-pocketing of the foregut to form the lungs, and the digestive system gives rise to the intestines, liver, gall bladder, and pancreas. The first fully functioning system is the cardiovascular system, which consists of the heart and a simple circulation that is continuously remodeled as the embryo develops.

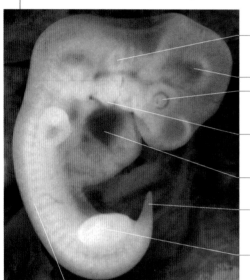

Ear
Marked by a shallow pit, this is the eventual site of the ear.

Brain
The rapid development that has occurred in the brain has caused the head to bend over.

Eye
The precursor to the lens is visible; the eyelid has still to develop and close the eye.

Pharyngeal arches
These five distinctive ridges give rise to many head and neck features in the fetus.

Heart
The dark area clearly shows the position of the heart.

Tail
This is not a tail in the true sense, but an extension of the skin that covers the spinal cord.

Limb bud
Signs of a developing leg are evident although it bears little resemblance to the real thing.

Somites
Bordering the neural tube, somites differentiate into skin, muscle, and vertebrae.

EARLY BODY STRUCTURES
This seven-week-old embryo is about midway through organogenesis. The development of body systems overshadows growth of the embryo.

THE DEVELOPMENT OF THE LUNGS

Lung development begins at day 50 and continues into early infancy. The rudimentary windpipe develops two branches that subdivide into successively finer tubes. The initial branching pattern is common to all embryos, but the final arrangement is unique. By the 18th week, 14 divisions have taken place to form the respiratory tree but the bronchioles are too large and their walls are too thick for gas exchange (respiration). Recognizable primitive alveoli, with walls thin enough for gas transfer (see pp.152–53), do not appear until 38 weeks.

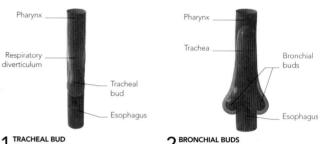

1 TRACHEAL BUD
The first sign of the developing windpipe or trachea is a pouch that grows outward and downward from the esophagus.

Labels: Pharynx, Respiratory diverticulum, Tracheal bud, Esophagus

2 BRONCHIAL BUDS
On day 56, after the trachea has lengthened sufficiently, it divides into two bronchial buds; each will form a lung.

Labels: Pharynx, Trachea, Bronchial buds, Esophagus

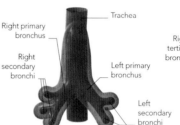

3 SECONDARY BRONCHI
The bronchial buds branch in a very specific manner. The right bud branches three times and the left branches twice.

Labels: Trachea, Right primary bronchus, Right secondary bronchi, Left primary bronchus, Left secondary bronchi

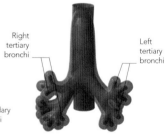

4 TERTIARY BRONCHI
By day 70, the third series of divisions is underway. The result is 10 lung segments on the right and eight segments on the left.

Labels: Right tertiary bronchi, Left tertiary bronchi

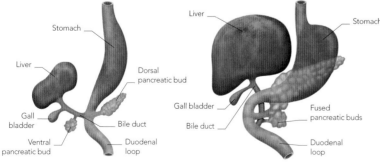

1 EMBRYO AT NINE WEEKS
The gut develops specialized structures that branch off the main tube. The early pancreas is made up of two separate buds.

Labels: Stomach, Liver, Gall bladder, Dorsal pancreatic bud, Bile duct, Ventral pancreatic bud, Duodenal loop

2 EMBRYO AT 10 WEEKS
The two separate pancreatic buds have fused, and the bile duct connecting the gall bladder to the duodenum has lengthened.

Labels: Liver, Stomach, Gall bladder, Bile duct, Fused pancreatic buds, Duodenal loop

THE DIGESTIVE SYSTEM

The digestive system starts as a simple tube joining mouth to anus. Gradually, portions specialize, with the stomach first to form, during the seventh week. During the ninth week, the bowel lengthens so much that it cannot be contained within the abdomen and it pushes out into the umbilical cord. Here it rotates 90° counterclockwise before returning to the abdominal cavity by the end of the 12th week. The small and large bowel reach their final positions by the 14th week. By 17 weeks, amniotic fluid enters the gut as the fetus makes regular swallowing motions. Although the gut cannot move until mid-pregnancy, villi in the intestines enable absorption of the fluid.

THE DEVELOPMENT OF THE HEART

The heart develops early, allowing nutrients to be distributed to support the embryo's development. It is the first system to work fully. The heart beats from day 50, and blood starts to circulate two to three days later. A cardiac bulge appears above the insertion point of the umbilical cord, and the heart is formed here from two thin-walled tubes as they fuse from top to bottom. The embryonic circulation continues as the final structure of the heart emerges. Looping and remodeling progress rapidly and are completed at the end of the 10th week. The fetal heart is lined with a special tissue called endocardium, and the muscular tissue of the heart (myocardium) is unique in its ability to contract spontaneously with a regular intrinsic rhythm.

DIVISION OF HEART CHAMBERS

The heart has left and right upper chambers (atria) that collect blood from the network of veins, and two lower chambers (ventricles) that pass blood out of the heart. The atria and ventricles are divided by walls (septa) that grow in toward the central crux of the heart (endocardial cushion). A one-way valve controls blood flow from each atrium to its corresponding ventricle. The ventricles are separated but the atria communicate via the foramen ovale, which allows oxygenated blood to pass through.

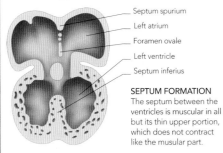

- Septum spurium
- Left atrium
- Foramen ovale
- Left ventricle
- Septum inferius

SEPTUM FORMATION
The septum between the ventricles is muscular in all but its thin upper portion, which does not contract like the musular part.

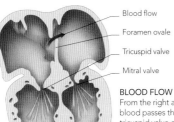

- Blood flow
- Foramen ovale
- Tricuspid valve
- Mitral valve

BLOOD FLOW
From the right atrium, blood passes through the tricuspid valve or through the foramen ovale. The left atrium passes blood into the left ventricle.

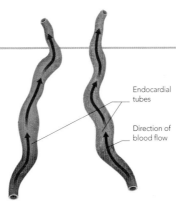

Endocardial tubes

Direction of blood flow

1 ENDOCARDIAL TUBES
Early on in development, two separate and parallel tubes direct blood toward the head of the embryo.

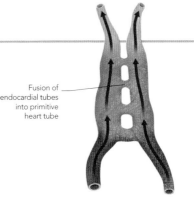

Fusion of endocardial tubes into primitive heart tube

2 PRIMITIVE HEART TUBE
The endocardial tubes merge from the base upward, eventually forming a single primitive heart tube by day 50.

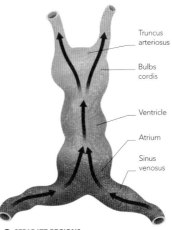

- Truncus arteriosus
- Bulbs cordis
- Ventricle
- Atrium
- Sinus venosus

3 SEPARATE REGIONS
Subtle constrictions demarcate separate portions while cardiac jelly and myocardium (beating cardiac muscle) surround the tube.

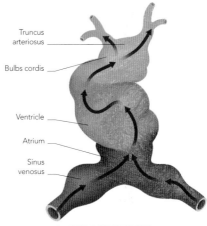

- Truncus arteriosus
- Bulbs cordis
- Ventricle
- Atrium
- Sinus venosus

4 BENDING OF HEART TUBE
On day 51, the beating heart tube elongates and loops to the right, forming a spiral. A basic circulation is established.

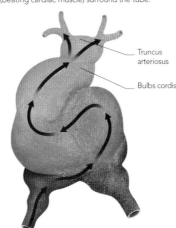

- Truncus arteriosus
- Bulbs cordis

5 S-FORMATION
By day 53, the heart tube has looped to form an S-shape, bringing the four chambers to their correct spatial orientation.

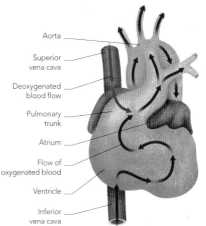

- Aorta
- Superior vena cava
- Deoxygenated blood flow
- Pulmonary trunk
- Atrium
- Flow of oxygenated blood
- Ventricle
- Inferior vena cava

6 FINAL POSITION OF CHAMBERS
The four chambers have achieved complete separation by day 84, and the heart valves are in place by day 91.

MONTH 3 | **WEEKS 9–12**

This month the embryo becomes a fetus. It is now recognizably human and moves vigorously. By the end of the first trimester the pregnancy is well established, and the risk of miscarriage much lower—many women choose this time to announce their pregnancy to the world.

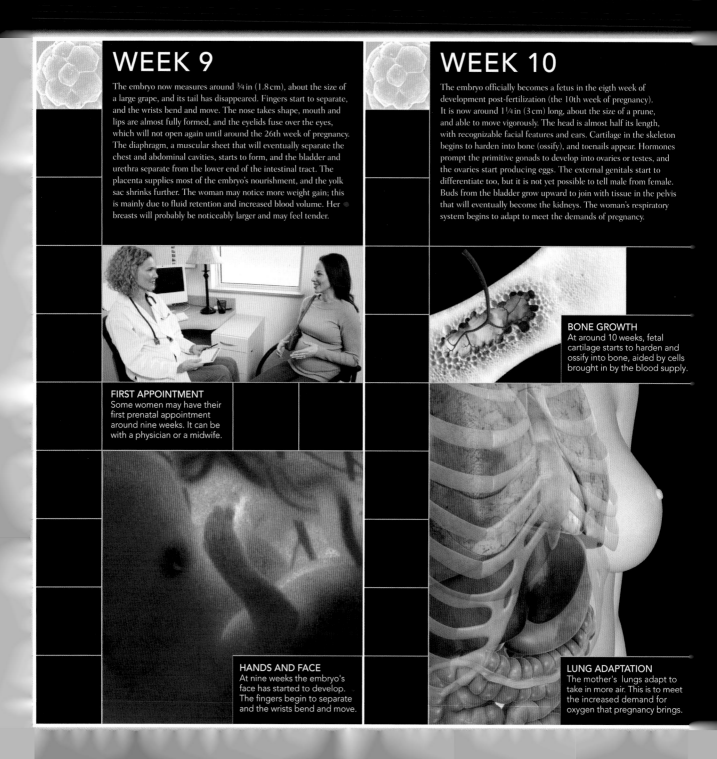

WEEK 9

The embryo now measures around ¾ in (1.8 cm), about the size of a large grape, and its tail has disappeared. Fingers start to separate, and the wrists bend and move. The nose takes shape, mouth and lips are almost fully formed, and the eyelids fuse over the eyes, which will not open again until around the 26th week of pregnancy. The diaphragm, a muscular sheet that will eventually separate the chest and abdominal cavities, starts to form, and the bladder and urethra separate from the lower end of the intestinal tract. The placenta supplies most of the embryo's nourishment, and the yolk sac shrinks further. The woman may notice more weight gain; this is mainly due to fluid retention and increased blood volume. Her breasts will probably be noticeably larger and may feel tender.

FIRST APPOINTMENT
Some women may have their first prenatal appointment around nine weeks. It can be with a physician or a midwife.

HANDS AND FACE
At nine weeks the embryo's face has started to develop. The fingers begin to separate and the wrists bend and move.

WEEK 10

The embryo officially becomes a fetus in the eigth week of development post-fertilization (the 10th week of pregnancy). It is now around 1¼ in (3 cm) long, about the size of a prune, and able to move vigorously. The head is almost half its length, with recognizable facial features and ears. Cartilage in the skeleton begins to harden into bone (ossify), and toenails appear. Hormones prompt the primitive gonads to develop into ovaries or testes, and the ovaries start producing eggs. The external genitals start to differentiate too, but it is not yet possible to tell male from female. Buds from the bladder grow upward to join with tissue in the pelvis that will eventually become the kidneys. The woman's respiratory system begins to adapt to meet the demands of pregnancy.

BONE GROWTH
At around 10 weeks, fetal cartilage starts to harden and ossify into bone, aided by cells brought in by the blood supply.

LUNG ADAPTATION
The mother's lungs adapt to take in more air. This is to meet the increased demand for oxygen that pregnancy brings.

WEEK 11

The fetus is now around 2 in (5 cm) long, about the size of a plum. It can open and close its mouth, enabling yawning and swallowing. Tiny teeth buds are forming within the jaw, fingers and toes start to lose their webbing, and the skin thickens and loses its previous transparency. The heart is beating faster, between 120 and 160 beats per minute, and blood is circulating rapidly around the fetal body. The woman's abdomen may protrude a little, and she may feel increasingly breathless on exertion due to the increased workload of the heart and lungs. The enlarging uterus now moves upward out of the pelvis, reducing pressure on the bladder, so urinary symptoms lessen, but existing varicose veins or hemorrhoids may swell up, or new ones may develop.

WEEK 12

The fetus is now on average 2¼ in (6 cm) long, about the size of a kiwi fruit. As brain cells multiply rapidly, the brain develops into two distinct halves (left and right hemispheres)—each one controls the opposite side of the body. Developed reflexes mean the fetus may move in response to pressure on the abdomen, suck a thumb or fist, and urinate. It begins to produce its own hormones, and the genitals may show the first outward signs of gender. Some women now appear pregnant and may have to adjust clothing to accommodate the bump. Hormonal changes can cause nipples and areolae to darken, although this will be more pronounced later on. Often any nausea now passes, appetite improves, and early pregnancy tiredness gives way to increased energy.

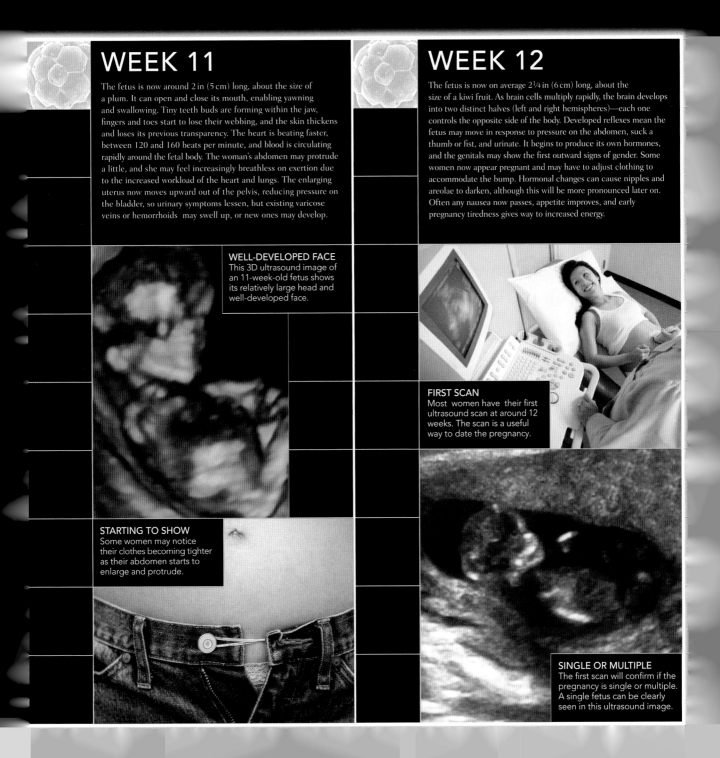

WELL-DEVELOPED FACE
This 3D ultrasound image of an 11-week-old fetus shows its relatively large head and well-developed face.

FIRST SCAN
Most women have their first ultrasound scan at around 12 weeks. The scan is a useful way to date the pregnancy.

STARTING TO SHOW
Some women may notice their clothes becoming tighter as their abdomen starts to enlarge and protrude.

SINGLE OR MULTIPLE
The first scan will confirm if the pregnancy is single or multiple. A single fetus can be clearly seen in this ultrasound image.

MONTH 3 | WEEKS 9–12
MOTHER AND FETUS

Early pregnancy symptoms, such as fatigue and nausea, usually peak this month, possibly due to high levels of the hormone hCG. During this month, the embryonic period ends and the fetal stage begins. The yolk sac shrinks, and its role diminishes as the placenta takes over. The placenta is now far larger than the fetus and can easily meet oxygen and nutrient demands and remove waste products and carbon dioxide. During this month, energy is directed toward completion of the basic organ structures. The face becomes more recognizable as the eyes form and the jaw and neck lengthen, bringing the ears toward their final positions. The neck is still relatively short, and the fetus adopts a curled position, with its head held close to its chest. Reflecting rapid brain growth, the head occupies half the total crown-to-rump length, and in just three weeks the fetus doubles in length.

MOTHER

🫀 66 beats per minute

🩸 105/68

💧 7½ pints (4.4 l)

1 fl oz

The volume of amniotic fluid in the 12th week of pregnancy is around 1 floz (27 ml). It will reach a peak of around 33¼ floz (946 ml) by week 34.

High levels of progesterone may lead to some women developing spots or acne.

The discomfort of early pregnancy often peaks, then improves, in month three.

MOTHER AT 12 WEEKS

Anticipating future demands, the mother breathes more deeply, absorbs nutrients more efficiently, and her circulation system delivers more blood to the placenta.

Stomach
At 12 weeks, hCG levels have reached their peak, which can lead to nausea and sickness in many women.

Bowel
Rising progesterone levels can slow bowel transit times, leading to constipation. A high-fiber diet and drinking plenty of water can reduce symptoms.

Uterus
The uterus has started to expand and now tips slightly forward. It can just be felt above the brim of the pelvis.

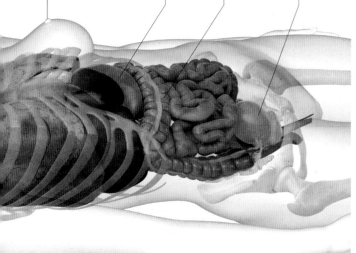

FETUS

🫀 175 beats per minute

📏 2¼ in (5.4 cm)

⚖️ ½ oz (14 g)

The embryonic period is complete by week 10, when the embryo becomes a fetus. The fetal period begins with early signs of organ growth.

The pregnancy can be accurately dated around week 12 by an ultrasound measurement of the length of the fetus, called the crown-rump length.

Ultrasound scanning will show the fetal heart beat and the limbs. Simple trunk and limb movements can also be seen at this stage. Swallowing starts and fluid can be seen in the stomach and bladder.

| 1 | 2 | 3 | 4 | 5 | 6 | 7 | 8 | 9 | 10 | 11 | 12 | 13 | 14 | 15 | 16 | 17 | 18 | 19 | 20 | 21 | 22 | 23 | 24 | 25 | 26 | 27 | 28 | 29 | 30 | 31 | 32 | 33 | 34 | 35 | 36 | 37 | 38 | 39 | 40 |

10 WEEKS

At 10 weeks, placental efficiency improves to support the demands of the rapidly growing fetus.
The fetus's bowel pushes through into the base of the umbilical cord.
It reenters the abdomen by 11–12 weeks, rotating as it moves back (see p.122).

Chorionic villi
Well-formed tertiary villi start to appear within the placenta to aid nutrient transfer.

Umbilical cord
Fetal movements encourage the umbilical cord to coil.

Head
The head represents 50 percent of the total fetal length. This is a reflection of the amount of brain development that needs to occur before other organs and body systems can mature.

Ear
The ear is quite low on the jaw line, but over the next two to three weeks it will ascend to its final position.

Neck
The neck is still shortened, which forces the head toward the chest and gives the fetus a curled-up appearance.

Legs
These are less developed than the arms; toes are not fully separated.

12 WEEKS

Head growth starts to slow down at 12 weeks. The neck lengthens, and the head lifts up and away from the chest—one effect of this change is that the fetus can now start swallowing. The kidneys begin to function by passing minute volumes of dilute urine into the amniotic fluid.

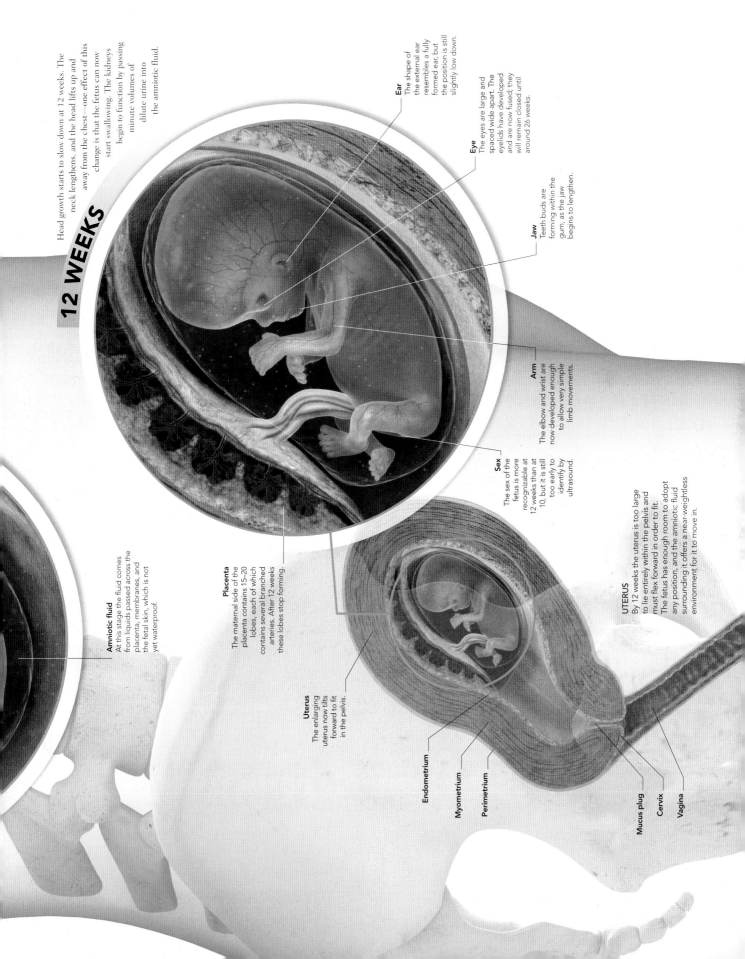

Ear
The shape of the external ear resembles a fully formed ear, but the position is still slightly low down.

Eye
The eyes are large and spaced wide apart. The eyelids have developed and are now fused; they will remain closed until around 26 weeks.

Jaw
Teeth buds are forming within the gum, as the jaw begins to lengthen.

Arm
The elbow and wrist are now developed enough to allow very simple limb movements.

Sex
The sex of the fetus is more recognizable at 12 weeks than at 10, but it is still too early to identify by ultrasound.

Amniotic fluid
At this stage the fluid comes from liquids passed across the placenta, membranes, and the fetal skin, which is not yet waterproof.

Placenta
The maternal side of the placenta contains 15–20 lobes, each of which contains several branched arteries. After 12 weeks these lobes stop forming.

Uterus
The enlarging uterus now tilts forward to fit in the pelvis.

Endometrium

Myometrium

Perimetrium

UTERUS
By 12 weeks the uterus is too large to lie entirely within the pelvis and must flex forward in order to fit. The fetus has enough room to adopt any position, and the amniotic fluid surrounding it offers a near-weightless environment for it to move in.

Mucus plug

Cervix

Vagina

MOTHER

EARLY PRENATAL CARE

At the first meeting, the health care professional offers information about pregnancy, healthcare services, and lifestyle choices, including screening tests and dietary information. The right to decline tests is also explained. This is the stage when questions should be asked and individual care plans discussed. Whether the prenatal care is hospital or community based, it will include regular meetings with a health care professional. Details of the hospital and care team involved are noted in the mother's medical records.

MEETING THE HEALTH CARE PROFESSIONAL
The first meeting with a health care professional should be well before 12 weeks, so there is plenty of time to discuss future pregnancy needs.

PRENATAL APPOINTMENTS (1ST AND 2ND TRIMESTERS)

At each prenatal appointment, a number of routine checks and tests are performed, to make sure everything is proceeding as it should and to identify whether additional care or medical attention is required.

TIMING	NATURE OF APPOINTMENT
11–14 weeks	The first ultrasound is carried out, which dates the pregnancy. Many hospitals offer the option to screen for Down syndrome at this stage.
16 weeks	The blood tests taken at the first visit are reviewed. Blood pressure is measured, and urine is checked for protein, which may indicate infection.
18–20 weeks	An ultrasound scan is done to assess placental and fetal development. If the placenta is lying low (see p.139) a further scan at 32 weeks is arranged.
20 weeks	Often you will be offered a review with the medical team to finalize the plan for your pregnancy or to discuss ultrasound results.
24 weeks	If it is a first pregnancy, this appointment is for routine checks, including blood pressure, and to measure growth of the uterus.

COMMON CONCERNS DURING PREGNANCY

Some women may be concerned about not feeling the baby move, but this can vary greatly (see p.138). Nausea and sickness are normal and may occur up to 20 weeks; heartburn can continue for longer. Some discomfort as the uterus enlarges and the ligaments and joints loosen is common, but the health care professional should be told if it is very painful. Vaginal discharge is normal, but it should not itch, smell, or be accompanied by any bleeding. The need to urinate may also become more frequent, but it should not hurt to do so.

UTERINE PAIN
Occasional discomfort is common at this early stage, but constant pain, bleeding, or fluid loss should always be investigated.

ADAPTATION OF LUNGS

The lungs adapt rapidly early on in pregnancy, in anticipation of increased oxygen demands. At first, this may make the mother feel short of breath, but the changes actually make the lungs work more efficiently. Deeper breathing increases oxygen absorption and removes more carbon dioxide. This is achieved through changes to the rib position and the elevation of the diaphragm, not through changes to the structure of the lungs. As the diaphragm is pushed up, the residual volume, which is not involved in gas exchange, is decreased in favor of the tidal volume, which is the amount of air that can be breathed in with a normal breath.

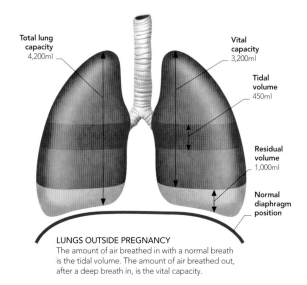

LUNGS OUTSIDE PREGNANCY
The amount of air breathed in with a normal breath is the tidal volume. The amount of air breathed out, after a deep breath in, is the vital capacity.

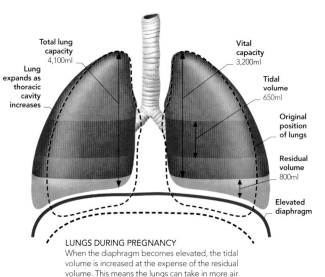

LUNGS DURING PREGNANCY
When the diaphragm becomes elevated, the tidal volume is increased at the expense of the residual volume. This means the lungs can take in more air.

PASSING ON IMMUNITY

Protection for the fetus, and for the baby once it is born, relies on immunity that is passed from the mother across the placenta. During pregnancy, most viral infections are fought by the mother's immune system. After birth, the antibody immunoglobulin type G (IgG), which crosses over the placenta from the mother during pregnancy, provides immunity for the baby. Breastfeeding allows immunoglobulin type A (IgA) to pass to the baby for additional protection. However, not all antibodies can pass over to the fetus. Immunoglobulin type M (IgM) antibodies, which are produced at the early stages of a viral attack, are too large to cross the placenta.

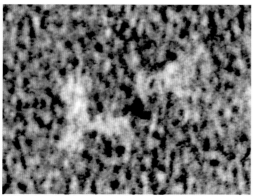

IMMUNOGLOBULIN G (IGG)
This color-enhanced electron micrograph shows the Y-shaped structure of IgG antibodies. They are the most abundant antibodies and are present in all body fluids. They are also the only ones able to cross the placenta.

PROTECTED FETUS
IgG antibodies provide the fetus with early immunity against disease, until it is able to boost this with its own antibodies. These are not usually produced until 20 weeks.

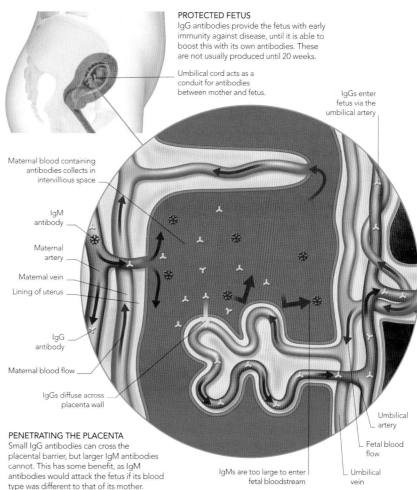

Umbilical cord acts as a conduit for antibodies between mother and fetus.

IgGs enter fetus via the umbilical artery

Maternal blood containing antibodies collects in intervillious space

IgM antibody

Maternal artery

Maternal vein

Lining of uterus

IgG antibody

Maternal blood flow

IgGs diffuse across placenta wall

Umbilical artery

Fetal blood flow

Umbilical vein

IgMs are too large to enter fetal bloodstream

PENETRATING THE PLACENTA
Small IgG antibodies can cross the placental barrier, but larger IgM antibodies cannot. This has some benefit, as IgM antibodies would attack the fetus if its blood type was different to that of its mother.

NASAL CONGESTION

Nasal congestion (pregnancy rhinitis) is now a recognized symptom during pregnancy, although it is still not clear why it occurs. It affects one in five women and, although it is often confused with hay fever, it is not an allergic reaction. Nasal congestion can develop at any time during pregnancy, but it settles down 1–2 weeks after delivery. Although no universally successful treatment exists, simple measures, such as raising the head of the bed, physical exercise, and saline washes, can help. A physician will be able to distinguish pregnancy rhinitis from a sinus infection, which may require antibiotic treatment.

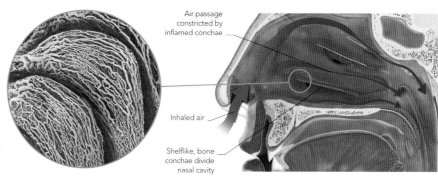

Air passage constricted by inflamed conchae

Inhaled air

Shelflike, bone conchae divide nasal cavity

NASAL CAPILLARIES
The lining of the nose has numerous capillaries that warm incoming air. Irritation of the lining can lead to engorgement of the capillaries, which can exacerbate pregnancy rhinitis.

CONSTRICTED AIR PASSAGE
Excess mucus production and irritation to the lining of the nasal passages both contribute to restricted airflow. To improve airflow, saline washes may be necessary.

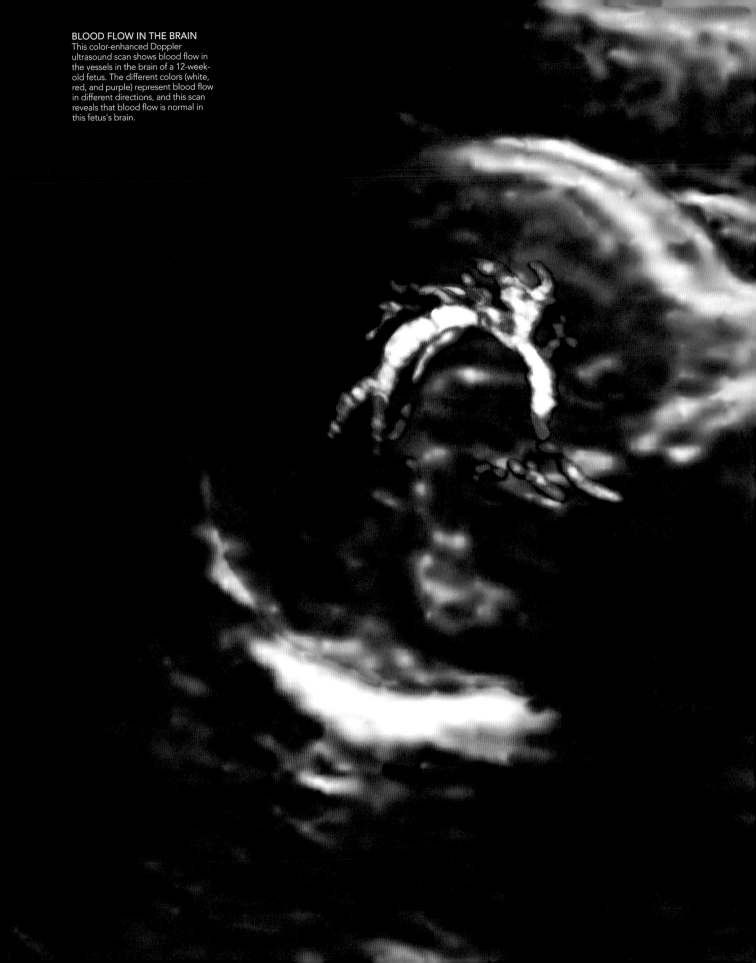

BLOOD FLOW IN THE BRAIN
This color-enhanced Doppler ultrasound scan shows blood flow in the vessels in the brain of a 12-week-old fetus. The different colors (white, red, and purple) represent blood flow in different directions, and this scan reveals that blood flow is normal in this fetus's brain.

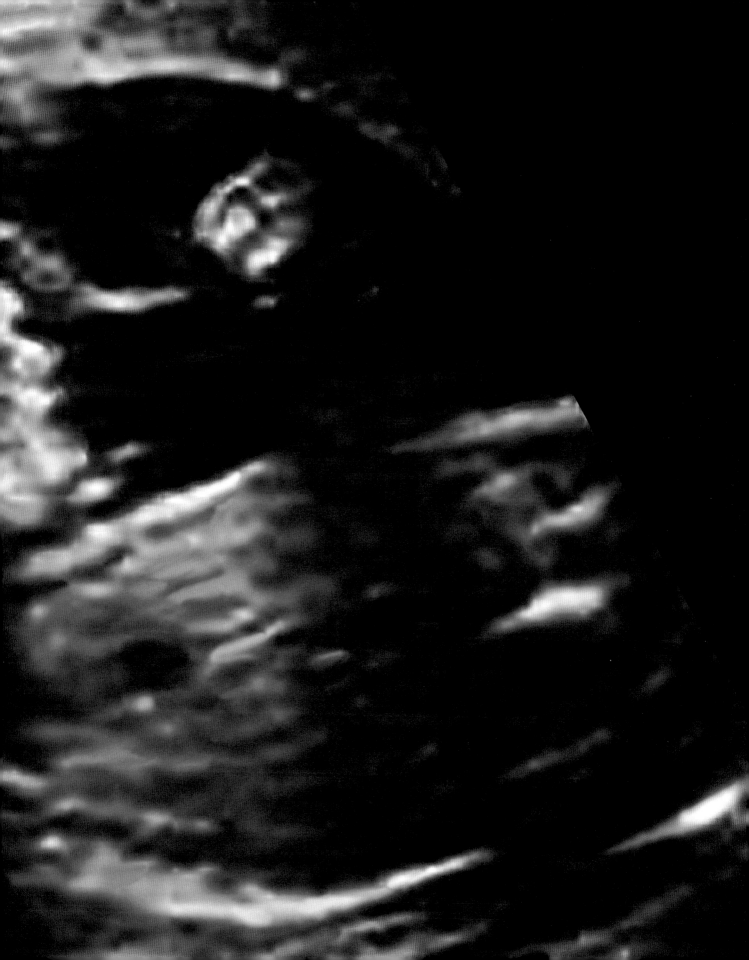

FETUS

THE DEVELOPING PLACENTA

The architecture of the placenta continues to mature as the surface area expands and the barrier between the maternal and fetal circulations thins. The walls of the mother's arteries have been invaded by fetal cells that weaken their structure, causing them to dilate, lowering resistance and releasing a continuous stream of blood into the intervillous space. On the fetal side, the villi branch into tertiary villi that float within pools of maternal blood. During the ninth week, these villi lengthen, reaching their maximum length by the 16th week. To meet the ever-increasing demands of the growing fetus, however, placental development continues well into the second half of pregnancy. To improve nutrient and gas transfer further, the villi walls start to thin out, and delicate side branches appear after 24 weeks.

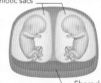

UMBILICAL CORD

The umbilical cord enables fetal blood to circulate to the placenta (along two coiled arteries) before returning with nutrients and oxygen to the fetus (within a single wide vein). Normally, arteries—and not veins—are the vessels that carry oxygen. However, veins take blood toward the heart and arteries away from it, hence the naming of the umbilical vessels. Stimulated by fetal movements, the cord gradually becomes coiled. This is a protective mechanism that, along with its jellylike covering (known as Wharton's jelly), prevents the cord from kinking.

FETAL LIFELINE
This photograph, taken inside the uterus, shows blood vessels within a section of coiled cord.

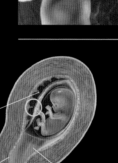

Villous chorion
Frondlike texture of chorion provides large surface area for gas exchange

Smooth chorion
Fronds erode as sac protrudes into uterine cavity

Uterine cavity

Mucus plug

Maternal blood vessel

Oxygen and nutrients diffuse into fetal bloodstream

Maternal blood pools in intervillous space

VITAL NOURISHMENT
Throughout pregnancy the fetus is entirely dependent on the umbilical cord for incoming nutrients and oxygen as well as for the removal of waste products and carbon dioxide.

Fetal waste passes back into maternal bloodstream

Umbilical vein carries oxygenated blood

Umbilical arteries carry deoxygenated blood

Blood flow to fetus

Blood flow from fetus

GAS EXCHANGE
Exchange between maternal and fetal blood occurs in the intervillous spaces. The villi, which are part of the fetus, project into these spaces, where maternal blood circulates. Gases pass from the maternal blood into the blood circulating in the villi, while waste travels in the opposite direction.

TWINS

Nonidentical twins (dizygotic) arise from two separate eggs so they may be of the same or different sex. This type accounts for 92 percent of all twins. More rarely, a single fertilized egg will divide into two identical, same-sex individuals (monozygotic). The timing of this division determines whether there will be one or two placentas (chorionicity) and one or two amniotic sacs (amnionicity).

CLEAVAGE AFTER 1–3 DAYS
If a single egg divides this early into two identical twins, each will exist separately from the other. Because they do not share the same placenta, they do not share a circulation. There is also no risk of them becoming entangled.

Separate amniotic sacs

Separate placentas

CLEAVAGE AFTER 4–8 DAYS
The twins are in separate sacs (diamniotic) so cannot become entwined but their circulations can mix across the joint placenta (monochorionic). If one donates more blood than it receives, problems can arise (see below).

Separate amniotic sacs

Shared placenta

CLEAVAGE AFTER 8–13 DAYS
The thin amniotic membrane separating the twins is absent (monoamniotic) and they share a placenta (monochorionic). Twins that share a placenta may have an unequal share of the circulation: twin–twin transfusion syndrome.

Shared amniotic sac

Shared placenta

CLEAVAGE AFTER 13–15 DAYS
Cleavage after 13–15 days produces conjoined twins. They may be joined at the head, chest, or abdomen. The complex circulation and sharing of organs at each level carries serious implications for separation.

Shared amniotic sac

Shared placenta

THE FIRST ULTRASOUND SCAN

The initial ultrasound scan is usually performed between 11 and 14 weeks of pregnancy. This is the best time to most accurately date the pregnancy and determine the expected delivery date. Dating is achieved by measuring the fetus from the top to tail, also called the crown-rump length. This measurement is identical in all fetuses of 11–14 weeks; major differences in fetal size only become apparent in the second half of pregnancy. In this first sighting of the fetus, both hands and feet can be seen, fluid in the stomach and bladder observed, and the heart beat recognized. If more than one fetus is present, the number of amniotic sacs and the number of placentas can be determined most accurately at this stage.

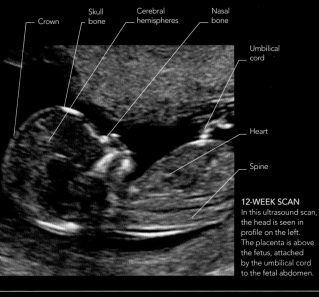

Crown · Skull bone · Cerebral hemispheres · Nasal bone · Umbilical cord · Heart · Spine

12-WEEK SCAN
In this ultrasound scan, the head is seen in profile on the left. The placenta is above the fetus, attached by the umbilical cord to the fetal abdomen.

NUCHAL TRANSLUCENCY SCAN

A nuchal translucency measurement is offered from the 11th week to 13 weeks and 6 days. It can identify pregnancies at increased risk of Down syndrome. Fetuses with Down syndrome may have an increase in the amount of fluid in the nuchal region, and this, along with the mother's age, will generate a risk estimate for Down syndrome. The scan identifies approximately 7 out of 10 fetuses with Down syndrome. Recently, blood hormone levels have been included to generate a more accurate estimate. A high risk result from the combined screening test will identify about 9 out of 10 fetuses with Down syndrome.

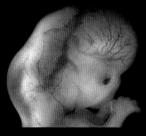

EXCESS NUCHAL FLUID
This fetus, photographed within the uterus, has excess nuchal fluid, which can be seen at the extreme left. All fetuses have some nuchal fluid, but the amount may be greater than normal for no apparent reason or in association with certain genetic conditions and some structural problems.

NORMAL SCAN
With the fetus in profile and clear of the amniotic membrane below, the widest part of the nuchal translucency (space between white plus signs) is carefully measured.

Normal nuchal fold
As seen here, the normal, narrow nuchal translucency is most commonly between $\frac{1}{16}$ and $\frac{1}{8}$in (1 and 3mm).

INCREASED NUCHAL FLUID SCAN
If there is a greater than normal measurement of nuchal fluid, the health care team will discuss possible implications with the parents.

Larger nuchal fold
In this fetus, the nuchal translucency exceeds $\frac{1}{8}$in (3.5mm).

CHORIONIC VILLUS SAMPLING (CVS)

If there is an increased risk of the fetus having a genetic or chromosomal disorder, the genetic material (karyotype) of the fetus can be checked by CVS after 10 weeks and up to 15 weeks, when an amniocentesis is often preferred. The genetic material in the placenta is almost always identical to that in the fetus. Under ultrasound guidance, the long, fine needle of a syringe is passed through the abdomen into the placenta, and a miniscule amount of placental tissue is removed and sent for analysis. The sample is always taken away from the cord insertion area. Sometimes the specimen has to be taken by passing a fine tube through the cervix and using gentle suction. CVS carries a chance of miscarriage that is 1–2 in 100; for amniocentesis the chance is 1 in 200.

ABDOMINAL PROCEDURE
Here, a needle is inserted through the abdominal wall (the transabdominal route), and cells are extracted away from the cord insertion. Ultrasound guidance ensures the safe and accurate positioning of the needle.

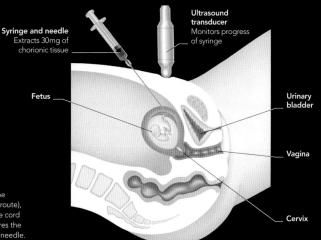

Syringe and needle
Extracts 30mg of chorionic tissue

Ultrasound transducer
Monitors progress of syringe

Fetus · Urinary bladder · Vagina · Cervix

FETUS

EARLY BRAIN DEVELOPMENT

The fetus's brain develops throughout pregnancy. By the third month, major changes have already occurred. The thalami are by now jointly the largest element of the brain, acting as a relay station for the hemispheres. Beneath the paired thalami is the hypothalamus, which controls organ functions such as heart rate. Under the hypothalamus is the third ventricle, a chamber filled with circulating cerebrospinal fluid (CSF) produced by the choroid plexus within each lateral ventricle. The cerebral hemispheres expand rapidly, although they are smooth at this stage, not achieving the familiar folded appearance until late in the second half of the pregnancy. This is only the start of brain development, which, unlike the other embryonic systems, undergoes major changes throughout pregnancy.

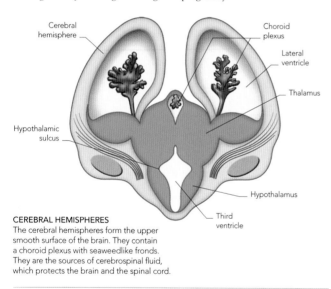

CEREBRAL HEMISPHERES
The cerebral hemispheres form the upper smooth surface of the brain. They contain a choroid plexus with seaweedlike fronds. They are the sources of cerebrospinal fluid, which protects the brain and the spinal cord.

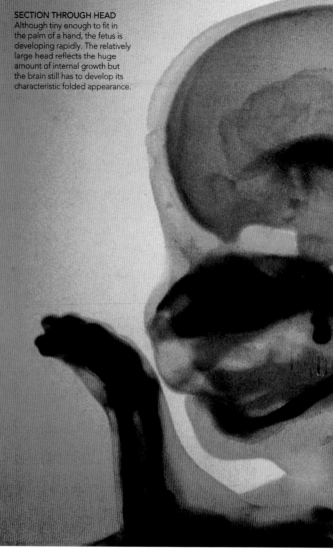

SECTION THROUGH HEAD
Although tiny enough to fit in the palm of a hand, the fetus is developing rapidly. The relatively large head reflects the huge amount of internal growth but the brain still has to develop its characteristic folded appearance.

FORMATION OF THE PITUITARY GLAND

The pituitary gland forms from two parts—a downward fold of neural tissue (the infundibulum) and an upward projection (Rathke's pouch) of an area close to the roof of the future mouth. Due to their different origins in the embryo, the anterior and posterior pituitary lobes function independently, each producing different hormones. The posterior pituitary lobe is attached to the hypothalamus by the pituitary stalk, from which it receives neurotransmitters. These structures regulate oxytocin and antidiuretic-hormone release. The anterior lobe secretes the neurotransmitter beta-endorphin as well as seven hormones that are regulated through feedback mechanisms: growth hormone, follicle-stimulating hormone, luteinizing hormone, prolactin, adrenocorticotropic hormone, thyroid-stimulating hormone, and melanocyte-stimulating hormone.

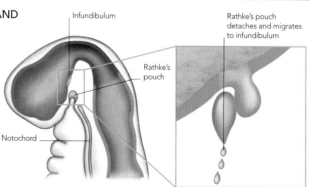

1 EMBRYOLOGICAL SITES
The pituitary gland has two lobes, each formed from separate primitive areas: Rathke's pouch and the infundibulum.

2 EARLY MIGRATION
During its upward journey, Rathke's pouch becomes detached from its original embryological position, at the back of the throat.

3 FINAL POSITION
The pituitary reaches its adult position when the lobes are joined and it is connected to the hypothalamus and cradled by bone.

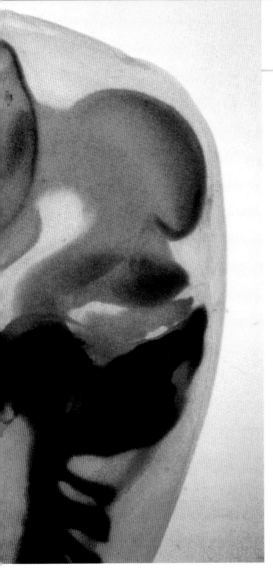

EAR DEVELOPMENT

The ear comprises three sections: an inner, middle, and outer ear (the visible or external part). The external ear develops from six small bumps on the skin (see p.150) and connects to the middle ear via the eardrum (tympanic membrane). Transmission along three tiny bones in the middle ear amplifies the sound over 20 times to the inner ear. These bones have Latin names that describe their shape: malleus (mallet), incus (anvil), and stapes (stirrup). Hair cells in the inner ear can change their length in response to sound. The movement is converted into nerve impulses and transmitted to the brain.

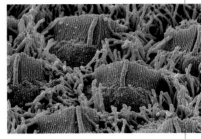

INNER EAR HAIR CELLS
This electron micrograph shows hair cells (pink) in the organ of Corti in the inner ear. They are fringed by microvilli (gray) in the fetus, which are reabsorbed by adulthood.

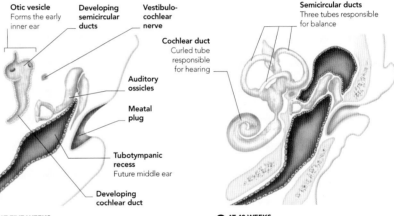

Otic vesicle
Forms the early inner ear

Developing semicircular ducts

Vestibulo-cochlear nerve

Cochlear duct
Curled tube responsible for hearing

Auditory ossicles

Meatal plug

Tubotympanic recess
Future middle ear

Developing cochlear duct

Semicircular ducts
Three tubes responsible for balance

1 AT FIVE WEEKS
The three sections of the ear—inner, middle, and outer—start out as completely separate elements that gradually merge together.

2 AT 40 WEEKS
The inner ear not only processes sound within the coiled cochlear but also judges head position and movement via three fluid-filled semicircular canals.

EYE DEVELOPMENT

During the sixth week, a shallow pit infolds to form the hollow lens. This is encircled by the optic cup, an outpocketing of the primitive forebrain. Over the next two weeks, lens fibers multiply and cause the lens to solidify. To accommodate this rapid growth, the optic stalk supplies blood to the lens (after birth there is no blood vessel). The eyes are open at this stage. During the sixth week, eyelids appear, then fuse by the eighth week and do not reopen until 26–27 weeks. Tear-forming lacrimal glands lubricate the eye, but do not fully function until six weeks after birth. The pigmented retina is very simple at this stage, but differentiates into distinct neural layers up to the time of birth. The optic stalk will become the optic nerve by eight weeks.

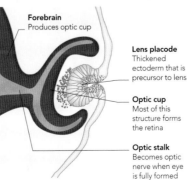

Forebrain
Produces optic cup

Lens placode
Thickened ectoderm that is precursor to lens

Optic cup
Most of this structure forms the retina

Optic stalk
Becomes optic nerve when eye is fully formed

Pigmented layer of retina
Nourishes retinal neural cells

Neural layer of retina
Formed from brain cells

Lens fibers
Cells on lens wall elongate to form lens fibers

Lens vesicle
Placode detaches, forming the solid, spherical lens

1 AT 46 DAYS
The structure is becoming eyelike. The optic cup appears to almost encircle the lens placode as it begins its separation from the skin surface to form the discrete lens.

2 AT 47 DAYS
The hollow lens vesicle close as the lens fibers multiply. The optic stalk, which was a hollow structure, now contains nerve fibers for its function as the optic nerve.

THE SKELETON

The skeleton protects and supports the developing fetus. It is initially composed of cartilage, but gradually ossifies into bone at varying rates so that the bones can expand to keep up with rapid fetal growth.

THE DEVELOPING SKELETON

The skeleton arises from the mesoderm cell layer. Bones form in two distinct ways. Most appear first as a soft cartilage framework that is later replaced with hard bone by the process of ossification. The flat bones of the skull miss the cartilaginous stage and arise directly as the mesoderm becomes ossified. In the bulk of the skeleton, cells called chondrocytes form the cartilaginous framework. The final shape of each bone results from a process of continuous bone formation, as calcium salts are laid down by cells known as osteoblasts; this is followed by remodeling through reabsorption of the bone matrix by osteoblasts.

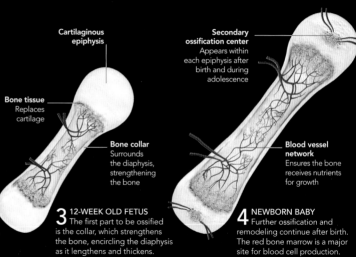

Maxilla

Flat bones
Facial bones and those of the frontal skull are known as the flat bones; bone tissue here forms without the presence of existing cartilage.

Mandible

Long bones
All limbs and girdles are known as long bones; bone tissue in these bones derives from a cartilage matrix.

17-WEEK OLD FETUS
The fetus's skeleton and joints have matured sufficiently for it to be able to perform a full range of movements. This is when the mother becomes aware of her fetus's movements.

Axial bones
The vertebral column and ribs are the axial bones; bone tissue derives from a cartilage matrix.

Rib

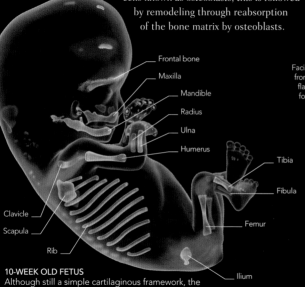

Frontal bone
Maxilla
Mandible
Radius
Ulna
Humerus
Tibia
Fibula
Clavicle
Scapula
Femur
Rib
Ilium

10-WEEK OLD FETUS
Although still a simple cartilaginous framework, the basic shape of each bone is complete. The bones anchor muscle attachments, allowing simple movement.

LONG BONES

All long bones, apart from the collarbone, form in the same way, through a process called ossification, by which osteoblasts deposit calcium salts. This process occurs in each bone at a different stage of pregnancy and some, such as the breastbone, are not fully ossified until after birth. During primary ossification, a central collar of bone forms around the shaft, while cartilage remains at the ends. Even after birth, when secondary ossification occurs, the tips of the bones remain cartilaginous. To allow for childhood growth, ossification in long bones is not complete until the age of 20.

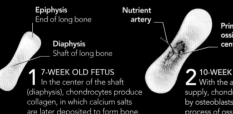

Cartilaginous epiphysis

Bone tissue
Replaces cartilage

Secondary ossification center
Appears within each epiphysis after birth and during adolescence

Bone collar
Surrounds the diaphysis, strengthening the bone

Blood vessel network
Ensures the bone receives nutrients for growth

Epiphysis
End of long bone

Nutrient artery

Diaphysis
Shaft of long bone

Primary ossification center

1 7-WEEK OLD FETUS
In the center of the shaft (diaphysis), chondrocytes produce collagen, in which calcium salts are later deposited to form bone.

2 10-WEEK OLD FETUS
With the arrival of a blood supply, chondrocytes are replaced by osteoblasts and the gradual process of ossification begins.

3 12-WEEK OLD FETUS
The first part to be ossified is the collar, which strengthens the bone, encircling the diaphysis as it lengthens and thickens.

4 NEWBORN BABY
Further ossification and remodeling continue after birth. The red bone marrow is a major site for blood cell production.

FLAT BONES

The flat bones of the face and skull arise from the direct conversion of mesoderm cells into osteoblasts, without an intermediate cartilage stage. This is called intramembranous ossification. The space between each pair of skull bones (fontanelle) remains open to allow for the increase in head size due to brain development. The spaces also allow the head to contract as it descends through the birth canal during labor.

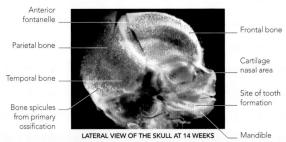

Anterior fontanelle
Parietal bone
Temporal bone
Bone spicules from primary ossification
Frontal bone
Cartilage nasal area
Site of tooth formation
Mandible

LATERAL VIEW OF THE SKULL AT 14 WEEKS

Phalanges
Ulna
Radius

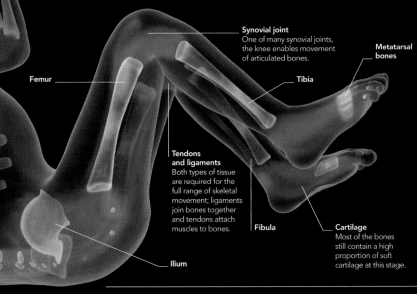

Femur

Synovial joint
One of many synovial joints, the knee enables movement of articulated bones.

Metatarsal bones

Tibia

Tendons and ligaments
Both types of tissue are required for the full range of skeletal movement; ligaments join bones together and tendons attach muscles to bones.

Fibula

Cartilage
Most of the bones still contain a high proportion of soft cartilage at this stage.

Ilium

SYNOVIAL JOINTS

Most of the joints in the body are synovial-type joints. The structure of a synovial joint allows a wide range of movement. In a synovial joint the ends of the bones are protected by cartilage and separated by a fluid-filled capsule. Movement (articulation) can occur without the hard bones coming into contact and rubbing against each other, which would erode their surfaces. By 15 weeks, all synovial joints have formed sufficiently to allow the fetus a full range of movement.

Connective tissue containing fibroblasts

1 UNDIFFERENTIATED STAGE Initial development involves portions of the soft cartilaginous bony framework transforming (differentiating) into connective tissue containing fibroblasts.

Cartilage
Dense connective tissue

2 TISSUE DIFFERENTIATION The fibroblasts form a dense layer of connective tissue—which becomes the joint—and stimulate further cartilage formation on either side of this region.

Articular cartilage (future joint lining)

3 FURTHER DIFFERENTIATION Articular cartilage forms but joint movement (articulation) cannot occur until the dense connective tissue transforms into the fluid-filled synovial joint.

Vacuole in connective tissue

4 SYNOVIAL CAVITY FORMS Vacuoles form in the dense connective tissue and join up to create the fluid-filled synovial cavity. Ligaments that connect the bones also start to appear.

Meniscus
Enclosed joint ligaments

5 COMPLETED JOINT The joint is now encased in a protective, ligamentous capsule, which means that the full range of movement at the joint is now possible.

Joint capsule | Synovial cavity

SPINAL DEVELOPMENT

The development of the spinal cord and vertebrae are closely linked. Each somite (see p.99) gives rise to a dermomyotome portion that forms the skin and underlying muscles of the trunk, and a sclerotome part that forms the vertebral column.

To enable the spinal nerves to emerge from the spinal cord, the somites undergo resegmentation, in which they divide into two parts to allow the spinal nerve to emerge. Later, they rejoin their neighboring halves and evolve into vertebrae.

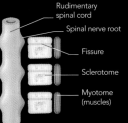

Rudimentary spinal cord
Spinal nerve root
Fissure
Sclerotome
Myotome (muscles)

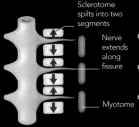

Sclerotome spilts into two segments
Nerve extends along fissure
Myotome

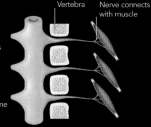

Vertebra
Nerve connects with muscle

1 SCLEROTOME FORMATION As the nerve roots emerge from the rudimentary spinal cord, each sclerotome begins to separate into two parts. A fissure appears at the area of division.

2 DIVIDING SCLEROTOME The fisure becomes a channel in the center of each sclerotome, through which the nerve roots emerge as they join a corresponding set of muscles (myotome).

3 FUSED VERTEBRAE The upper and lower parts of adjacent sclerotomes grow toward each other and fuse, becoming bony vertebrae. The spinal nerves join up with their assigned muscles.

MUSCLE DEVELOPMENT

There are three types of muscles in the body: cardiac muscle, skeletal (voluntary) muscle, and smooth (involuntary) muscle, such as the muscle of the gut. The skeletal muscles of the trunk, and also the limbs, diaphragm, and tongue, develop from the somites in a similar way to the bones of the vertebral column. Each somite has a myotome portion, from which the muscle arises. These portions are supplied by spinal nerves that allow the muscles to be controlled voluntarily. The process starts during the seventh week of pregnancy, when the muscle groups start to gradually emerge from the side of the future spine and extend around the trunk and down into the limb buds.

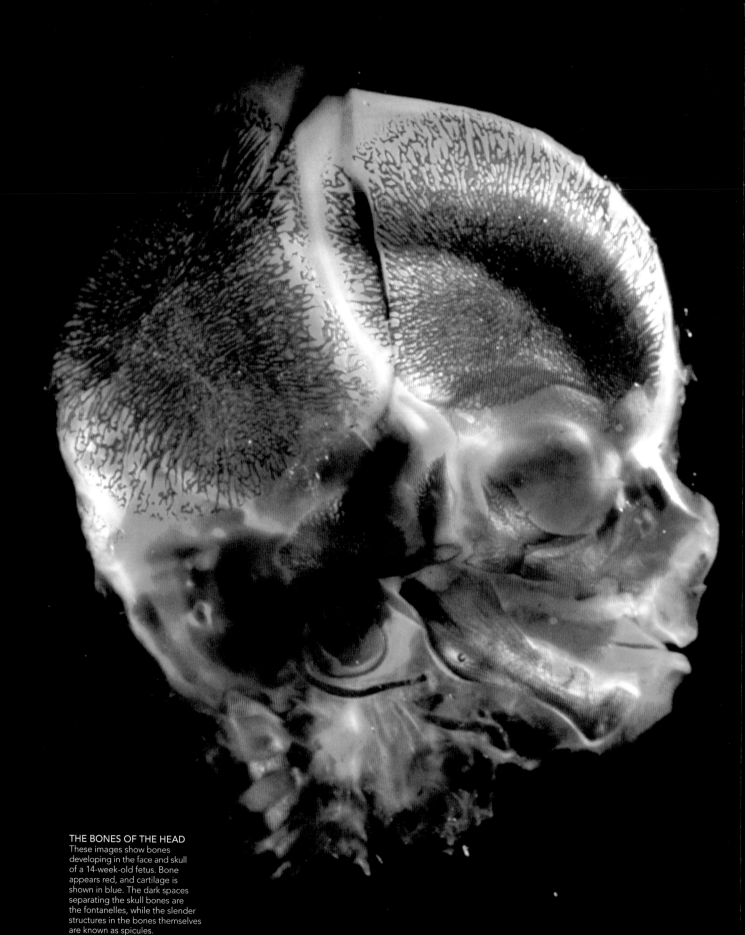

THE BONES OF THE HEAD
These images show bones developing in the face and skull of a 14-week-old fetus. Bone appears red, and cartilage is shown in blue. The dark spaces separating the skull bones are the fontanelles, while the slender structures in the bones themselves are known as spicules.

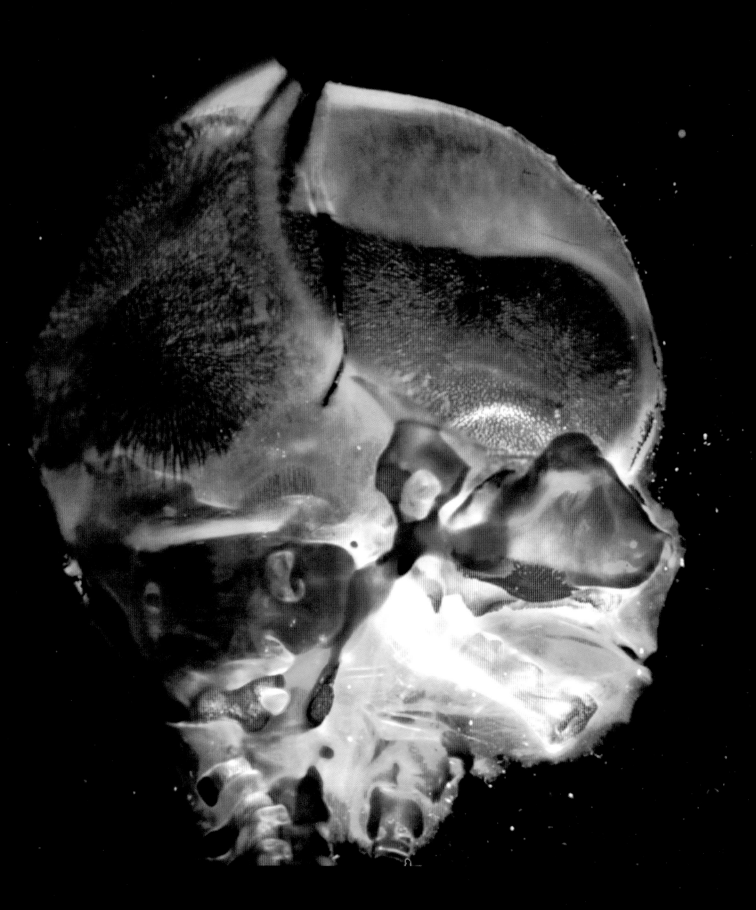

FETUS

LIMB FORMATION

By the 10th week of pregnancy, all the limb joints have formed and simple movements are possible. The joints can be bent (flexed) and extended, and the hands can be brought up to the face. Development of the upper limbs is slightly more advanced than that of the lower. Each limb starts as a bud and follows the same pattern of development, which involves a carefully sequenced program of cell growth and death. The bud gradually lengthens, and soft cartilaginous bones form within the tissues. This cartilaginous framework gradually hardens, each bone ossifying from the center outward (see pp.118–119). The blood vessels of the limbs are easily seen through the thin, transparent skin, which has virtually no fat layer at this stage.

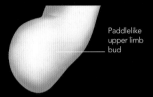

Paddlelike upper limb bud

1 HAND PLATE
The upper limb starts as a simple broad, short limb protruding from the surface at six weeks. Smooth paddle-shaped hand plates develop at the end of the bud.

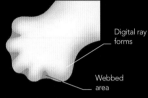

Digital ray forms

Webbed area

2 DIGITAL RAYS
Five short projections materialize at the edge of the hand plate and form the fingers. Development of the toes follows the same pattern about a week later.

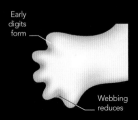

Early digits form

Webbing reduces

3 EARLY DIGITS
The projections lengthen, and the cells between the fingers die and disappear. The effect of this is to gradually reduce the webbing between each pair of fingers.

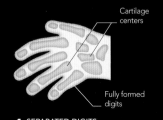

Cartilage centers

Fully formed digits

4 SEPARATED DIGITS
By the end of the eighth week all digits are distinct, but the overlying skin is thin and the ridges of each genetically determined fingerprint are not fully developed until 18 weeks.

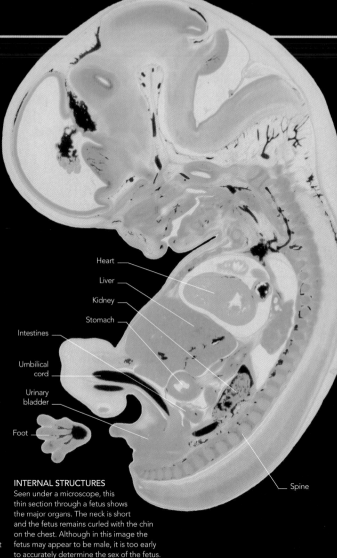

Heart
Liver
Kidney
Stomach
Intestines
Umbilical cord
Urinary bladder
Foot
Spine

INTERNAL STRUCTURES
Seen under a microscope, this thin section through a fetus shows the major organs. The neck is short and the fetus remains curled with the chin on the chest. Although in this image the fetus may appear to be male, it is too early to accurately determine the sex of the fetus.

DEVELOPING INTESTINES

The intestinal tube continues to elongate and differentiate into specialized sections (see p.104). The small bowel is too long to be accommodated within the embryo's abdomen, and it bulges into the base of the umbilical cord. Taking its blood supply with it, the intestine rotates in the umbilical cord and completes the rotation as it returns to the abdominal cavity. The large bowel then becomes fixed, holding the entire gut in place. This process starts at eight weeks and should be completed by 12 weeks. The gut is not yet functioning, and the embryo is unable to swallow amniotic fluid.

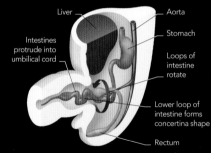

Liver
Aorta
Intestines protrude into umbilical cord
Stomach
Loops of intestine rotate
Lower loop of intestine forms concertina shape
Rectum

1 ROTATION OF INTESTINES
The simple intestinal tube rotates externally in the base of the umbilical cord through 90°, in a counterclockwise direction.

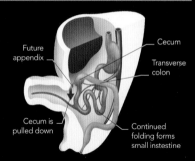

Future appendix
Cecum
Transverse colon
Cecum is pulled down
Continued folding forms small instestine

2 INTESTINES RETRACT INTO ABDOMEN
The intestinal loop retracts into the abdominal cavity, rotating 180° counterclockwise. The cecum is pulled down with it and forms the ascending colon.

THE URINARY SYSTEM

At first the bladder and lower bowel (rectum) open into a common opening called the cloaca. This separates into two, dividing the bowel from the bladder. A short ureteric bud grows upward on each side from the bladder to fuse with the primitive kidney at five weeks. Over the next four weeks, each kidney gradually ascends as it matures and lengthens the ureter on its journey upward. Branching of the ureter within the kidney forms the collecting system into which the filtered urine is passed. This process is not complete until 32 weeks, when approximately 2 million branches have formed.

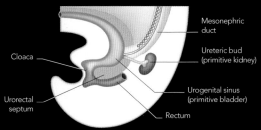

Cloaca

Urorectal septum

Mesonephric duct

Ureteric bud (primitive kidney)

Urogenital sinus (primitive bladder)

Rectum

1 DIVIDING CLOACA
The urorectal septum moves downward to the cloacal membrane to separate the bladder (and the unformed tube joining the bladder to the outside—the urethra) from the rectum.

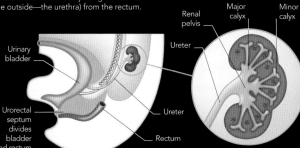

Urinary bladder

Urorectal septum divides bladder and rectum

Renal pelvis

Major calyx

Minor calyx

Ureter

Ureter

Rectum

2 BLADDER AND RECTUM FORMATION
The process of separation is complete by the seventh week. The rectum is not open—it is covered by a temporary thin membrane—but this will disappear over the next 10 days.

THE DEVELOPING KIDNEY
Branches of the ureter form the major calyces that divide into the minor calyces. These branch further to collect the urine from the kidney tissue.

THE LYMPHATIC SYSTEM

As liquid leaks out from the bloodstream to bathe the cells of the body, excess fluid (lymph) needs to be returned to the circulation. This is achieved by a series of sacs and, later in development, channels, which are together known as the lymphatic system. This system develops in parallel with the embryonic vascular system of blood vessels. The lymphatic system forms from a pair of upper lymph sacs during the fifth week, which remove lymph from the upper body. The following week, four lower sacs arise that drain the lower body. Further connections and modifications between these sacs result in the majority of the lymph draining into the upper body thoracic duct, which enters into the subclavian vein (a left-sided neck vein).

THE SEX ORGANS

In both males and females, the development of the urinary system is closely linked with the formation of the internal sex organs. Germ cells in the yolk sac move into the embryo during the sixth week to lie on the urogenital ridge close to the developing spine of the embryo. These cells stimulate formation of either the ovaries (female) or testes (male). Close by, a new pair of ducts form (mullerian ducts), which will disappear in the male but develop into the fallopian tubes, uterus, and upper part of the vagina in the female. The separation into male or female is governed by genes on the Y chromosome. Embryos without these genes develop as females and those with, become males.

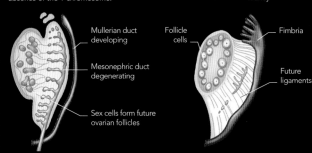

Sex cells migrate from umbilical cord to genital ridge

Liver

Somites

Mesonephric duct

Mesonephros

Mullerian duct

Genital ridge

Kidney

INDIFFERENT GONAD STAGE
The male and female gonads appear similar at this stage but the developmental pathway they will follow is predetermined by the presence or absence of the Y chromosome.

Mullerian duct developing

Mesonephric duct degenerating

Sex cells form future ovarian follicles

Follicle cells

Fimbria

Future ligaments

EARLY FEMALE SEX ORGANS
In the absence of a Y chromosome, the indifferent gonad defaults into the female state, forming an ovary containing millions of oocytes that remain inert until puberty.

DEVELOPING FEMALE SEX ORGANS
The upper part of the mullerian duct has formed the fimbrial end of the fallopian tube. Its lower portion will form the rest of the tube, the uterus, and upper vagina.

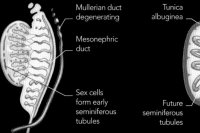

Mullerian duct degenerating

Mesonephric duct

Sex cells form early seminiferous tubules

Tunica albuginea

Vas deferens developing

Future seminiferous tubules

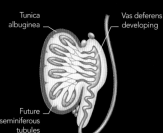

EARLY MALE SEX ORGANS
In each testis germ cells form Sertoli cells that nurture developing sperm. Leydig cells in the gonad produce testosterone to stimulate further male development.

DEVELOPING MALE SEX ORGANS
The mullerian duct is now a tiny remnant on top of the testis. The mesonephric duct links each testis to the urethra via the seminiferous tubules and vas deferens.

123

This 3D ultrasound scan shows a 13-week-old fetus touching its face with its hands. All of its joints are now present, allowing for a range of movements.

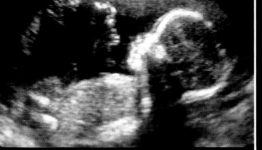

This 2D ultrasound scan shows a 20-week-old fetus in the uterus. A scan is usually performed at this time to check that the fetus is growing as expected.

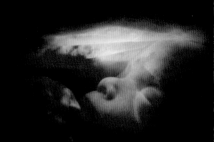

This photograph of a five-month-old fetus shows the developing facial features. The eyelids remain fused until the beginning of the third trimester.

TRIMESTER 2
MONTHS 4–6 | WEEKS 13–26

The second trimester of pregnancy is a time of continuing growth and development. All the body systems are in place, but the fetus is not yet capable of independent life.

The maternal discomforts of the first trimester, such as morning sickness and fatigue, start to settle at the beginning of the second trimester. Steadily increasing blood volume and a more dynamic circulation give the mother-to-be a healthy glow. The top, or fundus, of the uterus should rise above the pelvis in the fourth month, making the pregnancy obvious. The fundus of the uterus will continue to rise at a rate of approximately ⅜ in (1 cm) per week. This measurement gives a good estimation of the week of pregnancy, so that, for example, at 20 weeks of pregnancy, the fundal height will be around 8 in (20 cm). The first movement of the fetus felt by the mother-to-be is known as "quickening" and usually occurs during the fifth month, but it may be felt earlier if the woman has had a child before. Over the course of the second trimester, the fetus will more than triple in size, and its weight will increase by around 30 times. During the first half of the second trimester, the fetal brain and nervous system are still undergoing a critical period of development. The second half of the second trimester sees the rapid growth of the fetal body and limbs, while the head grows at a relatively slower rate. As a result, the proportions of the fetus look more like those of an adult by the end of the trimester.

TIMELINE

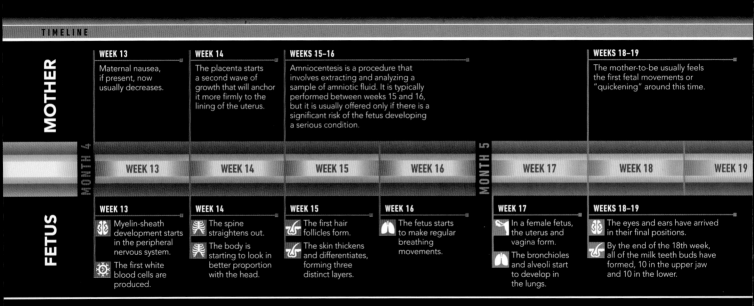

MOTHER

WEEK 13
Maternal nausea, if present, now usually decreases.

WEEK 14
The placenta starts a second wave of growth that will anchor it more firmly to the lining of the uterus.

WEEKS 15–16
Amniocentesis is a procedure that involves extracting and analyzing a sample of amniotic fluid. It is typically performed between weeks 15 and 16, but it is usually offered only if there is a significant risk of the fetus developing a serious condition.

WEEKS 18–19
The mother-to-be usually feels the first fetal movements or "quickening" around this time.

MONTH 4 | WEEK 13 | WEEK 14 | WEEK 15 | WEEK 16 | **MONTH 5** | WEEK 17 | WEEK 18 | WEEK 19

FETUS

WEEK 13
- Myelin-sheath development starts in the peripheral nervous system.
- The first white blood cells are produced.

WEEK 14
- The spine straightens out.
- The body is starting to look in better proportion with the head.

WEEK 15
- The first hair follicles form.
- The skin thickens and differentiates, forming three distinct layers.

WEEK 16
- The fetus starts to make regular breathing movements.

WEEK 17
- In a female fetus, the uterus and vagina form.
- The bronchioles and alveoli start to develop in the lungs.

WEEKS 18–19
- The eyes and ears have arrived in their final positions.
- By the end of the 18th week, all of the milk teeth buds have formed, 10 in the upper jaw and 10 in the lower.

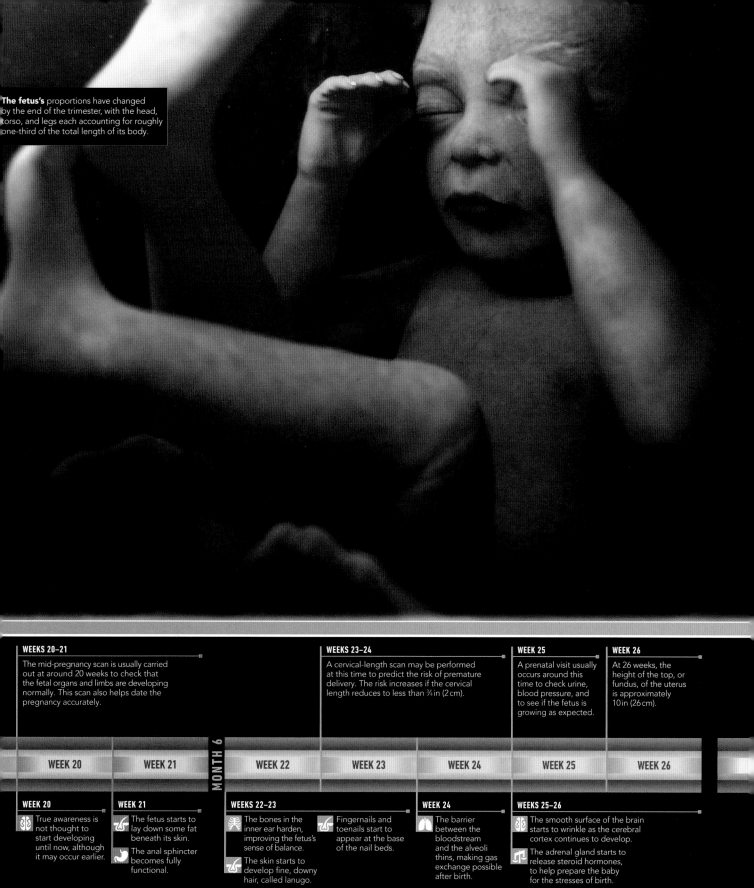

The fetus's proportions have changed by the end of the trimester, with the head, torso, and legs each accounting for roughly one-third of the total length of its body.

WEEKS 20–21

The mid-pregnancy scan is usually carried out at around 20 weeks to check that the fetal organs and limbs are developing normally. This scan also helps date the pregnancy accurately.

WEEKS 23–24

A cervical-length scan may be performed at this time to predict the risk of premature delivery. The risk increases if the cervical length reduces to less than ¾ in (2 cm).

WEEK 25

A prenatal visit usually occurs around this time to check urine, blood pressure, and to see if the fetus is growing as expected.

WEEK 26

At 26 weeks, the height of the top, or fundus, of the uterus is approximately 10 in (26 cm).

MONTH 6

| WEEK 20 | WEEK 21 | WEEK 22 | WEEK 23 | WEEK 24 | WEEK 25 | WEEK 26 |

WEEK 20

True awareness is not thought to start developing until now, although it may occur earlier.

WEEK 21

The fetus starts to lay down some fat beneath its skin.

The anal sphincter becomes fully functional.

WEEKS 22–23

The bones in the inner ear harden, improving the fetus's sense of balance.

The skin starts to develop fine, downy hair, called lanugo.

Fingernails and toenails start to appear at the base of the nail beds.

WEEK 24

The barrier between the bloodstream and the alveoli thins, making gas exchange possible after birth.

WEEKS 25–26

The smooth surface of the brain starts to wrinkle as the cerebral cortex continues to develop.

The adrenal gland starts to release steroid hormones, to help prepare the baby for the stresses of birth.

MONTH 4 | **WEEKS 13–16**

The fourth month marks the beginning of the second trimester. The uterus has expanded to reach the top of the pelvis and can be felt above the pubic bone. This means the pregnancy will soon start to show.

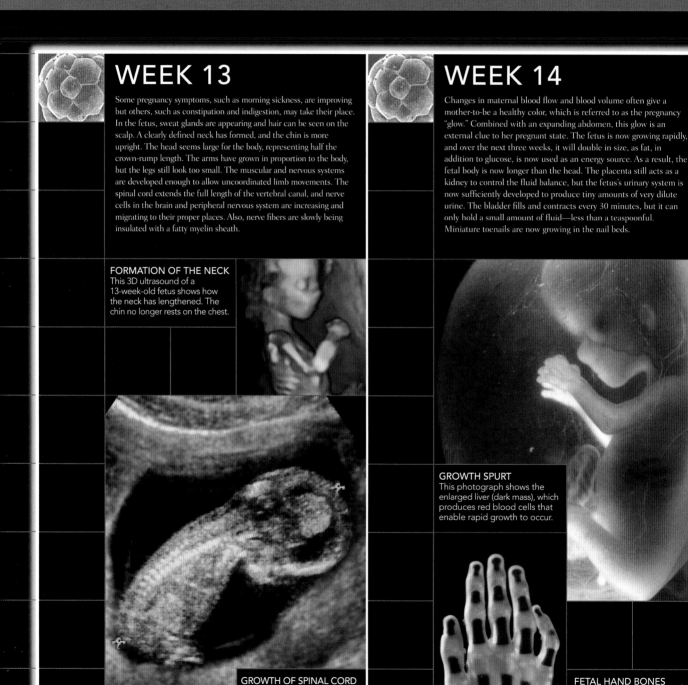

WEEK 13

Some pregnancy symptoms, such as morning sickness, are improving but others, such as constipation and indigestion, may take their place. In the fetus, sweat glands are appearing and hair can be seen on the scalp. A clearly defined neck has formed, and the chin is more upright. The head seems large for the body, representing half the crown–rump length. The arms have grown in proportion to the body, but the legs still look too small. The muscular and nervous systems are developed enough to allow uncoordinated limb movements. The spinal cord extends the full length of the vertebral canal, and nerve cells in the brain and peripheral nervous system are increasing and migrating to their proper places. Also, nerve fibers are slowly being insulated with a fatty myelin sheath.

FORMATION OF THE NECK
This 3D ultrasound of a 13-week-old fetus shows how the neck has lengthened. The chin no longer rests on the chest.

GROWTH OF SPINAL CORD
Vertebrae can be seen around the spinal cord in this ultrasound scan. The crown–rump length is indicated by the blue crosses.

WEEK 14

Changes in maternal blood flow and blood volume often give a mother-to-be a healthy color, which is referred to as the pregnancy "glow." Combined with an expanding abdomen, this glow is an external clue to her pregnant state. The fetus is now growing rapidly, and over the next three weeks, it will double in size, as fat, in addition to glucose, is now used as an energy source. As a result, the fetal body is now longer than the head. The placenta still acts as a kidney to control the fluid balance, but the fetus's urinary system is now sufficiently developed to produce tiny amounts of very dilute urine. The bladder fills and contracts every 30 minutes, but it can only hold a small amount of fluid—less than a teaspoonful. Miniature toenails are now growing in the nail beds.

GROWTH SPURT
This photograph shows the enlarged liver (dark mass), which produces red blood cells that enable rapid growth to occur.

FETAL HAND BONES
In this scan, the red areas show where hard bone is forming in the finger bones (phalanges) and hand bones (metacarpals).

WEEK 15

As fetal growth accelerates, amino acids are extracted from the mother's blood to build muscles and organs in the fetus. The fetus drinks amniotic fluid, which acquires flavors from the mother's diet. The lungs are expanding and producing small amounts of mucus. The external sex organs are now visible, and it may be possible to identify gender from an ultrasound scan. During this month, hundreds of thousands of eggs are forming within the ovaries of a female fetus. At the same time, the ovaries move down from the abdomen into the pelvis. The umbilical cord is thickening and lengthening as it carries more and more oxygenated blood, rich with nutrients, from the placenta to the fetus and returns deoxygenated blood and wastes to the mother's body.

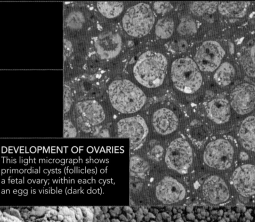

DEVELOPMENT OF OVARIES
This light micrograph shows primordial cysts (follicles) of a fetal ovary; within each cyst, an egg is visible (dark dot).

LINING OF THE AMNION
A scanning electron micrograph of the surface of the amniotic sac shows the cells that encircle the amniotic fluid.

WEEK 16

The fetus's face looks obviously human, with eyes in the correct forward-looking position and ears moving up toward their final position. The thyroid gland is descending from the base of the tongue into the neck. The fetus is now almost equal in size to the placenta, and a second wave of placental growth now anchors it more firmly to the uterus as blood flow to the fetus increases. Mothers-to-be are offered a number of screening tests, including amniocentesis, in which a sample of amniotic fluid is collected to analyze fetal cells. This procedure can be performed from 15 weeks, but is usually undertaken between 15 and 16 weeks. It is normally only offered to mothers with a higher than normal risk of a baby with chromosomal abnormalities such as Down syndrome.

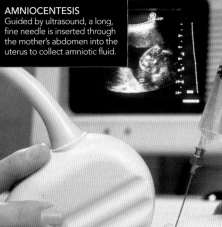

FETAL HEART MONITORING
A fetal heart rate monitor easily assesses how rapidly the fetal heart is beating. In this image, the rate is 165 beats per minute.

AMNIOCENTESIS
Guided by ultrasound, a long, fine needle is inserted through the mother's abdomen into the uterus to collect amniotic fluid.

MONTH 4 | WEEKS 13–16
MOTHER AND FETUS

The fourth month of pregnancy marks the beginning of the second trimester. Early symptoms of pregnancy, such as fatigue and nausea, have usually started to subside, the pregnancy is starting to show, and women often feel in the best of health and appear to be "blooming." A number of screening tests may be offered during this month to ascertain whether there is a risk of developmental abnormalities occurring in the fetus. If there is a high risk, amniocentesis can be performed at the end of this month to detect conditions such as Down syndrome. The fetus is still growing rapidly and fine, downy hair (lanugo) starts to cover the skin. It starts to produce small amounts of urine, which pass out of the urethra and into the amniotic fluid. The fetus's facial features continue to develop, and its proportions begin to look more like those of an adult.

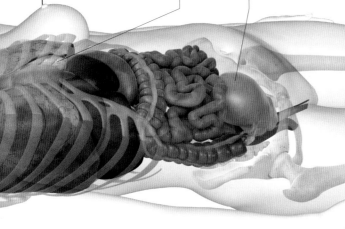

MOTHER AT 16 WEEKS
A fall in blood pressure and rising hormone levels are among the significant changes this month. The hormone changes are thought to take away the morning sickness that was a common feature of the first three months.

Blood volume and pressure
Blood volume increases significantly this month, while blood pressure falls slightly, after which it continues to rise until the birth.

Uterus expansion
The uterus starts to expand into the abdomen, causing the abdominal wall to stretch to accommodate it. This results in the first sign of a pregnancy "bump." Although they usually appear later in pregnancy, stretch marks may develop now as a result of this expansion.

MOTHER

- 68 beats per minute
- 104/66
- 8 pints (4.5l)

30%

Levels of human chorionic gonadotropin (hCG) in the blood fall by 30 percent this month.

Increasing blood flow to the skin causes the distinctive **pregnancy glow**.

The pregnancy **first starts to become obvious** during this month, depending on the mother's weight and build.

| 1 | 2 | 3 | 4 | 5 | 6 | 7 | 8 | 9 | 10 | 11 | 12 | 13 | 14 | 15 | 16 | 17 | 18 | 19 | 20 | 21 | 22 | 23 | 24 | 25 | 26 | 27 | 28 | 29 | 30 | 31 | 32 | 33 | 34 | 35 | 36 | 37 | 38 | 39 | 40 |

FETUS

- 158 beats per minute
- 4½in (12 cm)
- 3½oz (100 g)

100%

The fetus **doubles** in size during the fourth month of pregnancy.

30 minutes

The fetus's bladder empties **every 30 minutes,** contributing a small amount of urine to the amniotic fluid.

The **fetus's heartbeat** can be heard in the fourth month with a hand-held **doppler ultrasound machine.** The fetus's heart rate is over **twice as fast** as the mother's.

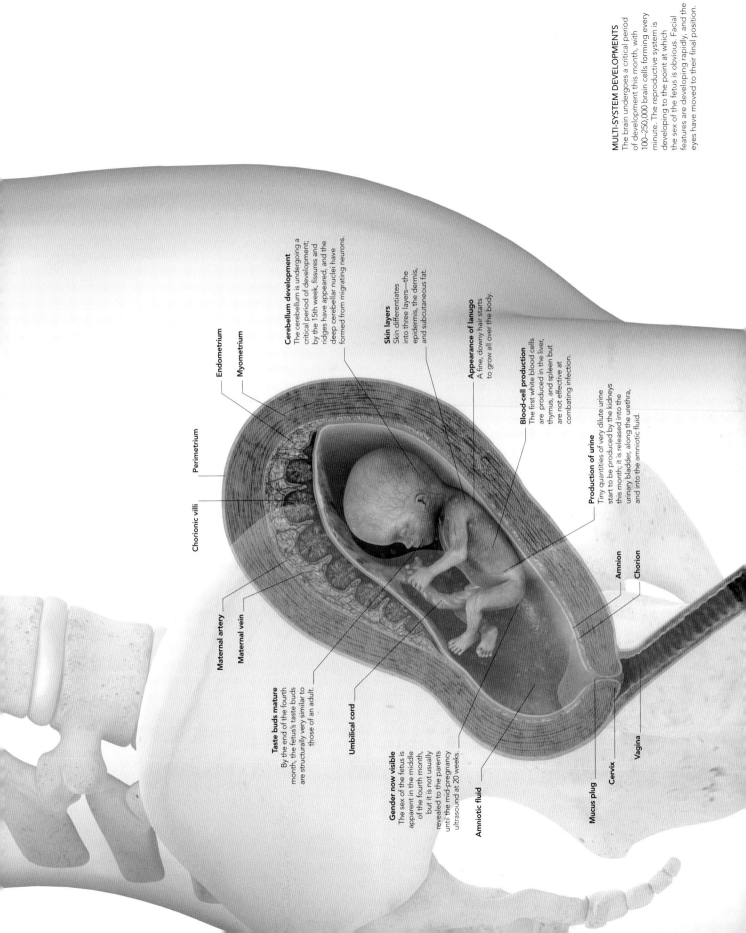

Endometrium

Myometrium

Perimetrium

Chorionic villi

Maternal artery

Maternal vein

Umbilical cord

Taste buds mature
By the end of the fourth month, the fetus's taste buds are structurally very similar to those of an adult.

Gender now visible
The sex of the fetus is apparent in the middle of the fourth month, but it is not usually revealed to the parents until the mid-pregnancy ultrasound at 20 weeks.

Amniotic fluid

Mucus plug

Cervix

Vagina

Amnion

Chorion

Cerebellum development
The cerebellum is undergoing a critical period of development; by the 15th week, fissures and ridges have appeared, and the deep cerebellar nuclei have formed from migrating neurons.

Skin layers
Skin differentiates into three layers—the epidermis, the dermis, and subcutaneous fat.

Appearance of lanugo
A fine, downy hair starts to grow all over the body.

Blood-cell production
The first white blood cells are produced in the liver, thymus, and spleen but are not effective at combating infection.

Production of urine
Tiny quantities of very dilute urine start to be produced by the kidneys this month; it is released into the urinary bladder, along the urethra, and into the amniotic fluid.

MULTI-SYSTEM DEVELOPMENTS
The brain undergoes a critical period of development this month, with 100–250,000 brain cells forming every minute. The reproductive system is developing to the point at which the sex of the fetus is obvious. Facial features are developing rapidly, and the eyes have moved to their final position.

MOTHER

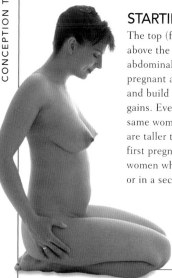

STARTING TO "SHOW"

The top (fundus) of the womb has now risen well above the pelvis, and it can be easily felt during an abdominal examination. Whether a woman looks pregnant at this stage depends partly on her height and build and partly on the amount of weight she gains. Every pregnancy is different, even for the same woman. In general, however, women who are taller than average, overweight, or in their first pregnancy do not tend to "show" as early as women who are shorter, of a more slender build, or in a second or subsequent pregnancy.

VISIBLE BUMP
The waistline has become noticeably thicker and the breasts are enlarging, although the pregnancy is not obvious and can easily be hidden beneath loose clothing.

MORNING SICKNESS SUBSIDES

Morning sickness, which affects around seven out of ten women, starts to ease after the first trimester and has usually disappeared by the 14th week of pregnancy. A small number of women continue to suffer throughout pregnancy. The exact cause is unknown, but it has been linked with low blood sugar, increased secretion of bile, and raised levels of some hormones, namely estrogen and human chorionic gonadotropin (hCG).

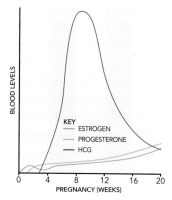

KEY
— ESTROGEN
— PROGESTERONE
— HCG

LINK WITH HORMONE LEVELS
Blood levels of the hormone, human chorionic gonadotropin (hCG), have fallen significantly by 12 weeks, which is possibly why morning sickness resolves at this time.

PREGNANCY "GLOW"

The healthy "bloom" of pregnancy begins around the fourth month and results from an increase in the volume of blood in circulation and the dilation of blood vessels. This diverts more blood to the skin to give her a glowing appearance. Dilation of blood vessels is due to the effects of progesterone, levels of which increase significantly during pregnancy. Although blood volume rises by 45 percent during pregnancy, the mass of red blood cells within the blood only increases by 20 percent. Most of the increased blood volume is due to fluid retention. This dilution causes hemoglobin levels to fall. At one time, this led to frequent diagnosis of anemia, and many pregnant women were treated with iron tablets. Doctors now realize blood dilution is natural in pregnancy, and iron tablets are no longer routinely prescribed.

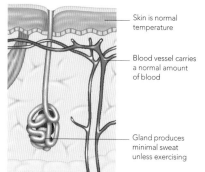

Skin is normal temperature

Blood vessel carries a normal amount of blood

Gland produces minimal sweat unless exercising

NORMAL WIDTH OF BLOOD VESSELS
Blood flow to the skin surface is normally determined by the temperature, exercise, and lifestyle factors such as alcohol intake.

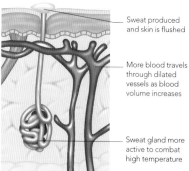

Sweat produced and skin is flushed

More blood travels through dilated vessels as blood volume increases

Sweat gland more active to combat high temperature

DILATED BLOOD VESSELS
During pregnancy, blood flow to the skin increases due to greater blood volume and dilation of blood vessels (vasodilation).

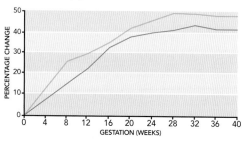

INCREASING BLOOD VOLUME
Total blood volume and amount pumped by the heart (cardiac output) increase from early pregnancy to peak around 32 weeks.

KEY
— CARDIAC OUTPUT
— TOTAL BLOOD VOLUME

BLOOD PRESSURE CHANGES

Blood pressure falls until the middle of the second trimester, where it starts to rise again. Posture has a significant impact on blood pressure. When a pregnant woman lies down, her growing uterus presses against the large vein situated behind the abdominal cavity. As a result of this, blood pressure is affected by whether the woman is sitting, lying on her back, or lying on her left side. It is therefore important that a woman's posture is the same each time her blood pressure is recorded so that values can be correctly compared.

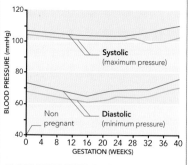

Systolic (maximum pressure)

Non pregnant

Diastolic (minimum pressure)

BLOOD PRESSURE READINGS
Both systolic (higher) and diastolic (lower) readings are consistently reduced when a woman lies on her back, compared to sitting down. Whatever the posture, blood pressure is measured with the arm strap level with the heart.

KEY
— SITTING
— LYING FLAT ON BACK

FETUS

SCREENING TESTS

A variety of screening tests is usually offered in the fourth month to assess fetal development. While some abnormalities can be identified by ultrasound scans, others can only be picked up through blood tests and more invasive procedures such as amniocentesis (see below). The decision to have a screening test is personal. Before deciding whether to go ahead, it is important to obtain as much information as possible about the risks and benefits. Discussing implications of a high-risk result should be part of the decision process. Genetic counselors, doctors, and other professionals can help parents make a decision.

ANALYZING A BLOOD SAMPLE
Down syndrome and several other fetal abnormalities can be detected by measuring the levels of different placental hormones in the blood.

SCREENING TESTS FOR DOWN SYNDROME

A number of tests are used to predict the risk of Down syndrome, all of which measure the levels of various hormones and proteins in the blood. A "false positive" result suggests a high risk for Down syndrome, which is shown to be incorrect following a diagnostic test.

METHOD OF SCREENING	TIMING (WEEKS)	DETECTION RATE (%)	FALSE POSITIVE RATE (%)
Quadruple test	15–20	76	5
Combined test	11–14	85	5
Integrated test	11–13 15–22	85	1
Noninvasive prenatal test	From 10 weeks	99	0.2

AMNIOCENTESIS

This is a procedure in which a small sample of amniotic fluid is removed from the uterus and then analyzed in a laboratory. A long, fine needle is inserted through the abdominal wall; an ultrasound transducer is used to make sure the needle is being directed into the right place. Approximately 4 teaspoons or ¾ fl oz (20 ml) is extracted from the sac surrounding the fetus. This fluid contains live fetal skin cells, the genetic material of which can be analyzed. Amniocentesis can be carried out from 15 weeks onward, but it is usually performed between 15 and 16 weeks. The procedure is generally only offered to women assessed as having a higher than normal risk of having a baby with chromosomal abnormalities such as Down syndrome (see p.237). Amniocentesis accurately identifies the number of chromosomes in fetal cells and can also determine whether the fetus is male or female. Later on in pregnancy, amniocentesis can predict fetal lung maturity and diagnose infections.

Ultrasound transducer
Ultrasound identifies the safest point of entry and helps guide the operator during collection of fluid.

Amniotic fluid
After collection, amniotic fluid is sent away for analysis. Depending on the test, results can take up to two weeks.

Syringe

Amniotic sac
The puncture quickly heals, and amniotic fluid is soon replenished.

Placenta

Uterus
The needle penetrates the muscular wall of the uterus.

Umbilical cord

Pubic bone

Bladder

Mucus plug

Vagina

Cervix

EXTRACTING AMNIOTIC FLUID
During this procedure, great care must be taken to make sure the collecting needle does not damage vital structures, including the placenta. Ultrasound scanning is used to guide the needle to an area where it is safe to take a sample of fluid.

DEVELOPMENT OF THE BRAIN

By the fourth month, the brain is the size of a kidney bean, making it fairly large compared with the rest of the body. Brain cells originate from cells that lined the central cavity of the neural tube. At this time, these cells multiply at the astonishing rate of 100,000—250,000 per minute, and migrate from the tube into the brain swellings. Every time the fetus moves, electrical messages pass from muscles to the developing brain. This helps stimulate development of the cerebellum (which controls posture and movement) and the motor cortex of the cerebral hemispheres, which are involved in the future initiation of voluntary muscle movements.

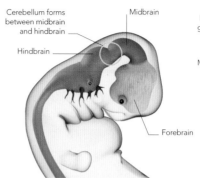

Cerebellum forms between midbrain and hindbrain

Hindbrain

Midbrain

Forebrain

LOCATION OF EMERGING CEREBELLUM

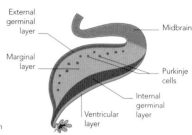

External germinal layer

Midbrain

Marginal layer

Purkinje cells

Internal germinal layer

Ventricular layer

1 LAYER DIFFERENTIATION
By 12 weeks, rapidly multiplying brain cells, including Purkinje cells, involved in regulating muscle movements, move to the surface to form the external germinal layer of gray matter.

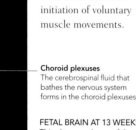

Primary fissure

Midbrain

Brain cells migrate outward from internal germinal layer

Internal germinal layer

Choroid plexus

Ventricular layer

2 FORMATION OF PRIMARY FISSURE
By 13 weeks, the cerebellum has started to fold in on itself, forming a large fissure. Developing brain cells continue to travel outward from the internal germinal layer.

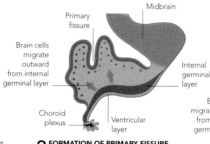

Gray matter

White matter

Brain cells migrate inward from external germinal layer

Deep cerebellar nuclei, formed by migrating neurons

Choroid plexus

3 DEVELOPMENT OF FISSURES AND RIDGES
By 15 weeks, the cerebellum has developed many more folds. Contained within these convolutions are a number of specialized cells that are involved in fetal movements.

Choroid plexuses
The cerebrospinal fluid that bathes the nervous system forms in the choroid plexuses.

FETAL BRAIN AT 13 WEEKS
This ultrasound scan of the brain of a 13-week fetus shows a choroid plexus in the left and right brain hemispheres. The dark areas above are the fluid-filled lateral ventricles.

THE PRODUCTION OF URINE

The fetal kidneys start to produce small amounts of urine around the beginning of the fourth month of development. The tiny bladder holds only a few milliliters of fluid, and dilute urine is regularly released into the amniotic fluid as shown below. As pregnancy progresses, urine is produced in larger quantities and becomes more concentrated. The fetus drinks and recycles this fluid.

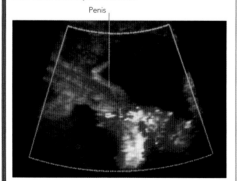

Penis

URINATING INTO THE AMNIOTIC FLUID
This Doppler ultrasound scan shows a male fetus (on the left) urinating (highlighted in blue, white, and red) through his penis into the amniotic fluid.

DEVELOPMENT OF THE URINARY SYSTEM

The urinary system starts to develop early in the fourth week of gestation in the lower pelvic region of the embryo. The kidneys start to form in the fifth week. Between this time and the fourth month, the kidneys shift position dramatically, from the pelvic region to the abdominal region. In the fourth month, the kidneys are capable of producing urine, which is expelled from the kidneys and released along the ureters to the urinary bladder before exiting via the urethra. In the female fetus, the entrance to the vagina and the urethral opening are part of the same structure until the sixth month.

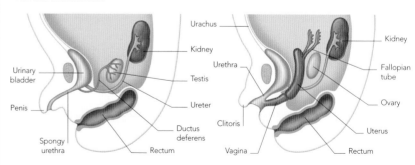

Urinary bladder

Penis

Spongy urethra

Urachus

Kidney

Testis

Ureter

Ductus deferens

Rectum

Urethra

Urachus

Kidney

Fallopian tube

Ovary

Uterus

Rectum

Clitoris

Vagina

MALE FETUS AT 14 WEEKS
In the male, development of the reproductive and urinary systems is closely linked because both systems share the same exit from the body via the penis.

FEMALE FETUS AT 14 WEEKS
In the female, the urinary and reproductive systems develop separately. The short urethra exits from the bladder at the urogenital sinus, in front of the vaginal plate.

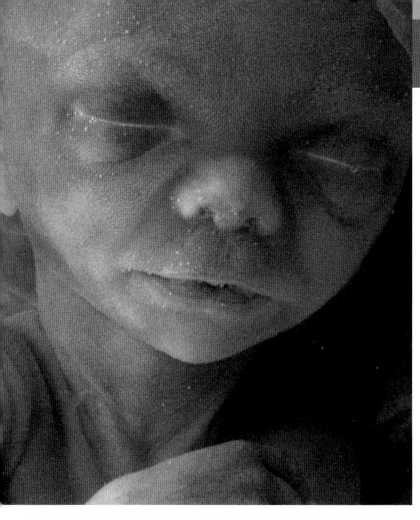

CHANGING APPEARANCE

Growth at this time is rapid, and the facial features of the fetus are developing quickly. Although the fetus still has a relatively large, bulging forehead, its eyes have moved from the sides of the head toward the front. This dramatically changes its facial appearance, and although the eyelids have not yet fully developed and remain fused, the fetus is beginning to look recognizably human. The outer ears have formed, and it has a button nose. The arms, wrists, hands, and fingers are developing more quickly than the lower legs, feet, and toes. The skin is thin and looks red as a result of the presence of many tiny blood vessels that are clearly visible.

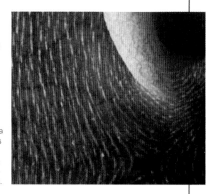

DEVELOPMENT OF FACIAL FEATURES
A photograph taken inside the uterus shows the fused eyelids of a fetus at four months. The umbilical cord floats behind.

LANUGO HAIR
The delicate skin of a four-month-old fetus is covered with fine lanugo hair, which is present even on the developing ear lobe.

THE FORMATION OF THE GENITALS

In early embryonic development, the male and female genitals look identical—this is known as the indifferent stage. The baby's sex cannot be easily identified until the fourth month of gestation. In a male, two ridges (the labioscrotal swellings) fuse along the midline to form the scrotum. A rounded bump (the genital tubercle) elongates to form the penis. In a female, the labioscrotal swellings are separate and form the labia (lips) surrounding the entrance to the vagina.

1 EARLY INDIFFERENT STAGE
The genital tubercle and labioscrotal swelling appear at around four weeks, and appear identical in males and females.

2 LATER INDIFFERENT STAGE
By six weeks, a division has formed that separates the developing anus from the urogenital membrane.

3 AT 14 WEEKS
Midway through the fourth month of gestation, the sex of the external genitals has become obvious. The urogenital membrane has fused in a boy, but forms the hymen in a girl.

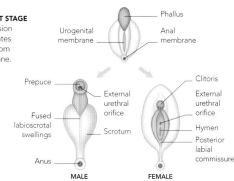

Urogenital fold — Genital tubercle
Labioscrotal swelling — Cloacal membrane

Urogenital membrane — Phallus
Anal membrane

Prepuce — External urethral orifice
Fused labioscrotal swellings — Scrotum
Anus

Clitoris — External urethral orifice
Hymen
Posterior labial commissure

MALE   FEMALE

THE FORMATION OF THE UTERUS

The uterus and cervix form from the fused ends of the mullerian ducts (see facing page). By the fourth month, the division between the two fused tubes has disappeared completely, leaving behind a hollow, muscular tube—the uterus. The vagina forms separately from a flat, circular collection of cells called the vaginal plate. These thicken and grow in a downward direction to form a solid cylinder. This structure starts to hollow out so that, by around 16 weeks' gestation, the vagina has fully formed.

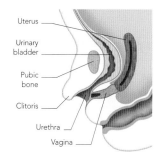

Uterus
Urinary bladder
Pubic bone
Clitoris
Urethra
Vagina

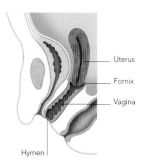

Uterus
Fornix
Vagina
Hymen

1 UTERUS AT 14 WEEKS
The uterus now forms a long tube, and the vagina is starting to hollow out. At 14 weeks, the lower part of the vagina opens into the urethral opening, but they soon develop separate entrances.

2 UTERUS OF A NEWBORN
The uterus is naturally slightly curved and tilts forward within the pelvis. The lower end of the vagina is protected by a thin, incomplete membrane called the hymen.

During the fifth month of pregnancy, the fetus grows rapidly in length and may double in weight. The growing uterus now makes the pregnancy more obvious, and the mother may become more aware of the life that is growing within her uterus.

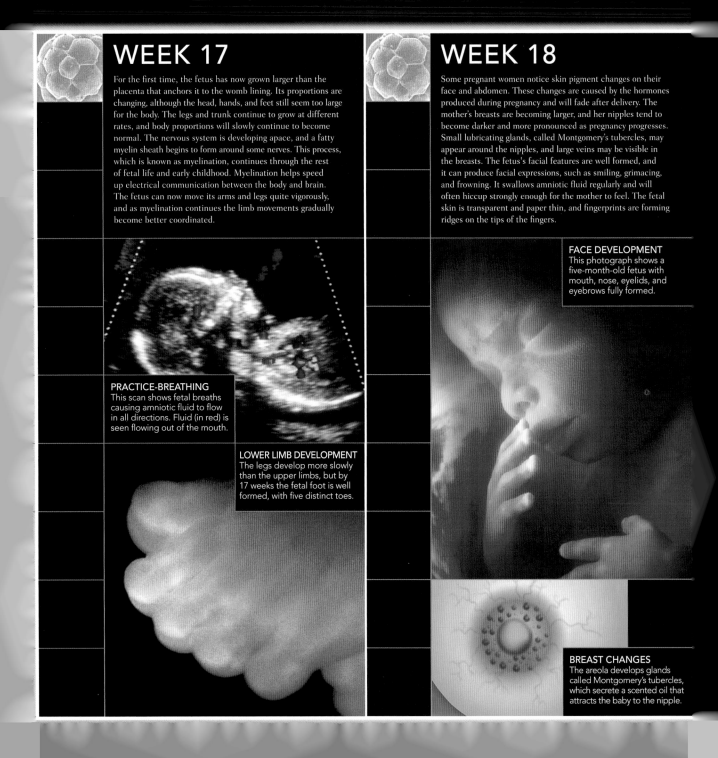

WEEK 17

For the first time, the fetus has now grown larger than the placenta that anchors it to the womb lining. Its proportions are changing, although the head, hands, and feet still seem too large for the body. The legs and trunk continue to grow at different rates, and body proportions will slowly continue to become normal. The nervous system is developing apace, and a fatty myelin sheath begins to form around some nerves. This process, which is known as myelination, continues through the rest of fetal life and early childhood. Myelination helps speed up electrical communication between the body and brain. The fetus can now move its arms and legs quite vigorously, and as myelination continues the limb movements gradually become better coordinated.

PRACTICE-BREATHING
This scan shows fetal breaths causing amniotic fluid to flow in all directions. Fluid (in red) is seen flowing out of the mouth.

LOWER LIMB DEVELOPMENT
The legs develop more slowly than the upper limbs, but by 17 weeks the fetal foot is well formed, with five distinct toes.

WEEK 18

Some pregnant women notice skin pigment changes on their face and abdomen. These changes are caused by the hormones produced during pregnancy and will fade after delivery. The mother's breasts are becoming larger, and her nipples tend to become darker and more pronounced as pregnancy progresses. Small lubricating glands, called Montgomery's tubercles, may appear around the nipples, and large veins may be visible in the breasts. The fetus's facial features are well formed, and it can produce facial expressions, such as smiling, grimacing, and frowning. It swallows amniotic fluid regularly and will often hiccup strongly enough for the mother to feel. The fetal skin is transparent and paper thin, and fingerprints are forming ridges on the tips of the fingers.

FACE DEVELOPMENT
This photograph shows a five-month-old fetus with mouth, nose, eyelids, and eyebrows fully formed.

BREAST CHANGES
The areola develops glands called Montgomery's tubercles, which secrete a scented oil that attracts the baby to the nipple.

WEEK 19

By the end of the 19th week, the full complement of milk teeth buds have formed—ten in the upper jaw and ten in the lower jaw. These tiny teeth buds lie dormant beneath the gums until some time after birth. The fetus's eyebrows and scalp hair are becoming visible, but the eyelids are still firmly fused together to protect the delicate developing eye beneath. The fetus continues growing rapidly, and the top of the uterus grows upward by around ⅜in (1 cm) per week. The top of the uterus is now almost level with the mother's belly button (navel or umbilicus). The cartilage that forms the blueprint for the fetal skeleton is starting to harden in some places to form areas of bone. This process, known as ossification, continues after birth to allow childhood growth.

WEEK 20

Between 18 and 20 weeks, a mid-pregnancy scan is carried out in order to check that the fetus's limbs and organs are developing correctly. The external genitals are now visible, and the sex of the fetus has become more obvious on ultrasound scans. In a female fetus, the ovaries have descended from the abdomen into the pelvis. In a male fetus, the testes are also descending, but they have not yet passed out of the body into the scrotum. The ability of the fetus to interact with its environment is increasing due to the development of the nervous system. Amazingly, the fetus can already detect a number of sounds and tastes, and nerve pathways that carry information about pain, temperature, and touch are starting to develop. The first tentative sparks of awareness are in place.

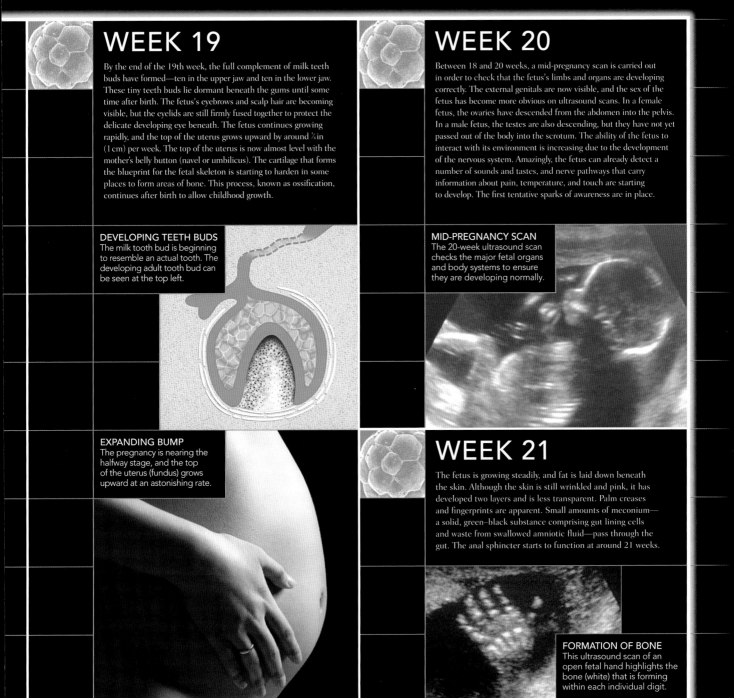

DEVELOPING TEETH BUDS
The milk tooth bud is beginning to resemble an actual tooth. The developing adult tooth bud can be seen at the top left.

MID-PREGNANCY SCAN
The 20-week ultrasound scan checks the major fetal organs and body systems to ensure they are developing normally.

EXPANDING BUMP
The pregnancy is nearing the halfway stage, and the top of the uterus (fundus) grows upward at an astonishing rate.

WEEK 21

The fetus is growing steadily, and fat is laid down beneath the skin. Although the skin is still wrinkled and pink, it has developed two layers and is less transparent. Palm creases and fingerprints are apparent. Small amounts of meconium—a solid, green–black substance comprising gut lining cells and waste from swallowed amniotic fluid—pass through the gut. The anal sphincter starts to function at around 21 weeks.

FORMATION OF BONE
This ultrasound scan of an open fetal hand highlights the bone (white) that is forming within each individual digit.

MONTH 5 | WEEKS 17–21
MOTHER AND FETUS

The first movement of the fetus felt by the mother—known as quickening—usually occurs this month. The mother may start to notice changes in skin pigmentation, such as the appearance of a dark line (linea nigra) running from the navel down to the pelvis and brown patches (chloasma) on the cheeks. Both of these pigment changes are thought to result from hormone changes and usually disappear or lighten after the birth. The mother's breasts are enlarging, and the nipples and surrounding areolae darken. The mid-pregnancy scan, usually performed at 20 weeks, detects any major fetal abnormalities, checks the position of the placenta, and reveals the sex of the baby. The fetus starts to move with increasing regularity and begins to hiccup. A fatty layer is starting to insulate the nerves, forming what makes the fetus's movements faster and better coordinated.

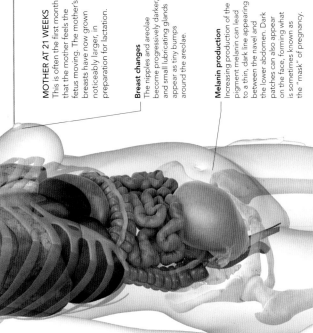

MOTHER AT 21 WEEKS
This is often the first month that the mother feels the fetus moving. The mother's breasts have now grown noticeably larger, in preparation for lactation.

Breast changes
The nipples and areolae become progressively darker, and small lubricating glands appear as tiny bumps around the areolae.

Melanin production
Increasing production of the pigment melanin can lead to a thin, dark line appearing between the navel and the lower abdomen. Dark patches can also appear on the face, forming what is sometimes known as the "mask" of pregnancy.

MOTHER

- 72 beats per minute
- 105/69
- 8 pints (4.6l)

20%
The mother's blood volume is now 20 percent higher than it was before pregnancy.

The **first movement** of the fetus felt by the mother is called **quickening** and usually occurs this month.

From the fifth month onward, the **top of the uterus (the fundus) rises** at a rate of ⅜in (1 cm) per week.

FETUS

- 150 beats per minute
- 10in (26cm)
- 12oz (350g)

50:50
For the first time in the pregnancy, the fetus is the same weight as the placenta.

90%
The fetus now has a water content of **90 percent.** This reduces to **70 percent** by birth and **60 percent** by adulthood.

The **mid-pregnancy ultrasound scan** checks that the fetus is **growing** as expected, and detects **major anomalies** or defects. The **sex of the fetus** is now apparent on this **scan.**

| 1 | 2 | 3 | 4 | 5 | 6 | 7 | 8 | 9 | 10 | 11 | 12 | 13 | 14 | 15 | 16 | 17 | 18 | 19 | 20 | 21 | 22 | 23 | 24 | 25 | 26 | 27 | 28 | 29 | 30 | 31 | 32 | 33 | 34 | 35 | 36 | 37 | 38 | 39 | 40 |

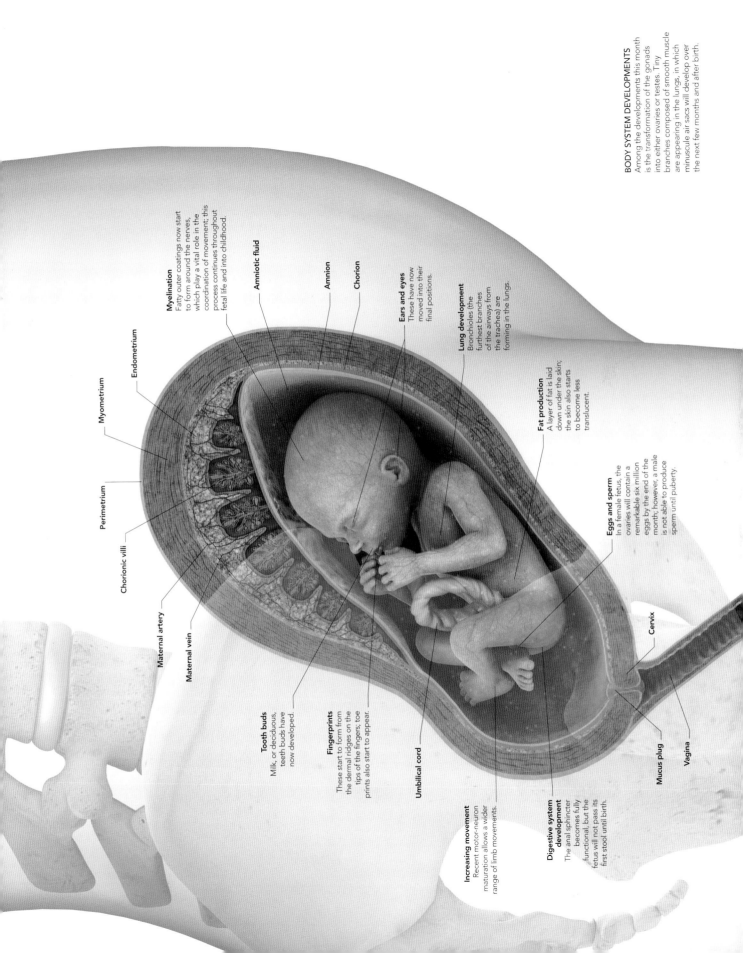

Myelination
Fatty outer coatings now start to form around the nerves, which play a vital role in the coordination of movement; this process continues throughout fetal life and into childhood.

Amniotic fluid

Amnion

Chorion

Ears and eyes
These have now moved into their final positions.

Lung development
Bronchioles (the furthest branches of the airways from the trachea) are forming in the lungs.

Fat production
A layer of fat is laid down under the skin; the skin also starts to become less translucent.

Eggs and sperm
In a female fetus, the ovaries will contain a remarkable six million eggs by the end of the month; however, a male is not able to produce sperm until puberty.

Endometrium

Myometrium

Perimetrium

Chorionic villi

Maternal artery

Maternal vein

Tooth buds
Milk, or deciduous, teeth buds have now developed.

Fingerprints
These start to form from the dermal ridges on the tips of the fingers; toe prints also start to appear.

Umbilical cord

Increasing movement
Recent motor-neuron maturation allows a wider range of limb movements.

Digestive system development
The anal sphincter becomes fully functional, but the fetus will not pass its first stool until birth.

Mucus plug

Cervix

Vagina

BODY SYSTEM DEVELOPMENTS
Among the developments this month is the transformation of the gonads into either ovaries or testes. Tiny branches composed of smooth muscle are appearing in the lungs, in which minuscule air sacs will develop over the next few months and after birth.

MONTH 5 | **KEY DEVELOPMENTS**

MOTHER

QUICKENING

The first movement of the fetus felt by a pregnant woman is known as quickening. This movement, likened to a fluttering sensation, typically occurs in the fifth month and was an important stage in pregnancy before ultrasound scanning was developed. Women who have not been pregnant before often mistake quickening for wind. Those experiencing their second or later pregnancy often feel quickening earlier than with their first baby. This is partly because they know what to expect and partly because the uterus is slightly thinner than it was before, allowing minor movements to be felt more easily.

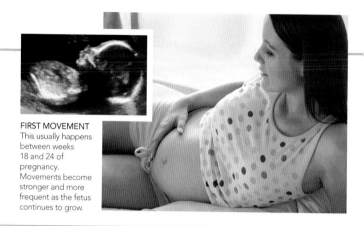

FIRST MOVEMENT
This usually happens between weeks 18 and 24 of pregnancy. Movements become stronger and more frequent as the fetus continues to grow.

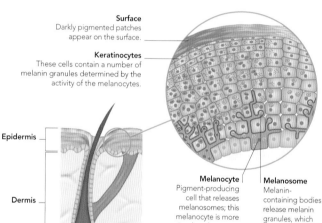

Surface
Darkly pigmented patches appear on the surface.

Keratinocytes
These cells contain a number of melanin granules determined by the activity of the melanocytes.

Epidermis

Dermis

Melanocyte
Pigment-producing cell that releases melanosomes; this melanocyte is more active than the one to the right, so the skin surface above it will generally be darker.

Melanosome
Melanin-containing bodies release melanin granules, which disperse within the skin cells (keratinocytes) above.

MELANIN PRODUCTION
Skin pigment changes are thought to result from increased stimulation of pigment cells (melanocytes), due to elevated levels of estrogen and progesterone. Because the pigment is not uniformly taken up into all skin cells, patchiness can result.

SKIN PIGMENT CHANGES

During pregnancy, hormone changes can affect skin pigmentation, which often becomes noticeable in the fifth month. The mother-to-be may develop a thin, dark line of melanin pigment running from the lower abdomen up to the navel and sometimes beyond. This is known as the linea nigra. A few women also develop irregular brown patches on the face. Known as chloasma, these patches can affect the upper cheeks, nose, forehead, or upper lip. Pigment changes usually fade and disappear after the baby is born.

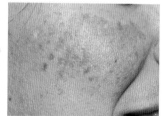

CHLOASMA
Brown patches of pigmentation on the face are sometimes described as the "mask of pregnancy."

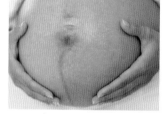

LINEA NIGRA
The term linea nigra simply means "black line" in Latin. It is thought to affect up to 75 percent of pregnant women.

BREAST CHANGES

The breasts start to change during early pregnancy in response to rising levels of estrogen. By the fifth month, they are usually noticeably larger. As well as enlarging, the breasts can become increasingly tender. The nipples and surrounding areolae darken and expand, and veins beneath the skin may look more pronounced. Small lubricating glands in the areolae, known as Montgomery's tubercles, are often visible as small bumps. In the second trimester, the breasts produce the first stage of milk, colostrum, which can leak from the nipples.

SIZE AND COLOR
The breasts continue to enlarge during pregnancy in preparation for nursing the baby. The nipples and areolae become progressively darker.

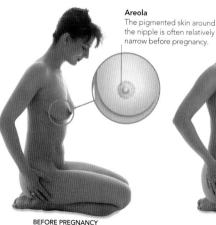

Areola
The pigmented skin around the nipple is often relatively narrow before pregnancy.

Secondary areola
A pale second areola may form; surrounding veins may also become more visible.

Nipple and areola
By the fifth month, the nipple and areola have become larger and darker.

Montgomery's tubercles
Tiny glands within the areola secrete a lubricating oil that attracts the baby to the nipple and may help prevent infection.

BEFORE PREGNANCY

FIFTH MONTH

FETUS

MID-PREGNANCY SCAN

By the fifth month of pregnancy, the fetus's organs and major bodily systems are well developed. The mid-pregnancy scan, usually carried out at 20 weeks, checks whether development is progressing normally and detects major structural abnormalities. Important checks include making sure that the fetus's heart has four chambers and is beating normally, as well as examining the abdomen to ensure that skin covers the internal organs. Because the fetus moves constantly, it is not always possible to check everything on one scan, so the mother may need to return for further checks.

Scan checks that skull bones are complete · Vertebrae examined to check for spina bifida · Right atrium · Right ventricle · Left ventricle · Left atrium

CHECKING THE VERTEBRAE
Examination of the position and width of the vertebrae can identify a number of developmental defects, including spina bifida.

HEART DEVELOPMENT
The heart is usually the first organ assessed to ensure that all four chambers have developed normally.

CONDITIONS A MID-PREGNANCY SCAN CAN DETECT

The mid-pregnancy scan is often referred to as a fetal anomaly scan because it can detect major problems with the fetus or the mother. This table shows that some developmental abnormalities are easier to identify than others.

NAME OF CONDITION	DETECTION RATE
Anencephaly (absence of the top of the head)	99%
Major limb abnormalities (missing or very short limbs)	90%
Spina bifida (open spinal cord)	90%
Major kidney problems (missing or abnormal kidneys)	85%
Cleft lip or palate (opening in the top lip or split in the roof of the mouth)	75%
Hydrocephalus (excess fluid in the brain)	60%
Major heart problems (defects of chambers, valves, or vessels)	25%

PLACENTA LOCATION

During the mid-pregnancy scan, the sonographer records whether the placenta is attached to the front or back wall of the uterus, at the top, or whether it is low-lying (positioned close to the cervix). As the uterus enlarges, a low-lying placenta usually rises up away from the cervix. However, women with a low-lying placenta may be offered another scan at 32 weeks to ensure that its position will not affect a vaginal delivery.

DETERMINING GENDER

The gender of the fetus is fixed as soon as the sperm fertilizes the egg. By the 12th week of pregnancy, the fetal reproductive system is fairly well developed, but the gender is not normally obvious until the mid-pregnancy scan at around 20 weeks. Within a female fetus, the ovaries already contain millions of eggs, and the vagina has started to develop as a hollow. Within a male fetus, the testes are anchored within the abdominal cavity but have not yet moved into the scrotum. The scrotal swelling forms a solid pouch at the base of the penis, which is often noticeable on the scan. The shape of the pelvic bones can help identify gender, too.

OLD WIVES' TALES
"Natural" methods of determining a baby's gender include dangling a gold ring over the abdomen; if the ring swings in a circle, the woman is thought to be carrying a boy, and if the ring swings back and forth, it is said to be a girl. Such methods are no more accurate than tossing a coin, however.

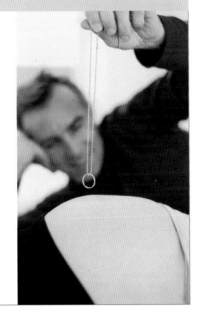

Placenta attached to top of uterus · Umbilical cord · Uterus · Lining of uterus · Mucus plug · Cervix

NORMAL

Placenta often attaches to side of uterus

NORMAL

Lower part of placenta is close to cervical opening

LOW-LYING

Placenta covers cervical opening

PLACENTA PREVIA

LOW-LYING PLACENTA
A placenta that lies over or within 1 in (2.5 cm) of the cervix is known as placenta previa (see p.228). If it stays in this position, delivery must be performed by cesarean section.

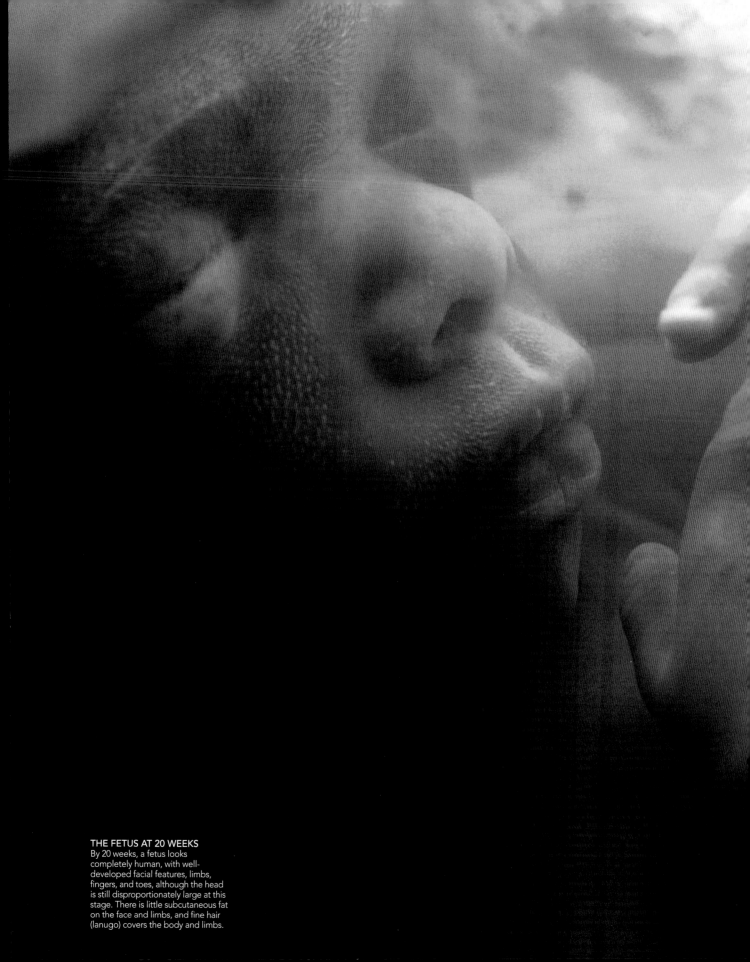

THE FETUS AT 20 WEEKS
By 20 weeks, a fetus looks
completely human, with well-
developed facial features, limbs,
fingers, and toes, although the head
is still disproportionately large at this
stage. There is little subcutaneous fat
on the face and limbs, and fine hair
(lanugo) covers the body and limbs.

FETUS

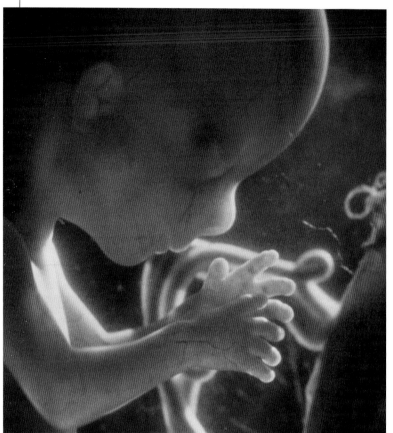

LONGER LIMBS
The arms are growing longer, as are the legs and body. Conversely, the hands and fingers still look large in proportion to the arms.

CHANGING PROPORTIONS

During the first trimester, the nervous system was undergoing a critical period of development. As a result, the brain and head grew quickly until they reached a size that comprised as much as half of the total length of the fetus's body. In the fifth month, the fetal trunk and limbs enter a rapid growth spurt, so the head begins to look more adultlike in terms of proportion to the body. From now until birth, the head grows relatively little compared with the huge growth that the body experiences during this time. Measurement of the head and thigh bones can be used to accurately date the pregnancy and assess the age of the fetus, but this information is usually gleaned from the first ultrasound (11—14 weeks) or mid-pregnancy ultrasound (20 weeks).

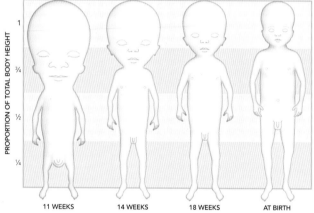

CHANGING RATES OF GROWTH
During the first trimester, the fetal head grows more quickly than the body. Relative growth of the head then slows, so that by the fifth month, fetal proportions look more like those of an adult.

INCREASING MOVEMENT

By the end of the fourth month the fetus's limbs are fully formed, and its joints can move. It can now make the complete range of movements that a full-term baby can make, such as yawning, sucking its thumb, and practicing breathing movements. It waves its arms and legs frequently and is startled by loud sounds. Although most movements are reflex actions, the continuing myelination of the nervous system (see facing page) causes some of these movements to become more coordinated. The fetus starts to make deliberate movements, such as touching its lips and sucking its thumb. Although it can move its eyes, the eyelids remain fused and do not open until the seventh month of pregnancy.

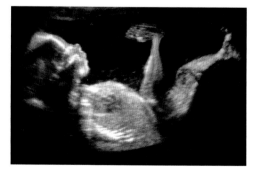

LIMB MOVEMENTS
This ultrasound scan shows a fetus flexing its muscles by waving its arms and legs. It kicks and punches the uterus. These movements are felt by the mother and often cause visible ripples on the abdomen.

HICCUPING

The fetus starts to hiccup in the middle of the second trimester; its hiccups increase in intensity and frequency over the course of the pregnancy. They occur when its diaphragm involuntarily contracts, causing a sudden rush of air that closes the opening between the vocal cords (glottis). This reflex may have adapted to prevent milk from entering the lungs of modern newborns during suckling, but this is not certain.

MYELINATION

By the fifth month, some of the nerve axons linking the fetal limbs to the spinal cord are developing a fatty outer coating. This process is known as myelination—the nerves are electrically insulated so they can carry messages without affecting neighboring nerve cells. After myelination, messages pass more easily from the brain to the body (and from the body to the brain).

As a result, fetal movements become faster and more coordinated rather than being slow and jerky. Myelination continues throughout fetal life and early childhood.

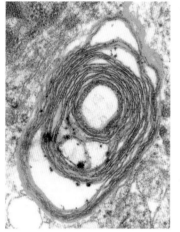

MYELIN AROUND NERVE AXON
An electron micrograph showing rings of myelin sheath (blue) around a nerve axon, similar to insulation tape around an electric wire.

SENSORY STIMULI
A 20-week-old fetus explores its left ear with one hand and grasps its forearm in the other. The brain stimulation received from exploring the environment contributes toward the fetus's growing awareness.

AWARENESS OF SURROUNDINGS

Exactly when a fetus becomes aware of its surroundings is unclear. The first connections between brain cells (synapses) form during the 12th week of pregnancy, but it is thought that true awareness does not start until around the 20th week. Different types of awareness develop, such as "quiet" awareness, when the fetus is awake but seems to rest, and "active" awareness, when it is awake and moves, often quite vigorously. The fetus reacts to sounds within its mother's body, and noises in its external environment. As myelination and brain development progress, the fetus's awareness of its own body and its movements will increase.

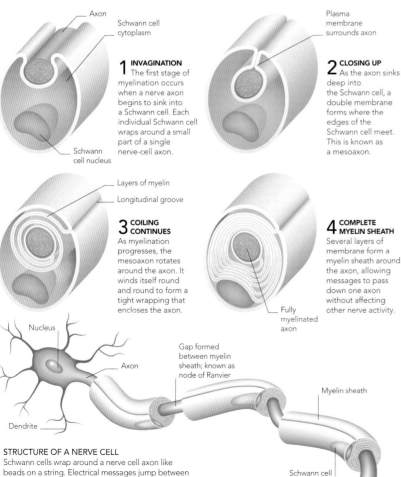

1 INVAGINATION The first stage of myelination occurs when a nerve axon begins to sink into a Schwann cell. Each individual Schwann cell wraps around a small part of a single nerve-cell axon.

2 CLOSING UP As the axon sinks deep into the Schwann cell, a double membrane forms where the edges of the Schwann cell meet. This is known as a mesoaxon.

3 COILING CONTINUES As myelination progresses, the mesoaxon rotates around the axon. It winds itself round and round to form a tight wrapping that encloses the axon.

4 COMPLETE MYELIN SHEATH Several layers of membrane form a myelin sheath around the axon, allowing messages to pass down one axon without affecting other nerve activity.

STRUCTURE OF A NERVE CELL
Schwann cells wrap around a nerve cell axon like beads on a string. Electrical messages jump between nodes to speed up nerve cell transmission.

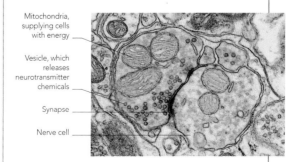

THE JUNCTION BETWEEN NERVE CELLS
This electron micrograph shows a synapse that forms a link between nerve cells (green). Electrical signals are carried across through the action of neurotransmitters (red dots).

The sixth month brings the mother-to-be toward the end of the second trimester. The uterus and breasts are growing larger, and the amount of blood the heart pumps every minute increases. Most women gain around 18 oz (500 g) per week at this stage.

WEEK 22

The bones within the fetus's inner ear are beginning to harden, and the coiled cochlear membrane is sufficiently developed to process low-frequency sounds. Over the coming weeks, the fetus starts to be aware of higher sound frequencies, too. The nervous system is now developed enough for the fetus to start recognizing the sounds inside the uterus, such as the mother's breathing, heart beat, stomach and intestinal rumbles, and her voice. It may be noticeable that the fetus becomes increasingly responsive to sounds, and it will develop a startle reaction to loud noises. As the nervous system develops, the fetus becomes able to make much more sophisticated movements, such as kicking and turning somersaults, and the mother will be aware of this increased internal activity.

WEEK 23

The fetal skin cells now start to accumulate a tough, protective protein called keratin, the thickest layer of which is on the palms and soles of the feet. The skin is very wrinkled and is covered in greasy vernix and fine lanugo hair, which protect the fetus in its aquatic environment and may have an insulating effect. The nails start to appear at the base of the nail beds, and eyelids and eyebrows are developing. Small blood vessels appear in the lungs. The barrier between these capillaries and the future air sacs is thinning to allow the exchange of gases when the baby is born. Specialized lung lining cells (pneumocytes) are appearing. These will produce a substance called surfactant, which reduces surface tension so the small air sacs can expand more easily after birth.

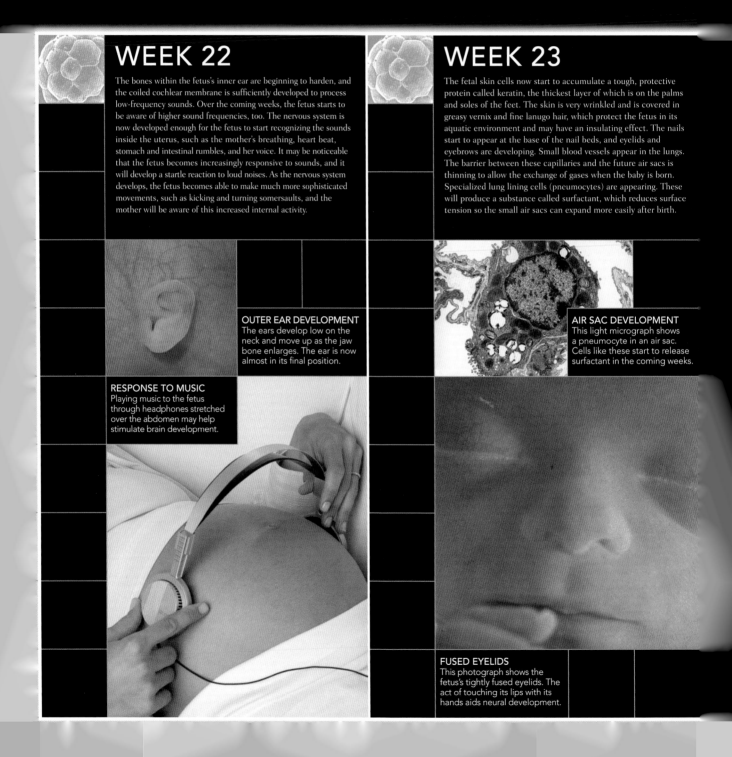

OUTER EAR DEVELOPMENT
The ears develop low on the neck and move up as the jaw bone enlarges. The ear is now almost in its final position.

AIR SAC DEVELOPMENT
This light micrograph shows a pneumocyte in an air sac. Cells like these start to release surfactant in the coming weeks.

RESPONSE TO MUSIC
Playing music to the fetus through headphones stretched over the abdomen may help stimulate brain development.

FUSED EYELIDS
This photograph shows the fetus's tightly fused eyelids. The act of touching its lips with its hands aids neural development.

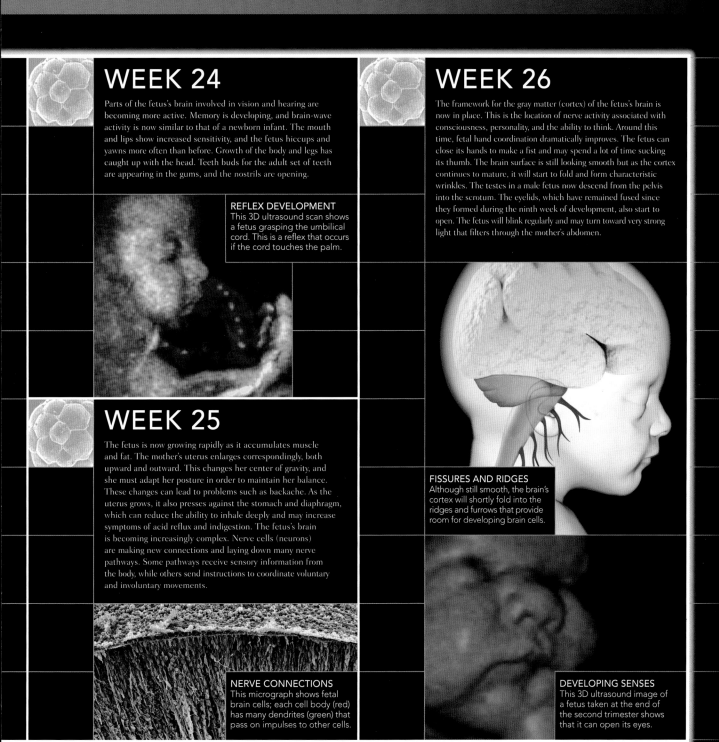

WEEK 24

Parts of the fetus's brain involved in vision and hearing are becoming more active. Memory is developing, and brain-wave activity is now similar to that of a newborn infant. The mouth and lips show increased sensitivity, and the fetus hiccups and yawns more often than before. Growth of the body and legs has caught up with the head. Teeth buds for the adult set of teeth are appearing in the gums, and the nostrils are opening.

REFLEX DEVELOPMENT
This 3D ultrasound scan shows a fetus grasping the umbilical cord. This is a reflex that occurs if the cord touches the palm.

WEEK 25

The fetus is now growing rapidly as it accumulates muscle and fat. The mother's uterus enlarges correspondingly, both upward and outward. This changes her center of gravity, and she must adapt her posture in order to maintain her balance. These changes can lead to problems such as backache. As the uterus grows, it also presses against the stomach and diaphragm, which can reduce the ability to inhale deeply and may increase symptoms of acid reflux and indigestion. The fetus's brain is becoming increasingly complex. Nerve cells (neurons) are making new connections and laying down many nerve pathways. Some pathways receive sensory information from the body, while others send instructions to coordinate voluntary and involuntary movements.

NERVE CONNECTIONS
This micrograph shows fetal brain cells; each cell body (red) has many dendrites (green) that pass on impulses to other cells.

WEEK 26

The framework for the gray matter (cortex) of the fetus's brain is now in place. This is the location of nerve activity associated with consciousness, personality, and the ability to think. Around this time, fetal hand coordination dramatically improves. The fetus can close its hands to make a fist and may spend a lot of time sucking its thumb. The brain surface is still looking smooth but as the cortex continues to mature, it will start to fold and form characteristic wrinkles. The testes in a male fetus now descend from the pelvis into the scrotum. The eyelids, which have remained fused since they formed during the ninth week of development, also start to open. The fetus will blink regularly and may turn toward very strong light that filters through the mother's abdomen.

FISSURES AND RIDGES
Although still smooth, the brain's cortex will shortly fold into the ridges and furrows that provide room for developing brain cells.

DEVELOPING SENSES
This 3D ultrasound image of a fetus taken at the end of the second trimester shows that it can open its eyes.

MONTH 6 | WEEKS 22–26
MOTHER AND FETUS

As the second trimester draws to a close, most women feel well and exhibit a healthy glow. However, stretchmarks may start to appear during this month around the abdomen, and libido can decrease. A vaginal ultrasound may be performed to measure the length of the cervix and predict the risk of a premature birth; this scan is often offered if the woman has suffered a late miscarriage in a previous pregnancy. The development of the fetus's body systems has reached a point where it can now start using the energy and nutrients supplied via the placenta to lay down some fat. This causes its weight to increase rapidly. Red blood cells, which were previously only produced in the liver, are now also being created in the marrow of the long bones. If born prematurely toward the end of this month, a fetus has a moderate chance of survival with intensive neonatal care.

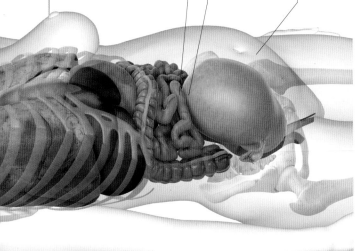

MOTHER

- 72 beats per minute
- 105/70
- 8½ pints (4.8l)

MOTHER AT 26 WEEKS
The expanding uterus starts to cause a reduction in lung volume, which can cause breathlessness. Other discomforts, such as constipation, may also occur.

Constipation
The growing uterus can exert pressure on the digestive system, leading to constipation.

Fundal height
The height of the uterus above the pubic bone gives a good indication of the duration of pregnancy. At 24 weeks, the height is around 9½in (24cm); the uterus now expands upward at a rate of around ⅜in (1 cm) per week.

Stretchmarks
The expanding uterus causes the abdominal wall to stretch, leading to the rapid thinning of collagen and elastin fibers in the skin, resulting in stretchmarks.

50%
Levels of progesterone rise by 50 percent this month. Estrogen levels are also steadily increasing.

A **cervical-length scan** may be performed to detect premature opening of the cervix.

Most fetuses now settle into a more **regular cycle** of rest interspersed with periods of **vigorous movement.**

STATISTICS

| 1 | 2 | 3 | 4 | 5 | 6 | 7 | 8 | 9 | 10 | 11 | 12 | 13 | 14 | 15 | 16 | 17 | 18 | 19 | 20 | 21 | 22 | 23 | 24 | 25 | 26 | 27 | 28 | 29 | 30 | 31 | 32 | 33 | 34 | 35 | 36 | 37 | 38 | 39 | 40 |

FETUS

- 150 beats per minute
- 14in (36cm)
- 26oz (750g)

One-third
The head, trunk, and legs now each comprise **a third of the overall length** of the fetus.

12%
A fetus's bones contain as little as 12 percent calcium compared to the 90 percent calcium in adult bones.

65%
Premature babies born at 26 weeks have a 65 percent chance of survival, compared with only a 25 percent chance at 24 weeks.

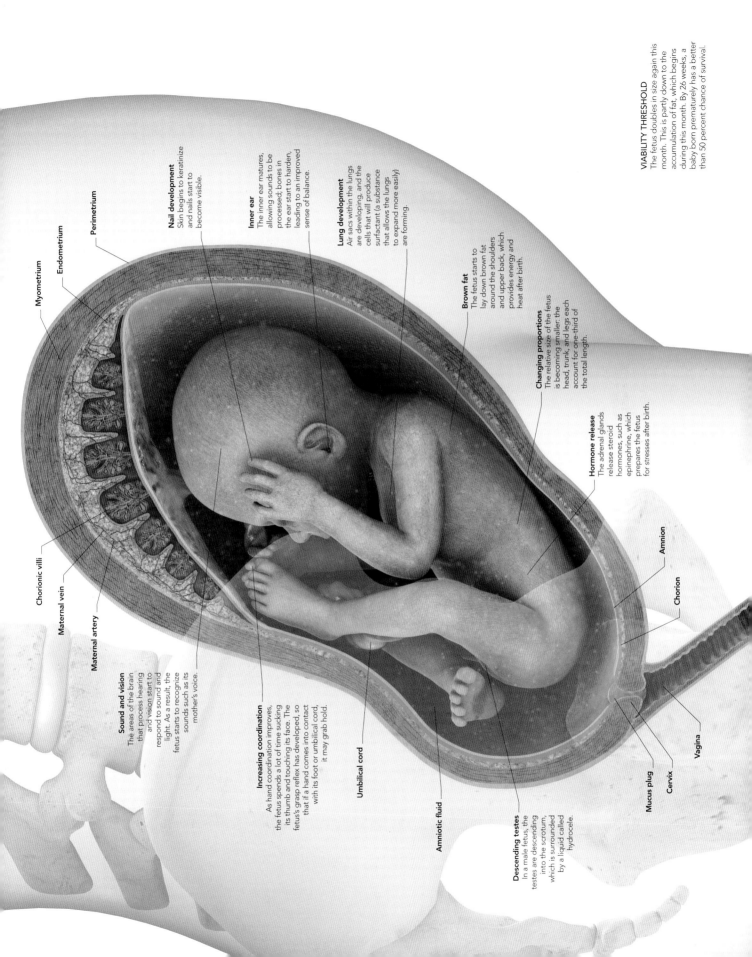

Myometrium

Endometrium

Perimetrium

Chorionic villi

Maternal vein

Maternal artery

Nail development
Skin begins to keratinize and nails start to become visible.

Inner ear
The inner ear matures, allowing sounds to be processed; bones in the ear start to harden, leading to an improved sense of balance.

Lung development
Air sacs within the lungs are developing, and the cells that will produce surfactant (a substance that allows the lungs to expand more easily) are forming.

Brown fat
The fetus starts to lay down brown fat around the shoulders and upper back, which provides energy and heat after birth.

Changing proportions
The relative size of the fetus is becoming smaller: the head, trunk, and legs each account for one-third of the total length.

Hormone release
The adrenal glands release steroid hormones, such as epinephrine, which prepares the fetus for stresses after birth.

Amnion

Chorion

VIABILITY THRESHOLD
The fetus doubles in size again this month. This is partly down to the accumulation of fat, which begins during this month. By 26 weeks, a baby born prematurely has a better than 50 percent chance of survival.

Sound and vision
The areas of the brain that process hearing and vision start to respond to sound and light. As a result, the fetus starts to recognize sounds such as its mother's voice.

Increasing coordination
As hand coordination improves, the fetus spends a lot of time sucking its thumb and touching its face. The fetus's grasp reflex has developed, so that if a hand comes into contact with its foot or umbilical cord, it may grab hold.

Umbilical cord

Amniotic fluid

Descending testes
In a male fetus, the testes are descending into the scrotum, which is surrounded by a liquid called hydrocele.

Mucus plug

Cervix

Vagina

MOTHER

STRETCHMARKS

Stretchmarks are creaselike ruptures in the skin. Also known as striae gravidarum, they are common in pregnancy. Their appearance is partly related to rapid weight gain and expansion of the abdominal wall, and partly to the effects of the hormone progesterone. Stretchmarks are initially a red–purple color and fade over time to a silver-gray. For unknown reasons, some women escape them altogether, even after multiple pregnancies. Being overweight at the start of pregnancy increases the risk of stretchmarks. Moisturizing massages and plenty of essential fatty acids in the diet may help to reduce their formation.

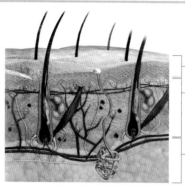

Epidermis
The outer exposed layer of the skin—known as the epidermis—remains intact over the surface of the stretch marks.

Dermis
Support tissues in the deeper layer of skin, the dermis, become stretched and thinned. This causes painless tears that appear as stretchmarks on the surface.

Subcutaneous fat
During pregnancy, increased amounts of fat are laid down beneath the dermis, contributing to skin stretching.

CAUSE OF STRIAE
Stretchmarks occur when collagen and elastin fibers in the dermis are rapidly thinned and stretched.

AFFECTED AREAS
Stretchmarks can form anywhere, but are most common on the abdomen, hips, thighs, and breasts.

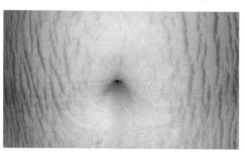

CHANGES TO LIBIDO DURING PREGNANCY

Sex drive can go up, down, or remain the same during pregnancy; despite the predictable hormone changes that they experience, all women have different libidos. Psychological influences play a large part, as does increased blood flow to the genital area, increased lubrication, and the fact that orgasm is often easier to achieve and more intense during pregnancy. Low sex drive may be linked to physical exhaustion, especially during the last three months. During this time, blood levels of the hormone prolactin—which tends to lower libido—start to increase in preparation for lactation. Low sex drive is not inevitable because high levels of estrogen and progesterone help reduce the effects of prolactin.

CERVICAL-LENGTH SCAN

If there is a risk of premature delivery, a vaginal ultrasound scan may be carried out to measure cervical length. A lubricated ultrasound probe is gently inserted into the vagina to assess whether the cervix is becoming shorter and softer than normal. The cervical length is measured and the shape formed by the upper cervical canal (internal os) is also examined. A tight internal os, which forms a T-shape, is less likely to be associated with premature delivery. As the cervix shortens and the internal os beings to open, it will form a Y-shape, then a V-shape, and finally a U-shape. This funneling allows the amniotic membranes to bulge through and greatly increases the chance of premature delivery.

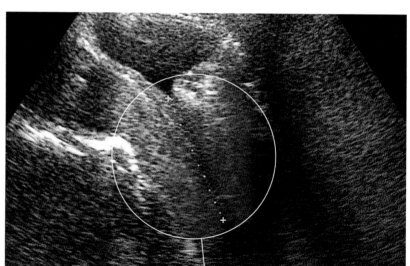

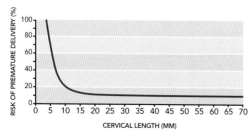

RISK OF PREMATURE DELIVERY
This graph shows the relationship between cervical length and the risk of premature delivery during the 23rd week of pregnancy. If cervical length falls below approximately ¾ in (2 cm), the risk increases.

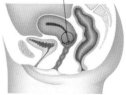

LOCATION OF THE CERVIX

NORMAL LENGTH
This ultrasound scan shows the cervix during the fifth month of pregnancy. The scan shows a cervical length of over 1 in (2.5 cm), which is normal. The risk of premature delivery in this pregnancy as a result of an incompetent cervix—a cervix that cannot retain a baby beyond a certain stage—is low. The fetus is not visible in this view.

FETUS

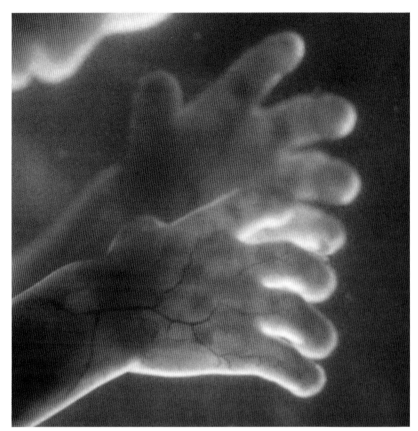

THE DEVELOPMENT OF DIGITS

The fetal fingers and toes are fully developed by the sixth month of pregnancy. The nail beds have formed, and the nail plates start to appear. The epidermal ridges that form creases on the palms of the hands and soles of the feet were laid down during early embryonic life; these are now becoming more visible as the fetal skin thickens and appears more opaque. These creases are determined by the genes a fetus inherits. The fingerprints on the fleshy pads of the fingertips are also becoming more obvious. Fingerprints are unique to each individual, and the whorls they form are thought to reflect nutrition and placental blood flow during the early stages of development. Some research suggests that the prints may help predict the individuals who will develop high blood pressure in later life.

FINGERS
This image of a six-month-old fetus shows how well the hands and fingers are developed. The nail beds are laid down, and the nail plates are starting to grow, and on the underside the fleshy pads are beginning to reveal their unique fingerprint. patterns.

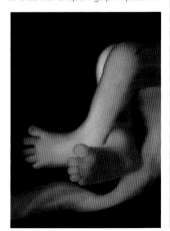

SPREADING TOES
This computer-generated image of a fetus's legs at 22 weeks of pregnancy shows it can already spread its toes.

RED-BLOOD-CELL PRODUCTION

Red blood cells, which carry oxygen around the fetus's body, are the most abundant of all cells in the body. During embryonic life, red blood cells are produced first in the yolk sac, and from the third or fourth month of pregnancy by the developing liver and spleen. In the sixth month of pregnancy, the red bone marrow within the hollow spaces of the fetal long bones starts to take over this role. Substances produced in the fetal kidneys and the placenta regulate this process.

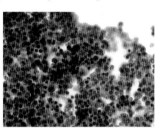

FETAL BONE MARROW
This light micrograph shows a section of fetal bone marrow, containing many red blood cells. These cells have differentiated from embryonic stem cells (see p.99).

HEART RATE

The fetal heart rate at this stage is around 140–150 beats per minute. This varies in a predictable pattern during the day and night. As with the mother's heart rate and blood pressure, the lowest fetal heart rate and blood pressure occur in the early hours of the morning (around 4 am); they rise again just before waking to reach a natural peak in mid-morning. Although it is often said that higher heart rates occur in male fetuses, research involving 10,000 fetal heart rates shows this is not the case. Fetal heart rate does, however, vary with the stage of pregnancy.

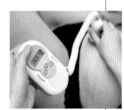

DOPPLER
The fetal heart beat is detected via the mother's abdomen with a Doppler fetal-heart-rate monitor. The rate is displayed on the screen.

FETAL HEART RATE
The chart below shows the fetal heart rate at different stages of pregnancy. It peaks early on and gently fluctuates until birth.

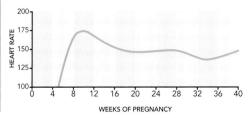

WEEKS OF PREGNANCY

FETUS

THE DEVELOPMENT OF HEARING

Hearing is one of the first senses to develop. The inner, middle, and external ear develop separately, from three different parts of the embryo, but work together to detect sound. The complex outer ear shell (auricle) develops from six tiny bumps (auricular hillocks) that are visible in a six-week-old embryo. These slowly enlarge and fuse to form the folded auricle. At first, the ear forms on the lower neck. As the jaw bone develops, different parts of the fetal head grow at different rates, so the ears appear to move up until they are level with the eyes. The external ear's shape helps collect and funnel sound waves into the auditory canal. They pass through the eardrum (tympanic membrane) and are transmitted across the three tiny bones (ossicles) in the middle ear to reach the inner ear. Here, sound vibrations are converted into nerve signals that are sent to the brain for processing.

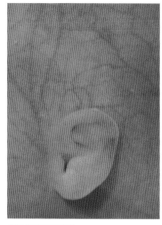

EAR AT 22 WEEKS
By 22 weeks' gestation the ear is almost completely formed. The auricle has risen up to its correct anatomical position, halfway up the head and level with the eyes.

RECOGNIZING THE MOTHER'S VOICE

A fetus recognizes its mother's voice above all others. In part this is because her voice is heard most often, but it is also because her body is a good conductor of sound and vibration. The sound is conducted to the fetus through the mother's body tissues internally and also externally through the air to reach the outer abdomen. Her voice is one of the first things it learns to recognize as it becomes more aware of its surroundings. The powerful soothing effect of the mother's voice also provides comfort after birth.

EFFECT ON HEART RATE
Studies carried out on newborn babies have shown that their heart rate slows whenever they hear their mother speaking.

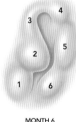

THE AURICULAR HILLOCKS
During early embryonic life, the outer ear starts to form from six auricular hillocks. These enlarge and fuse and eventually form the folded auricle.

MONTH 1

MONTH 6

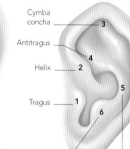

Cymba concha
Antitragus
Helix
Tragus
Concha
Antihelix

NEWBORN

WHAT A FETUS HEARS
The uterus is full of noises, such as the sound of the mother's heartbeat and the gurgling of her intestines. This means that the fetus is exposed to around 70 decibels of sound in the uterus—around the same sound level as an average conversation.

The level of sound to which a fetus is accustomed

SOURCE OF NOISE	
WHISPER	
QUIET ROOM	
NOISE LEVEL INSIDE THE UTERUS	
BUSY STREET	
LOUD MUSIC	
JET ENGINE	

0   20   40   60   80   100   120   140   160
LOUDNESS OF NOISE OR SOUND (DECIBELS)

REFLEX DEVELOPMENT

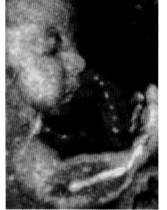

GRASP REFLEX
This color enhanced 3D ultrasound scan of a 24-week-old fetus shows it playing with its umbilical cord (purple).

A baby is born with over 70 primitive reflexes that give protection in the early days of life. They are programmed into the nervous system during early development as nerve connections are laid down. Some reflexes, such as the rooting and suckling reflexes, help with feeding. Others, such as the grasp reflex, are survival instincts that help stabilize the body. The grasp reflex develops at around 10 weeks' gestation— the fetus can close its fingers but only in an incomplete way. By six months, a true but weak grasp reflex is evident.

2 Each sensory nerve impulse is sent directly to the spinal cord by sensory neurons (the brain is not involved in reflex actions).

3 The cell bodies of motor neurons in the spinal cord initiate their own impulse back to the muscles.

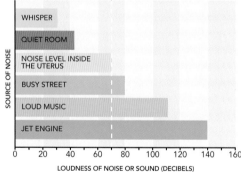

Brain
Spinal cord

1 Stimulus is applied.

4 The two sets of muscles involved in grasping—one in the forearm and one in the hand— are activated.

MECHANICS OF THE GRASP REFLEX
A baby will grasp tightly if the palm is stroked with a finger. A rapid sequence of nervous activity initiated by the spinal cord is responsible for this action.

KEY
— SENSORY NERVE
— MOTOR NERVE

PREMATURE BIRTH

A singleton baby born before 37 weeks' gestation is classed as preterm. If born prematurely, a fetus of 24 weeks' gestation has a moderate chance of survival in a neonatal intensive care unit. Resuscitation will be needed, and the infant will receive 24-hour care from health professionals to ensure that he or she remains warm and receives the correct amount of oxygen and nourishment. Where possible, the mother is encouraged to express her milk, which is then fed to her baby through a feeding tube that passes through the baby's nose and down into the tiny stomach. Constant monitoring is vital because babies who are born early can develop breathing problems and are also at increased risk of infection. Their body systems, including their lungs and immune function, are not fully mature. At six months' gestation, the baby will look tiny with wrinkled skin and very little subcutaneous fat. The fetal liver has difficulty processing the red blood pigment—bilirubin—so jaundice will develop, making the skin an orange-yellow color. This is treated with a special "blue" light that converts the pigment into a form that can be excreted in the baby's urine and bowel motions. The length of treatment depends on the baby's birth weight, age, and the level of bilirubin in the blood. As soon as the baby's condition allows, parents are encouraged to actively help with their baby's care. Skin-to-skin contact is recommended, which provides comfort and helps the mother bond with the baby.

LUNG UNDER-DEVELOPMENT

Many babies born before 34 weeks' gestation have some degree of breathing difficulty. This is mainly due to the lack of surfactant—a chemical secreted by specific cells in the air sacs (alveoli) in the lungs that prevents these air sacs from collapsing.

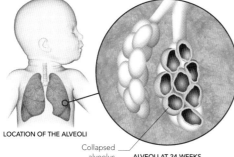

LOCATION OF THE ALVEOLI

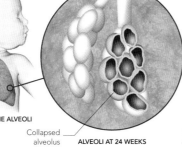

Collapsed alveolus — **ALVEOLI AT 24 WEEKS**

Normal alveolus — **ALVEOLI AT BIRTH**

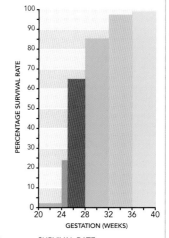

SURVIVAL RATE
The rate of survival increases the longer the baby remains in the uterus. At 24 weeks, a baby has a 24 percent chance of survival. By 28 weeks' gestation, this rises to 86 percent. The youngest premature baby to reach adulthood was born in Canada at 21 weeks and 5 days.

Treating jaundice
Blue light illuminates the infant to treat the jaundice that causes orange skin. A shield protects the eyes.

Ventilator
A ventilator supplies varying amounts of oxygen to the lungs according to the needs of the baby. Low positive pressure helps keep the tiny alveoli open.

Heart rate monitor
The baby's heart is closely monitored to ensure that it is working properly. A typical heart rate is 140–150 beats per minute.

LIFE SUPPORT AND MONITORING
This image shows a 24-week-old baby in a neonatal intensive care unit. The tubes and sensors monitor the infant's well-being and deliver oxygen, milk, and drugs.

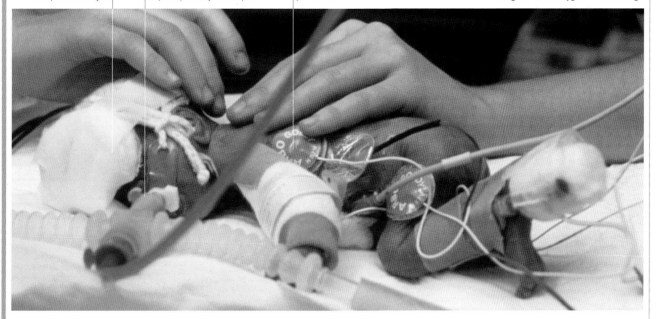

THE FORMATION OF THE RESPIRATORY SYSTEM

The respiratory system undergoes regular stages of development, with a critical period occurring late in pregnancy. It does not perform its key function—breathing—until after birth, and is filled with fluid until then.

In the uterus, the fetus receives oxygen via the mother's blood circulation in the placenta. After birth, the baby must immediately start to breathe on its own, both to obtain oxygen from the surrounding air and to exhale waste carbon-dioxide gas. The main airway in the lower respiratory tract is the trachea, which starts to develop in the fifth week. In the same week, it branches to form the left and right main bronchi. The right lung eventually develops three lobes, while the left lung develops two, leaving extra space for the heart. Development of the lungs is not usually complete until around 36 weeks' gestation. Babies born prematurely may therefore need treatment to help overcome breathing difficulties during the first few days or weeks of life.

Surfactant
Released to help air sacs expand and contract

Alveolar type II cell
Secretes surfactant; contains fine, hairlike structures on the surface

SURFACTANT PRODUCTION
Surfactant is a chemical produced by a specific type of cell in the air sacs of the lungs. It lowers surface tension so the air sacs can expand and contract easily. This image shows surfactant (green) being released by these alveolar cells.

THE UPPER RESPIRATORY SYSTEM

The mouth, nose, and throat develop at the same time as the lower airways and lungs, but from a different part of the embryo. At five weeks' gestation, a thickening at the front of the head folds inward to form two nasal pits. This creates a ridge of tissue that becomes compressed by the developing upper jaw, forming the structure of the nose. The mouth takes shape as the upper and lower jaw arches grow in from each side and fuse.

ORAL AND NASAL CAVITIES
The cavities of the nose and mouth are initially separated by the palate. As development progresses, the two airways meet at the back of the throat.

Nasal cavity
Brain
Rupturing oronasal membrane
Pharynx
Olfactory bulb
Olfactory nerves
Primary palate
Oral cavity
Heart
Oral cavity
Nasal conchae
Tongue
Secondary palate

6 WEEKS
12 WEEKS

Trachea
The main airway, this is also known as the windpipe.

Developing cartilage
Rings of cartilage help keep the larger airways open.

4 WEEKS LUNG BUD STAGE
The respiratory system develops from a tiny lung bud that branches off from the foregut. The base of the bud eventually becomes the trachea and larynx. The lower end branches to form left and right bronchial buds, which will become the left and right main bronchi. These continue branching to form secondary and tertiary bronchial buds.

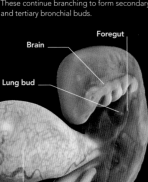

Foregut
Brain
Lung bud
Yolk sac
Umbilical cord

LUNG BUD FORMING IN A 4-WEEK-OLD FETUS

Tertiary buds
Secondary bronchial buds divide to form tertiary buds.

Right main bronchus
This divides to form three secondary bronchial buds.

LUNGS AT 6 WEEKS

LUNGS AT 7 WEEKS

LUNGS AT 5 WEEKS

First branching
The lung bud branches to form left and right main bronchi.

Continual branching
The bronchial buds branch many times over the next few weeks

Left main bronchus
This divides to form two secondary bronchial buds.

LUNGS AT 16 WEEKS

Epithelium
This will soon differentiate to form two types of cell.

Connective tissue cell

Capillary
These will gradually move closer to the alveoli.

5–7 WEEKS PSEUDOGLANDULAR STAGE
The developing respiratory system keeps dividing to form progressively more and increasingly smaller tubes. After the secondary and tertiary bronchi have formed, they will divide another 14 times to produce bronchioles by around the 24th week of gestation. These divisions determine the position, size, and shape of the lung lobes and lobules. At this early stage of development, the tiniest tubes are known as terminal bronchioles

16 WEEKS
CANALICULAR STAGE
The terminal bronchioles divide to form canal-like respiratory bronchioles. These develop rounded protrusions at the ends known as terminal sacs. Blood vessels are developing nearby.

Respiratory bronchiole
The furthest branch of the respiratory tree at this point

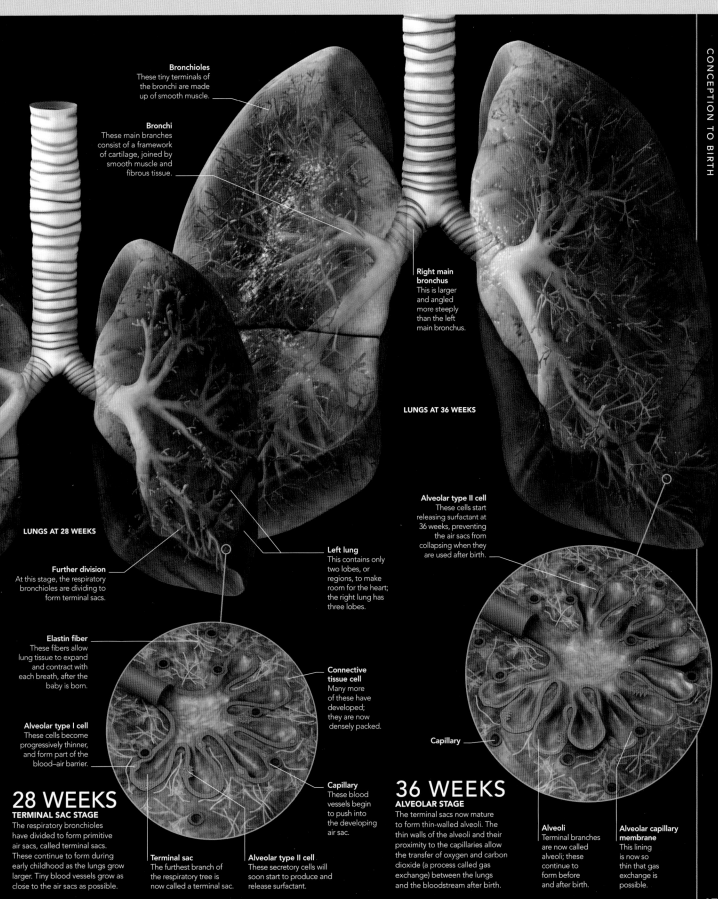

Bronchioles
These tiny terminals of the bronchi are made up of smooth muscle.

Bronchi
These main branches consist of a framework of cartilage, joined by smooth muscle and fibrous tissue.

Right main bronchus
This is larger and angled more steeply than the left main bronchus.

LUNGS AT 36 WEEKS

LUNGS AT 28 WEEKS

Further division
At this stage, the respiratory bronchioles are dividing to form terminal sacs.

Alveolar type II cell
These cells start releasing surfactant at 36 weeks, preventing the air sacs from collapsing when they are used after birth.

Elastin fiber
These fibers allow lung tissue to expand and contract with each breath, after the baby is born.

Connective tissue cell
Many more of these have developed; they are now densely packed.

Left lung
This contains only two lobes, or regions, to make room for the heart; the right lung has three lobes.

Alveolar type I cell
These cells become progressively thinner, and form part of the blood–air barrier.

Capillary

Capillary
These blood vessels begin to push into the developing air sac.

28 WEEKS
TERMINAL SAC STAGE
The respiratory bronchioles have divided to form primitive air sacs, called terminal sacs. These continue to form during early childhood as the lungs grow larger. Tiny blood vessels grow as close to the air sacs as possible.

Terminal sac
The furthest branch of the respiratory tree is now called a terminal sac.

Alveolar type II cell
These secretory cells will soon start to produce and release surfactant.

36 WEEKS
ALVEOLAR STAGE
The terminal sacs now mature to form thin-walled alveoli. The thin walls of the alveoli and their proximity to the capillaries allow the transfer of oxygen and carbon dioxide (a process called gas exchange) between the lungs and the bloodstream after birth.

Alveoli
Terminal branches are now called alveoli; these continue to form before and after birth.

Alveolar capillary membrane
This lining is now so thin that gas exchange is possible.

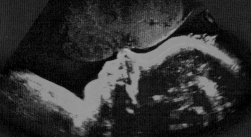

This 2D ultrasound shows a 33-week-old fetus in the uterus. Space is getting increasingly tight, and its nose can be seen pressed against the placenta.

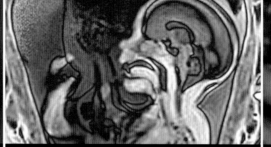

This MRI scan shows full-term twin fetuses. Twins are usually delivered earlier than single babies, at around 37 weeks of pregnancy.

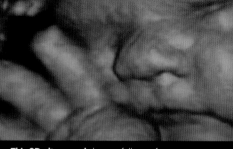

This 3D ultrasound shows a full-term fetus rubbing its eye. Its eyes have now opened and it is now sensitive to light, although it cannot yet focus.

TRIMESTER 3
MONTHS 7–9 | WEEKS 27–40

The third trimester of pregnancy is a time of maturation and rapid growth. By 40 weeks, a fetus's organs will have developed to the point at which it is capable of independent life.

During the third trimester, the important fetal developments include the laying down of fat, and the maturation of the body systems so that they can function fully on their own after birth. The respiratory system has to undergo a particularly dramatic transformation to enable breathing for the first time. To assist this, special cells in the lining of the air sacs (alveoli) produce a substance called surfactant; this lowers surface tension, allowing the lungs to inflate easily. The fetal brain continues to expand during these last three months, so the head circumference increases from around 11 in (28 cm) to 15 in (38 cm). At the same time, the fetus's total body length increases from roughly 15 in (38 cm) to

19 in (48 cm), and its weight rises from an average of 3 lb (1.4 kg) to 7½ lb (3.4 kg). The final 10 weeks are a remarkable period of growth, with the fetus gaining half of its final, full-term weight. By the end of this trimester, the fetus is fully formed and may have settled into a head-down position ready for birth. The mother-to-be may suffer from back pain in the last three months, as postural changes lead to increased strain on muscles and ligaments. Fatigue can also be a problem, mainly due to the added weight of the fetus. The breasts start to produce a creamy pre-milk called colostrum, which will nourish the baby in the days after the birth.

TIMELINE

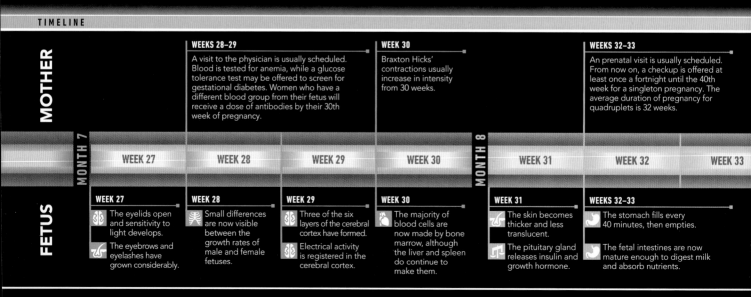

MOTHER

WEEKS 28–29
A visit to the physician is usually scheduled. Blood is tested for anemia, while a glucose tolerance test may be offered to screen for gestational diabetes. Women who have a different blood group from their fetus will receive a dose of antibodies by their 30th week of pregnancy.

WEEK 30
Braxton Hicks' contractions usually increase in intensity from 30 weeks.

WEEKS 32–33
An prenatal visit is usually scheduled. From now on, a checkup is offered at least once a fortnight until the 40th week for a singleton pregnancy. The average duration of pregnancy for quadruplets is 32 weeks.

MONTH 7 | WEEK 27 | WEEK 28 | WEEK 29 | WEEK 30 | **MONTH 8** | WEEK 31 | WEEK 32 | WEEK 33

FETUS

WEEK 27
The eyelids open and sensitivity to light develops.
The eyebrows and eyelashes have grown considerably.

WEEK 28
Small differences are now visible between the growth rates of male and female fetuses.

WEEK 29
Three of the six layers of the cerebral cortex have formed.
Electrical activity is registered in the cerebral cortex.

WEEK 30
The majority of blood cells are now made by bone marrow, although the liver and spleen do continue to make them.

WEEK 31
The skin becomes thicker and less translucent.
The pituitary gland releases insulin and growth hormone.

WEEKS 32–33
The stomach fills every 40 minutes, then empties.
The fetal intestines are now mature enough to digest milk and absorb nutrients.

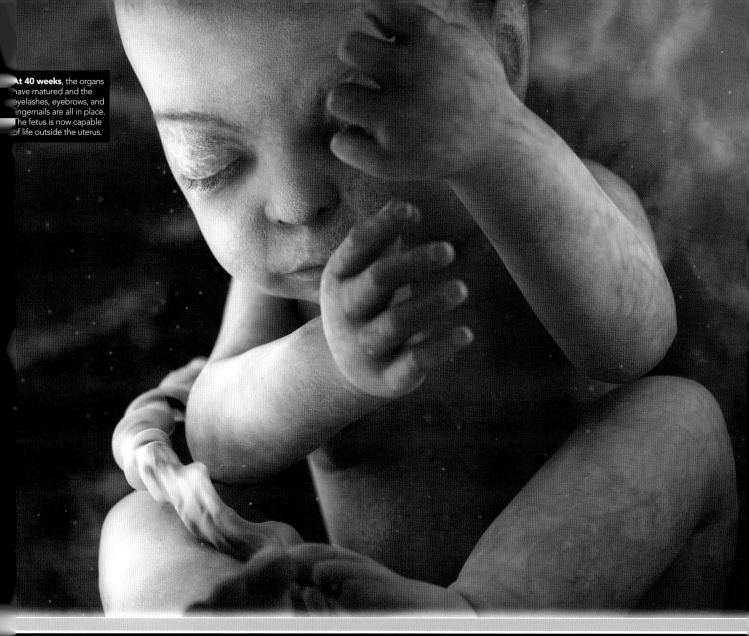

At **40 weeks**, the organs have matured and the eyelashes, eyebrows, and fingernails are all in place. The fetus is now capable of life outside the uterus.

WEEK 34

A visit to the physician is usually arranged for this week to discuss the birth plan and to receive vitamin K injections if required.

WEEK 36

Tests may be offered to check placental function, fetal growth, heart rate, and general well-being.

WEEKS 37–38

Examination of the mother's abdomen shows whether the fetus is in a head-down position. If in a breech position, there is still time for the fetus to turn by itself. The optimal time for the birth of twins is considered to be 37 weeks.

WEEK 39

The breasts are producing colostrum and preparing for lactation.

WEEK 40

A prenatal visit is scheduled if the baby has not already been delivered. If birth has not occurred by 42 weeks, labor is induced.

WEEK 34	WEEK 35	MONTH 9	WEEK 36	WEEK 37	WEEK 38	WEEK 39	WEEK 40

WEEK 34

The suckling reflex develops.

WEEK 35

The lungs now consistently produce surfactant, allowing the alveoli to expand and collapse more easily when breathing air after birth.

WEEKS 36–37

Most of the lanugo hair has been shed and replaced by fine vellus hair.

Ossification is occurring in the humerus, femur, and tibia.

Urine is now more concentrated due to the developing kidneys.

WEEK 38

The fingernails now reach the ends of the fingers.

The eyes can move but cannot focus.

WEEKS 39–40

The liver is now mature enough to take over all the metabolic functions performed by the placenta.

In male fetuses, the testes have usually descended into the scrotum by now.

MONTH 7 | WEEKS 27–30

The mother-to-be now enters the third trimester. If the fetus is born prematurely, it is now capable of independent life and has a good chance of surviving with special care. Most development now focuses on the maturing of the brain, lungs, and digestive system.

WEEK 27

From now on, natural differences in the growth rate between male and female fetuses cause boys to be slightly bigger and heavier at birth than girls. This difference is not usually noticeable to the mother while the baby is still in the womb. The fetus is now regularly swallowing, yawning, and making practice-breathing movements. It starts to develop a regular pattern of resting and sleeping that alternates with periods of wakefulness and activity. The protective layer of grease (vernix) covering the skin thickens. This coincides with the kidneys maturing. They start to produce small quantities of urine that pass into the amniotic fluid, and the vernix protects the fetus's delicate skin from irritation. The eyebrows and eyelashes are growing, and scalp hair is growing longer.

STRETCHING FACE MUSCLES
This 3D ultrasound image of a fetus at the beginning of the third trimester shows its mouth stretched wide open in a yawn.

UPSIDE-DOWN POSITION
This MRI shows that the uterus's shape helps the fetus into a head-down position, but its position may change frequently.

WEEK 28

Despite massive fetal growth, there is still plenty of space in the uterus, and the fetus may turn somersaults, spending some time in a head-down position and some time with its head upright. As a result of all this exercise, the mother-to-be may feel kicks in several different parts of her abdomen. Skin creases are becoming visible on the fetal hands, which have taken on a chubby appearance, and the minute fingernails are perfectly formed. Within the upper and lower gums, the teeth buds have now formed separate layers of enamel and dentine. During a prenatal visit, hemoglobin levels in the mother's blood will be checked to detect signs of anemia. A glucose tolerance test may also be carried out in order to check for maternal gestational diabetes.

GLUCOSE TEST
Urinalysis sticks make it easy for healthcare professionals to screen urine for glucose, hidden blood, protein, and infection.

TOOTH FORMATION
Distinct layers now appear in each tooth bud: an outer layer of enamel (white), dentin (cream), and pulp (red).

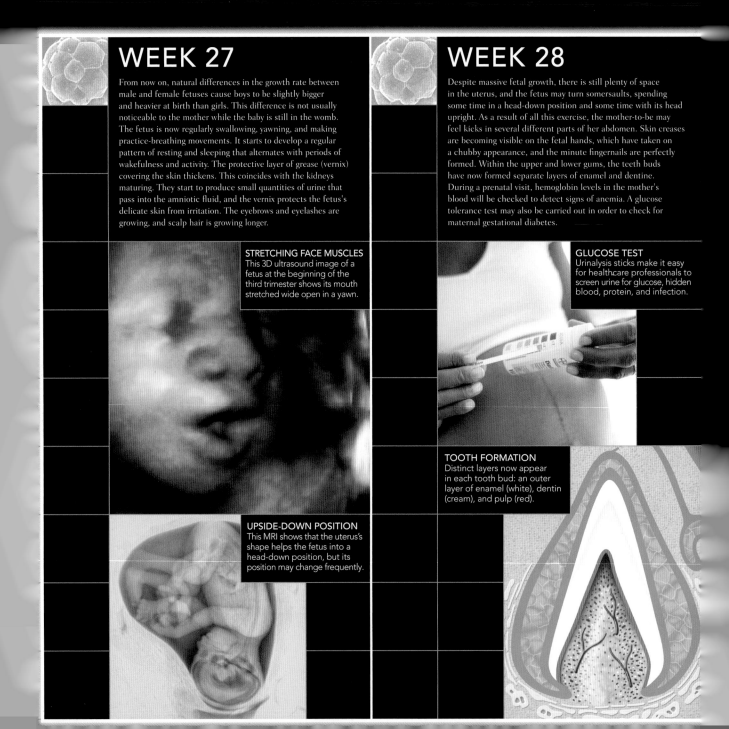

WEEK 29

The surface of the fetus's brain is becoming increasingly folded to expand its surface area in order to accommodate the many millions of nerve cells that are being formed. More nerves are gaining the fatty myelin sheath that helps insulate them from one another; this speeds up the development of the fetus's movements. The amniotic sac, which envelops the fetus, and the amniotic fluid it contains are now fully developed. The two layers of the amniotic sac—the inner amnion and the outer chorion—slide over one another to reduce friction as the fetus twists and turns in the womb. Even up to the final weeks of pregnancy, when the fetus has reached its maximum size, the amniotic sac remains amazingly flexible and continues to stretch as the fetus grows.

WEEK 30

The fetus is now beginning to look increasingly rounded, plump, and well-nourished, and over the final 10 weeks of pregnancy, it will double its weight. The fetus now has a regular sleep–wake pattern and spends approximately half the time resting quietly. Women with a Rhesus-negative blood group will receive a injection of anti-D antibodies by the 30th week of pregnancy; another dose is given shortly after the birth. This helps neutralize an immune response if the mother is carrying a fetus with a Rhesus-positive blood group. It will reduce the chances of the mother producing her own anti-D antibodies, which might cause problems if she becomes pregnant with another Rhesus-positive fetus at some time in the future.

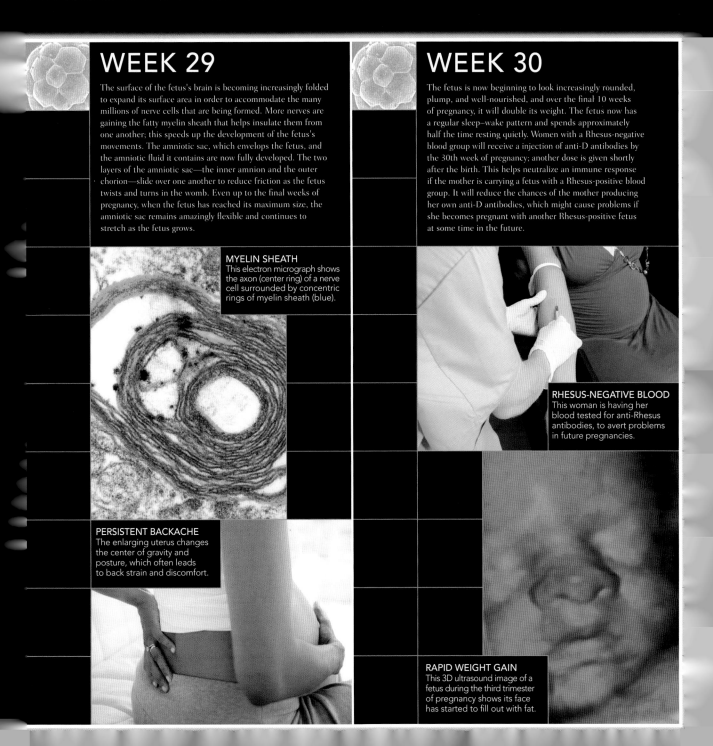

MYELIN SHEATH
This electron micrograph shows the axon (center ring) of a nerve cell surrounded by concentric rings of myelin sheath (blue).

PERSISTENT BACKACHE
The enlarging uterus changes the center of gravity and posture, which often leads to back strain and discomfort.

RHESUS-NEGATIVE BLOOD
This woman is having her blood tested for anti-Rhesus antibodies, to avert problems in future pregnancies.

RAPID WEIGHT GAIN
This 3D ultrasound image of a fetus during the third trimester of pregnancy shows its face has started to fill out with fat.

MONTH 7 | WEEKS 27–30
MOTHER AND FETUS

The seventh month marks the start of the final trimester. In the first week of this month, the fetus's eyelids usually separate and it begins to blink. Nutrients are increasingly diverted toward producing muscle and fat, and so the fetus continues the growth spurt that started at the end of the previous month. The fetus's kidneys now produce urine in increasing quantities, which is frequently released into the amniotic fluid. Its skin is covered in a protective layer of grease called vernix that, among other functions, helps the fetus descend through the birth canal when the time is right. The mother may undergo a glucose tolerance test to check for maternal gestational diabetes. If blood tests in early pregnancy revealed the mother to have a Rhesus-negative blood group, she will usually receive her first dose of anti-D antibodies in the middle of this month.

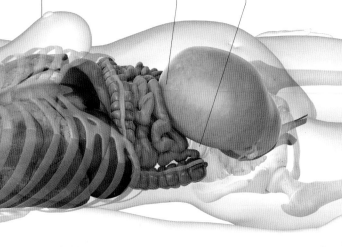

MOTHER AT 30 WEEKS
The mother may start to suffer from backache as her changing center of gravity puts increasing strain on ligaments and muscles. She may also notice that the fetus startles when hearing a loud noise.

Minor contractions
Minor contractions stemming from the fundus of the uterus become more noticeable toward the end of the month.

Changing center of gravity
As the uterus enlarges, the mother's center of gravity moves forward, and her posture changes. This accentuates the lumbar curve in the lower back and can lead to backache.

MOTHER

- 72 beats per minute
- 106/70
- 9 pints (5.1l)

40%
The tidal volume—the amount of air inhaled and exhaled in one breathing motion—has risen 40 percent since pregnancy began.

Braxton Hicks' contractions begin to **increase in intensity** from the 30th week.

The mother has typically gained **15lb (7kg)** by the end of the seventh month.

STATISTICS

| 1 | 2 | 3 | 4 | 5 | 6 | 7 | 8 | 9 | 10 | 11 | 12 | 13 | 14 | 15 | 16 | 17 | 18 | 19 | 20 | 21 | 22 | 23 | 24 | 25 | 26 | 27 | 28 | 29 | 30 | 31 | 32 | 33 | 34 | 35 | 36 | 37 | 38 | 39 | 40 |

FETUS

- 150 beats per minute
- 15½in (40cm)
- 2¾lb (1.3kg)

33%
In one-third of pregnancies, the fetus lies in a breech (bottom-down) position in the 30th week, but only 3 percent of fetuses remain in this position until birth.

10%
Twins put on the same amount of weight as single fetuses until the 28th week, after which there is a 10 percent reduction in their relative rate of growth.

In the 30th week, the fetus only spends a **tenth** of its time **awake**.

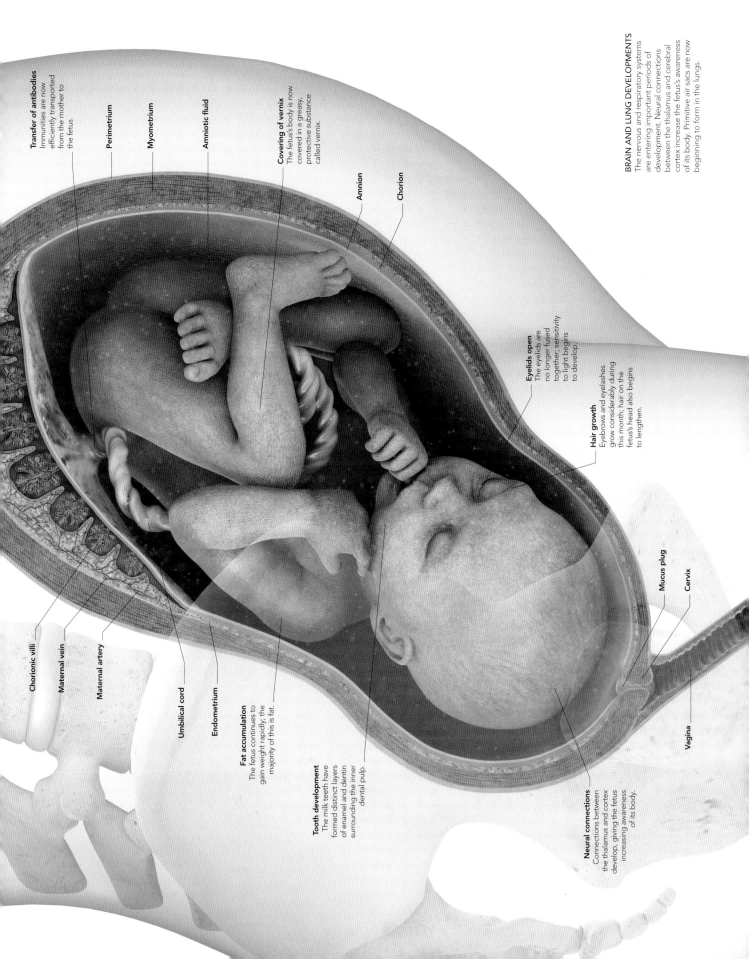

Transfer of antibodies
Immunities are now efficiently transported from the mother to the fetus.

Perimetrium

Myometrium

Amniotic fluid

Covering of vernix
The fetus's body is now covered in a greasy, protective substance called vernix.

Amnion

Chorion

Eyelids open
The eyelids are no longer fused together; sensitivity to light begins to develop.

Hair growth
Eyebrows and eyelashes grow considerably during this month; hair on the fetus's head also begins to lengthen.

Mucus plug

Cervix

Vagina

Neural connections
Connections between the thalamus and cortex develop, giving the fetus increasing awareness of its body.

Tooth development
The milk teeth have formed distinct layers of enamel and dentin surrounding the inner dental pulp.

Fat accumulation
The fetus continues to gain weight rapidly; the majority of this is fat.

Endometrium

Umbilical cord

Maternal artery

Maternal vein

Chorionic villi

BRAIN AND LUNG DEVELOPMENTS
The nervous and respiratory systems are entering important periods of development. Neural connections between the thalamus and cerebral cortex increase the fetus's awareness of its body. Primitive air sacs are now beginning to form in the lungs.

MOTHER

THE CHANGING CENTER OF GRAVITY

During the third trimester, the increasing volume and weight of the uterus moves the pregnant woman's center of gravity forward. To counteract this and maintain stability, it is natural for the mother-to-be to lean backward. However, this causes the long muscles running down the spine to work harder to pull the shoulders back and lift the abdomen. As the shoulders pull back, the head naturally moves forward. These changes in posture can lead to back, shoulder, and neck aches.

FLEXIBLE VERTEBRAE

The vertebral bones interlock in a series of sliding joints to produce four gentle curves that provide strength, flexibility, and stability. These curves are known as the cervical, thoracic, lumbar, and sacral curves. As the center of gravity changes during pregnancy, it is natural to lean back, placing more strain on the five vertebrae that make up the lumbar curve.

PRENATAL CLASSES

Prenatal classes provide important information that helps a pregnant woman, and her partner, prepare for childbirth emotionally and physically. Classes usually cover what happens to the baby and the mother during birth, positions to adopt during labor, and possible interventions, such as cesarean section, delivery by vacuum extraction, and delivery by forceps. Breathing exercises and relaxation techniques are taught, and the different methods of pain relief are discussed.

BIRTH EDUCATION

Prenatal classes are a relaxed way of preparing for labor, birth, and the first few months after the birth.

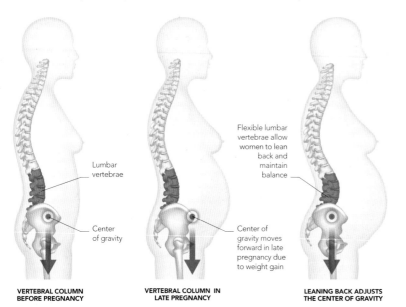

Lumbar vertebrae

Center of gravity

Flexible lumbar vertebrae allow women to lean back and maintain balance

Center of gravity moves forward in late pregnancy due to weight gain

VERTEBRAL COLUMN BEFORE PREGNANCY

VERTEBRAL COLUMN IN LATE PREGNANCY

LEANING BACK ADJUSTS THE CENTER OF GRAVITY

PRENATAL APPOINTMENTS

Regular tests in the third trimester include checking the mother's blood pressure, the height of the uterus, and the position of the fetus. Urine is tested for protein, glucose, the presence of blood, and other signs of infection. Blood is checked for signs of anemia, and a glucose tolerance test may be performed. If the mother is Rhesus negative and her partner is Rhesus positive (see p.230), anti-Rhesus antibody levels are checked regularly. Injections may be needed if the antibody level is too high.

CHECKING GLUCOSE LEVELS

If glucose is found in the mother-to-be's urine, a glucose tolerance test will be needed to check for gestational diabetes.

BACKACHE DURING PREGNANCY

The change in posture during late pregnancy places extra strain on the muscles, ligaments, and joints in the lower back, causing pain. Other factors that increase back pain include reduced levels of exercise, diminishing muscle tone in the core abdominal region, and the increased secretion of the hormone relaxin, which softens ligaments as delivery approaches, often leading to inflammation and pain in many joints in the body. Back problems can also occur during pregnancy through lifting heavy objects with the back hunched and without bending the knees.

LOCALIZED PAIN AREAS

In addition to discomfort from strained joints and ligaments, surrounding muscles may go into spasm, causing pain and tenderness over a larger area.

Inflammation of sacroiliac joint
Causes persistent pain in middle and lower back

Pressure on vertebrae
Causes pain around coccyx

Pubic joint strain
Leads to pain in front of pelvis

THIRD TRIMESTER HOSPITAL APPOINTMENTS

WEEK 28	Hospital visit to test for gestational diabetes and anemia; injection may be given if Rhesus incompatible
WEEK 34	Hospital visit to discuss birth plan; second injection may be given if blood is Rhesus incompatible
WEEK 41	Hospital visit organized to discuss possible induction of labor
WEEK 41 + 3 days	Hospital visit to maternity day care unit; ultrasound performed to assess fetal well-being

FETUS

THE DESCENT OF THE TESTES

The testes develop within the abdominal cavity of a developing male embryo, near the kidneys. They become attached to a ligament on each side, known as the gubernaculum. Between the 28th and 35th weeks of pregnancy, each gubernaculum becomes shorter and thicker. This acts as a guide, pulling the testes downward through the inguinal canal and into the scrotum. Moving outside the abdomen into the scrotum helps keep the testes cool, which improves the quality of sperm when it is produced from puberty onward.

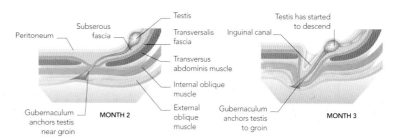

MONTH 2

Peritoneum
Subserous fascia
Testis
Transversalis fascia
Transversus abdominis muscle
Internal oblique muscle
External oblique muscle
Gubernaculum anchors testis near groin

MONTH 3

Inguinal canal
Testis has started to descend
Gubernaculum anchors testis to groin

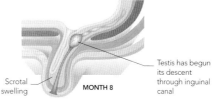

MONTH 8

Scrotal swelling
Testis has begun its descent through inguinal canal

MONTH 9

Gubernaculum starts to disintegrate after testis has descended
Testis has now descended into the scrotum

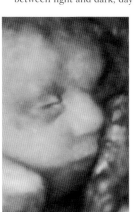

Scrotum

THE FINAL DESCENT
The testes should move into the scrotum before birth. In 1 percent of full-term and 10 percent of premature boys, one testis remains undescended.

HOW AND WHERE THE TESTES MOVE
The testes descend from the abdominal cavity through the inguinal canal, a narrow tunnel that passes over the pelvic bone and into the scrotum. Once the testes are in position in the scrotum, the gubernaculum withers away on either side.

THE DEVELOPMENT OF THE EYE

The eyelids, which have been fused since the end of the first trimester, begin to separate at the beginning of the seventh month, allowing the fetus to open its eyes and start to blink. All the layers of the retina at the back of each eye have now developed, including the light-sensitive cells known as rods and cones. A small amount of light passes through the mother's abdominal wall to stimulate the fetal rods, which detect shades of black, gray, and white in the dim conditions. The fetus can recognize the difference between light and dark, day and night, and see the outline of its hands, knees, and umbilical cord. Color vision, which arises as a result of stimulation of the cone cells, is not thought to develop until after birth.

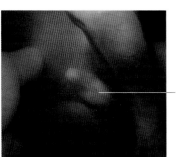

Suspensory ligament of lens
Iris
Cornea
Eyelids fused
Eyelid
Lens
Neural layer of retina
Inner layer of retina
Hyaloid artery

17 WEEKS

Scleral venous sinus
Iris
Cornea
Suspensory ligament of lens
Ciliary body
Choroid
Optic nerve
Hyaloid artery

26 WEEKS

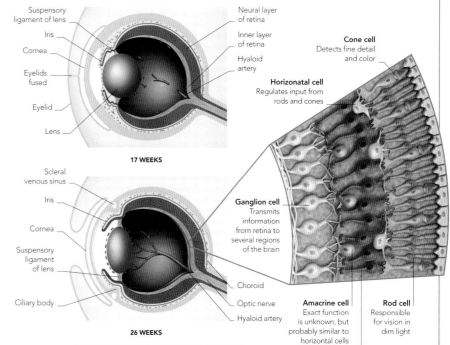

Cone cell
Detects fine detail and color

Horizonatal cell
Regulates input from rods and cones

Ganglion cell
Transmits information from retina to several regions of the brain

Amacrine cell
Exact function is unknown, but probably similar to horizontal cells

Rod cell
Responsible for vision in dim light

Bipolar cell
Transfers information from ganglion cells to rods and cones

EYES START TO OPEN
This 3D ultrasound of a fetus in the seventh month shows the eyelids beginning to separate. Sensitivity to light develops, and the fetus will turn toward bright light.

ANATOMY OF THE EYE AT 17 AND 26 WEEKS
A number of developments take place between 17 and 26 weeks. The lens becomes less spherical and more ovoid, the eyelids separate, and the ciliary body forms, which allows the lens to move and change shape.

FETUS

THE FORMATION OF TEETH

The first set of 20 milk (or deciduous) teeth start developing around eight weeks into pregnancy. Buds form from the band of tissue (dental lamina) that runs along both jaws. The lamina guides the buds into position and disintegrates. The tooth buds then fold inward to form a bell-shaped structure. Cells of the inner enamel epithelium deposit hard enamel on the developing tooth's surface, while dental papilla underneath produces the softer dentin and pulp. In the seventh month, enamel and dentin have formed separate layers. The permanent tooth buds form during the third month, but then lie dormant until around six years of age.

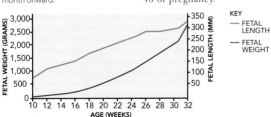

Enamel

Dentin

Dental pulp

PROTECTIVE LAYER
A thick layer of hard enamel (red) protects the softer dentin (pink) and pulp (yellow) beneath.

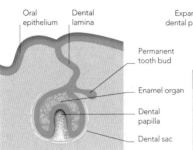

Oral epithelium — Dental lamina

Permanent tooth bud

Enamel organ

Dental papilla

Dental sac

1 EARLY BELL STAGE
By 10 weeks, the milk teeth start to form within a dental sac. The permanent tooth bud begins to develop beside it.

Expanding dental papilla — Disintegrating dental lamina — Developing permanent tooth bud

2 LATE BELL STAGE
By 14 weeks, the dental lamina connecting the tooth to the gum surface is no longer needed and starts to break down.

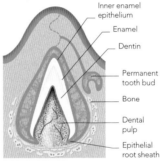

Inner enamel epithelium

Enamel

Dentin

Permanent tooth bud

Bone

Dental pulp

Epithelial root sheath

3 ENAMEL AND DENTIN
By the seventh month of pregnancy, the milk teeth have distinct layers of enamel and dentin surrounding the inner dental pulp.

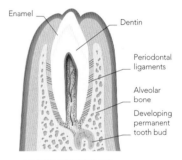

Enamel — Dentin

Periodontal ligaments

Alveolar bone

Developing permanent tooth bud

4 EARLY ERUPTION STAGE
The tooth bulges onto the surface of the gum until the crown breaks through. Eruption of the milk teeth occurs between six months and two years after birth.

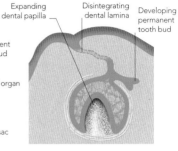

MUSCLE AND FAT ACCUMULATION

The length of the fetus increases steadily throughout pregnancy. This allows the age of the fetus to be assessed with relative accuracy using measurements taken during ultrasound scans. The rate at which the fetus gains weight increases slowly at first, but starts to accelerate in the seventh month. Muscle and fat are being laid down, and the fetus starts a growth spurt, doubling in weight between week 30 and week 40 of pregnancy.

GROWTH SURGE
Fetal length increases steadily throughout pregnancy, but most weight gain occurs from the seventh month onward.

KEY
— FETAL LENGTH
— FETAL WEIGHT

FETAL WEIGHT (GRAMS): 3,000 / 2,500 / 2,000 / 1,500 / 1,000 / 500 / 0

FETAL LENGTH (MM): 350 / 300 / 250 / 200 / 150 / 100 / 50

AGE (WEEKS): 10 12 14 16 18 20 22 24 26 28 30 32

VERNIX

Vernix caseosa is a white, greasy substance that forms a coating on the fetus's skin. It starts to appear around the 20th week of gestation, and by the seventh month it covers most of the fetus's body. Made up of fetal skin oil (sebum), skin cells, and lanugo (fine hair), vernix helps moisturize and protect the skin from constant exposure to amniotic fluid, which, due to the development of the kidneys in the third trimester, contains more concentrated fetal urine. Vernix also helps lubricate the baby as it passes down the birth canal during labor.

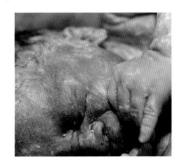

PROTECTIVE COVERING
A thick, slippery layer of vernix may still be present at birth. The term vernix caseosa means "cheesy varnish" in Latin.

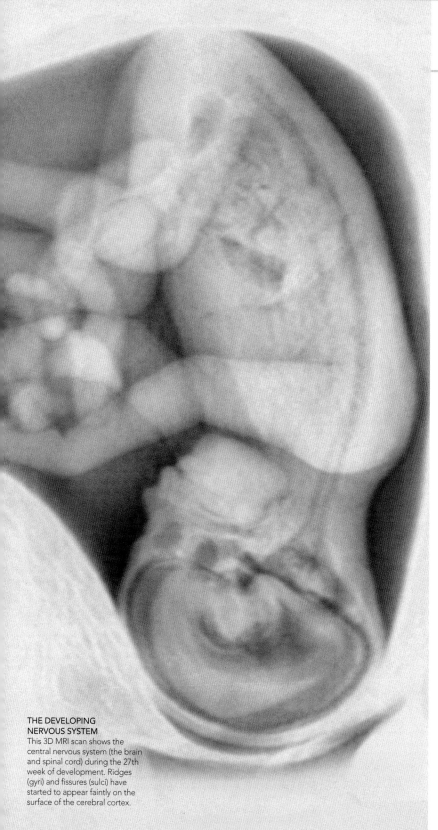

THE DEVELOPING NERVOUS SYSTEM
This 3D MRI scan shows the central nervous system (the brain and spinal cord) during the 27th week of development. Ridges (gyri) and fissures (sulci) have started to appear faintly on the surface of the cerebral cortex.

THE BIRTH OF CONSCIOUSNESS

Consciousness is roughly defined as sensory awareness of the body, awareness of the self, and awareness of the world. The fetus starts to develop one of these constituent parts—awareness of the body—by the seventh month, because it can now react to smell, touch, and sound. The other constituent parts only start to develop after birth. In the seventh month, the number of connections (synapses) between brain cells is increasing, and the nerve activity associated with consciousness, personality, and the ability to think are developing. Many different nerve pathways are being laid down between the brain and the body. Some pathways receive sensory information from the body, while others send instructions that help coordinate voluntary and involuntary movements. Much of the information coming into the brain passes through the thalamus, where it is processed and sent on to the correct part of the cerebral cortex for analysis. The thalamus is also involved in the regulation of consciousness, alertness, and awareness.

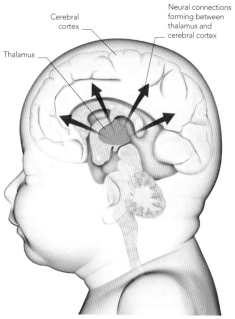

Cerebral cortex

Neural connections forming between thalamus and cerebral cortex

Thalamus

DEVELOPING NEURAL NETWORKS
This illustration shows the fetal brain at 28 weeks. At this time, connections form between the thalamus (green area) and the cerebral cortex. One of the roles of the thalamus is to process sensory signals. The connections that are forming allow these signals to be relayed from the thalamus to the relevant part of the cortex.

MONTH 8 | WEEKS 31–35

During the eighth month of pregnancy, the fetus gains weight at a dramatic rate. All the body systems are maturing in readiness for delivery in the near future. The mother-to-be may develop an urge to clean, tidy, and "ready the nest," but it is important to find time for rest and relaxation.

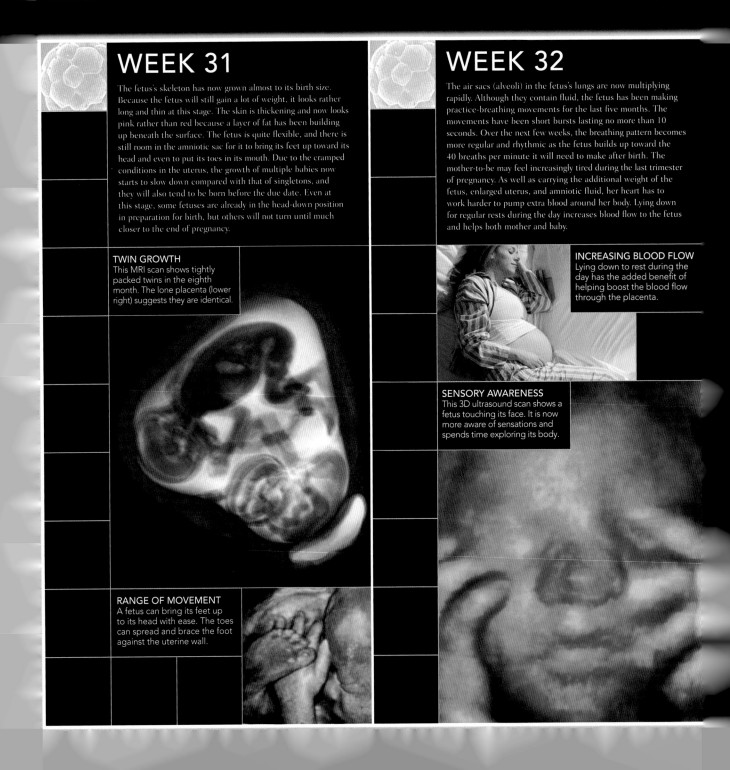

WEEK 31

The fetus's skeleton has now grown almost to its birth size. Because the fetus will still gain a lot of weight, it looks rather long and thin at this stage. The skin is thickening and now looks pink rather than red because a layer of fat has been building up beneath the surface. The fetus is quite flexible, and there is still room in the amniotic sac for it to bring its feet up toward its head and even to put its toes in its mouth. Due to the cramped conditions in the uterus, the growth of multiple babies now starts to slow down compared with that of singletons, and they will also tend to be born before the due date. Even at this stage, some fetuses are already in the head-down position in preparation for birth, but others will not turn until much closer to the end of pregnancy.

TWIN GROWTH
This MRI scan shows tightly packed twins in the eighth month. The lone placenta (lower right) suggests they are identical.

RANGE OF MOVEMENT
A fetus can bring its feet up to its head with ease. The toes can spread and brace the foot against the uterine wall.

WEEK 32

The air sacs (alveoli) in the fetus's lungs are now multiplying rapidly. Although they contain fluid, the fetus has been making practice-breathing movements for the last five months. The movements have been short bursts lasting no more than 10 seconds. Over the next few weeks, the breathing pattern becomes more regular and rhythmic as the fetus builds up toward the 40 breaths per minute it will need to make after birth. The mother-to-be may feel increasingly tired during the last trimester of pregnancy. As well as carrying the additional weight of the fetus, enlarged uterus, and amniotic fluid, her heart has to work harder to pump extra blood around her body. Lying down for regular rests during the day increases blood flow to the fetus and helps both mother and baby.

INCREASING BLOOD FLOW
Lying down to rest during the day has the added benefit of helping boost the blood flow through the placenta.

SENSORY AWARENESS
This 3D ultrasound scan shows a fetus touching its face. It is now more aware of sensations and spends time exploring its body.

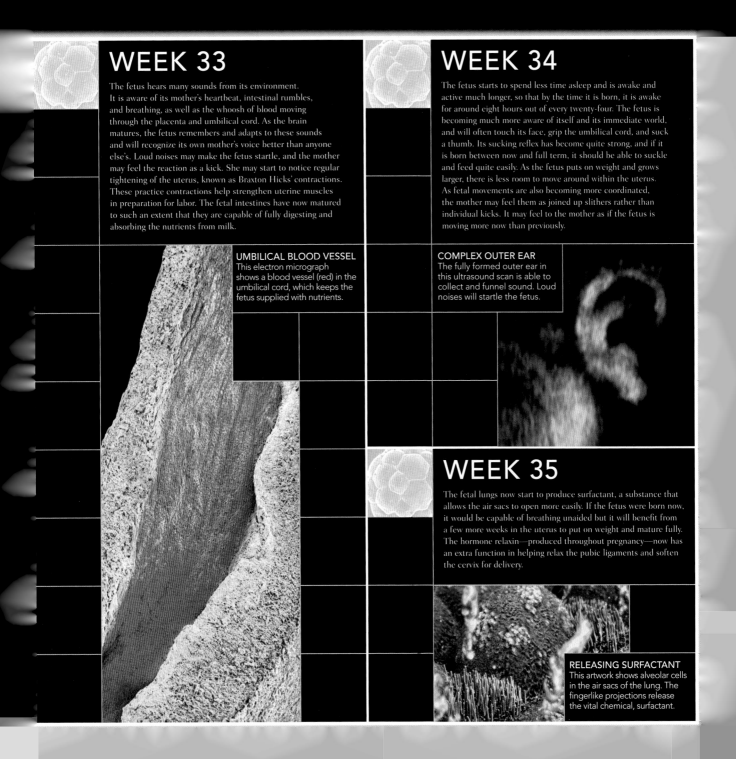

WEEK 33

The fetus hears many sounds from its environment. It is aware of its mother's heartbeat, intestinal rumbles, and breathing, as well as the whoosh of blood moving through the placenta and umbilical cord. As the brain matures, the fetus remembers and adapts to these sounds and will recognize its own mother's voice better than anyone else's. Loud noises may make the fetus startle, and the mother may feel the reaction as a kick. She may start to notice regular tightening of the uterus, known as Braxton Hicks' contractions. These practice contractions help strengthen uterine muscles in preparation for labor. The fetal intestines have now matured to such an extent that they are capable of fully digesting and absorbing the nutrients from milk.

UMBILICAL BLOOD VESSEL
This electron micrograph shows a blood vessel (red) in the umbilical cord, which keeps the fetus supplied with nutrients.

WEEK 34

The fetus starts to spend less time asleep and is awake and active much longer, so that by the time it is born, it is awake for around eight hours out of every twenty-four. The fetus is becoming much more aware of itself and its immediate world, and will often touch its face, grip the umbilical cord, and suck a thumb. Its sucking reflex has become quite strong, and if it is born between now and full term, it should be able to suckle and feed quite easily. As the fetus puts on weight and grows larger, there is less room to move around within the uterus. As fetal movements are also becoming more coordinated, the mother may feel them as joined up slithers rather than individual kicks. It may feel to the mother as if the fetus is moving more now than previously.

COMPLEX OUTER EAR
The fully formed outer ear in this ultrasound scan is able to collect and funnel sound. Loud noises will startle the fetus.

WEEK 35

The fetal lungs now start to produce surfactant, a substance that allows the air sacs to open more easily. If the fetus were born now, it would be capable of breathing unaided but it will benefit from a few more weeks in the uterus to put on weight and mature fully. The hormone relaxin—produced throughout pregnancy—now has an extra function in helping relax the pubic ligaments and soften the cervix for delivery.

RELEASING SURFACTANT
This artwork shows alveolar cells in the air sacs of the lung. The fingerlike projections release the vital chemical, surfactant.

MONTH 8 | WEEKS 31–35
MOTHER AND FETUS

The continuing accumulation of fat stores this month proves to be vital because it provides a baby with energy in the first few days after birth, before milk is produced. The fetus now starts to spend less time asleep and more time awake. Practice breathing occurs as the chest wall moves consistently, which prepares the lungs and respiratory control centers in the brain for the baby's first breath after birth. The mother's levels of the hormone relaxin increase, loosening the pubic ligaments and softening the cervix in readiness for birth. The enlarging uterus presses down on the pelvic floor, compressing the bladder and increasing the mother's urge to urinate. Most expectant mothers now start to feel increasingly tired. Prenatal visits usually increase in frequency to monitor both mother and fetus as pregnancy approaches its final stage.

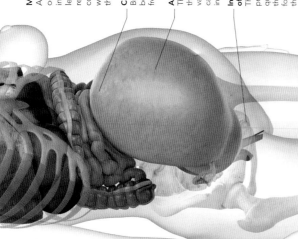

MOTHER AT 35 WEEKS
A number of changes occur in the mother's body, including rising hormone levels and the increasing regularity of Braxton Hicks' contractions, both of which begin to prepare the body for labor.

Continuing contractions
Braxton Hicks' contractions become stronger and more frequent during this month.

Additional weight
The increasing weight of the fetus, compounded by various hormonal changes, can make the mother increasingly tired.

Increasing production of relaxin
The hormone relaxin is produced in increasing quantities, which softens the joints in preparation for the baby's descent through the birth canal.

MOTHER

🤍 **74 beats per minute**

🩸 **109/73**

💧 **9¼ pints (5.25 l)**

1.5 pints
The amount of amniotic fluid in the uterus has now reached 1.5 pints (800 ml). It starts to decrease in the ninth month.

Over 40%
The mother's total blood volume is now over 40 percent greater than it was prior to pregnancy.

STATISTICS

FETUS

🤍 **144 beats per minute**

📏 **18 in (46 cm)**

⚖️ **5¼ lb (2.4 kg)**

0.88 pints
The fetus swallows around 0.88 pints (500 ml) of amniotic fluid a day. Most of this is urinated back into the amniotic fluid.

In a male fetus, the **testes** are starting the final **descent** through the inguinal canal and into the **scrotum**.

Specific cells within the **air sacs** (alveoli) in the lungs start to release **surfactant** in the **35th week**. This allows the air sacs to **inflate and deflate without collapsing**, which is vital when the baby starts **breathing** after birth.

Perimetrium

Myometrium

Endometrium

Chorionic villi

Maternal artery

Maternal vein

Amniotic fluid

| 1 | 2 | 3 | 4 | 5 | 6 | 7 | 8 | 9 | 10 | 11 | 12 | 13 | 14 | 15 | 16 | 17 | 18 | 19 | 20 | 21 | 22 | 23 | 24 | 25 | 26 | 27 | 28 | 29 | 30 | 31 | 32 | 33 | 34 | 35 | 36 | 37 | 38 | 39 | 40 |

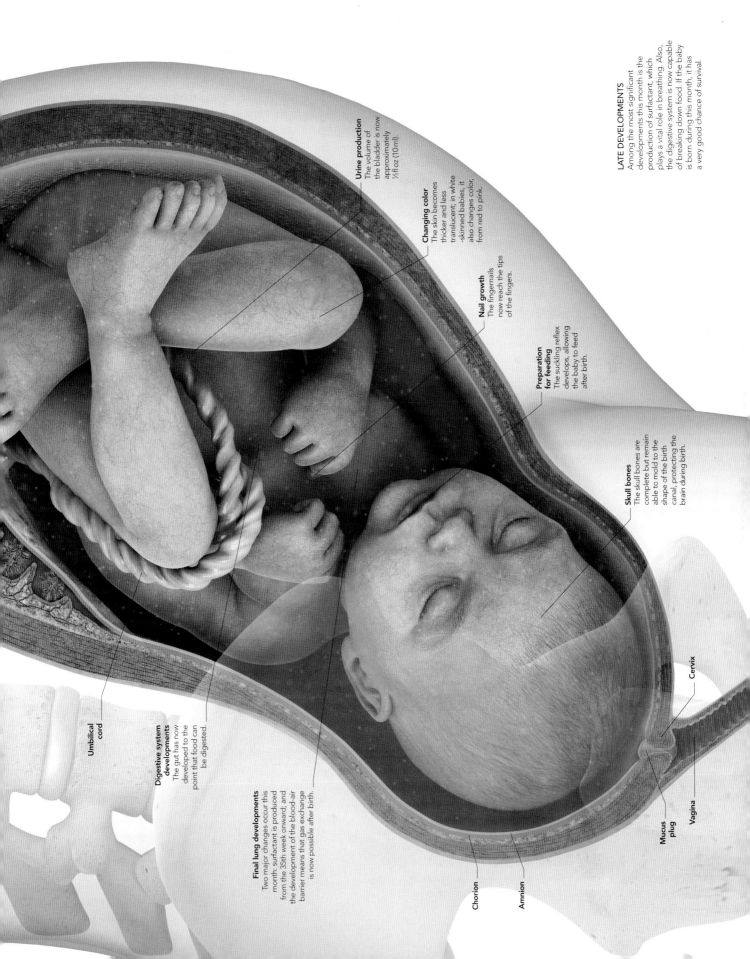

Urine production
The volume of the bladder is now approximately ½ fl oz (10ml).

Changing color
The skin becomes thicker and less translucent; in white -skinned babies, it also changes color, from red to pink.

Nail growth
The fingernails now reach the tips of the fingers.

Preparation for feeding
The sucking reflex develops, allowing the baby to feed after birth.

Skull bones
The skull bones are complete but remain able to mold to the shape of the birth canal, protecting the brain during birth.

Cervix

Mucus plug

Vagina

Amnion

Chorion

Final lung developments
Two major changes occur this month: surfactant is produced from the 35th week onward; and the development of the blood-air barrier means that gas exchange is now possible after birth.

Digestive system developments
The gut has now developed to the point that food can be digested.

Umbilical cord

LATE DEVELOPMENTS
Among the most significant developments this month is the production of surfactant, which plays a vital role in breathing. Also, the digestive system is now capable of breaking down food. If the baby is born during this month, it has a very good chance of survival.

MOTHER

BRAXTON HICKS' CONTRACTIONS

The uterus contracts regularly throughout pregnancy. Known as Braxton Hicks' contractions, these "practice" contractions become more noticeable from the eighth month onward and they are sometimes mistaken for labor. These contractions are felt as tightening sensations and may last for a minute or more. However, they do not produce the cervical dilation that occurs during labor. They squeeze the fetus and are thought to be an important stimulus for its developing senses, and to tone the uterine muscle in preparation for labor.

UTERINE ACTIVITY IN PREGNANCY
These charts show Braxton Hicks' contractions appearing as regular increases in uterine pressure (measured in millimeters of mercury). These contractions become increasingly intense in the eighth month, but are still minor compared with "true" labor.

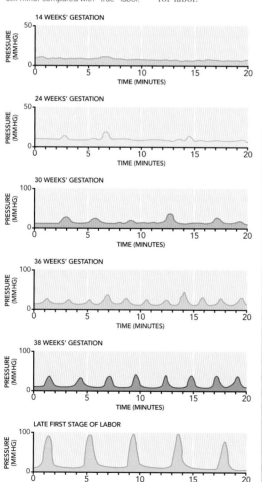

14 WEEKS' GESTATION

24 WEEKS' GESTATION

30 WEEKS' GESTATION

36 WEEKS' GESTATION

38 WEEKS' GESTATION

LATE FIRST STAGE OF LABOR

RELAXIN IN LATE PREGNANCY

Relaxin is a hormone that softens the pelvic joints and ligaments—as well as other ligaments in the body—in preparation for childbirth. Although these changes can lead to the backache and pelvic pain that are often experienced in late pregnancy, relaxin also makes the bones of the mother's pelvis more flexible, which allows the birth canal to widen enough for the fetal head to pass through. In addition, relaxin may aid in the development of blood vessels in the uterus and placenta; it is also thought to relax the uterus, allowing it to stretch as pregnancy progresses.

WHERE RELAXIN IS PRODUCED
Relaxin is produced in the breasts, ovaries, placenta, chorion, and decidua.

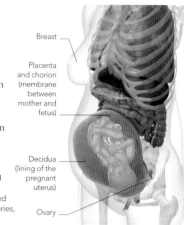

Breast

Placenta and chorion (membrane between mother and fetus)

Decidua (lining of the pregnant uterus)

Ovary

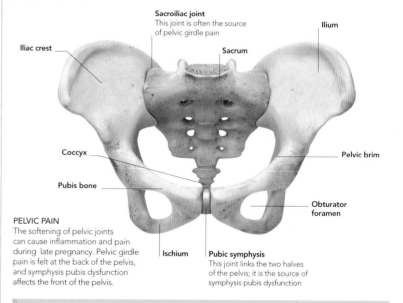

Sacroiliac joint
This joint is often the source of pelvic girdle pain

Ilium

Iliac crest

Sacrum

Coccyx

Pelvic brim

Pubis bone

Obturator foramen

PELVIC PAIN
The softening of pelvic joints can cause inflammation and pain during late pregnancy. Pelvic girdle pain is felt at the back of the pelvis, and symphysis pubis dysfunction affects the front of the pelvis.

Ischium

Pubic symphysis
This joint links the two halves of the pelvis; it is the source of symphysis pubis dysfunction

INCREASING FATIGUE

A pregnant woman often feels increasingly tired toward the end of pregnancy. This is partly because of the extra weight she has to carry and partly because of the various hormonal changes that are taking place within her body. Exceptional fatigue can also be a sign of iron deficiency (anemia). It is for this reason that prenatal clinics perform blood tests to screen for anemia at various stages of pregnancy.

BENEFITS OF REST
Sitting or lying down increases blood flow to the uterus and is therefore beneficial to both mother and fetus.

FETUS

RAPID GROWTH

As the placenta matures, it approaches its peak efficiency, allowing for maximum transfer of oxygen, glucose, and other vital nutrients to the fetus. As much as 70 percent of these nutrients are destined for the rapidly growing fetal brain. The fetal body is now almost fully developed, and it is able to divert precious energy resources toward laying down stores of body fat. The fetus starts to look better nourished as the wrinkles in its skin begin to fill out, and as it grows, the fetus is starting to get cramped inside the uterus.

MUSCLE FORMATION
This color-enhanced MRI scan of an 8-month-old fetus in the uterus shows that the fetal musculature (pink areas) is well formed.

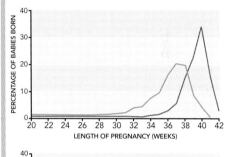

TWIN PREGNANCY

When twins share a uterus they also share maternal resources, including nutrients and space. As a result of this competition, their growth now starts to slow compared with single babies (or "singletons"), and they tend to be born earlier. On average, a twin pregnancy lasts for 38 weeks, while a singleton pregnancy lasts around 40 weeks. As a result of being born earlier, twins usually weigh less than single babies.

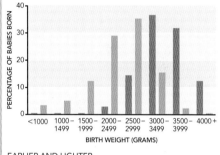

EARLIER AND LIGHTER
The top graph shows that twin babies are usually born a couple of weeks earlier than singletons. The bottom graph shows that twins are born around 2½lb (1 kg) lighter.

KEY
■ SINGLE BABIES
■ TWINS

"PRACTICE" BREATHING

The air sacs (alveoli) within the fetal lungs are almost fully formed, and the fetus now spends around half its time "practice" breathing—so called because the fetus is preparing to breathe oxygen, which will take place only after birth. During "practice" breathing, the amniotic fluid does not actually enter the fetus's lungs, but the accompanying movements of the diaphragm and chest wall are vital for stimulating normal lung development.

Red areas show amniotic fluid being expelled

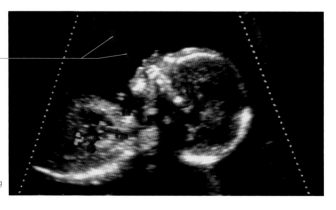

EARLY BREATHING
This colored Doppler ultrasound scan shows a fetus at about 17 weeks "practice" breathing amniotic fluid. The red patches show fluid coming out of the fetus's mouth.

The fetus is now fully formed and may already have settled in a head-down position ready for birth. During the last few weeks of pregnancy, the fetus lays down increasing amounts of fat as a reserve for the less protected life outside the uterus.

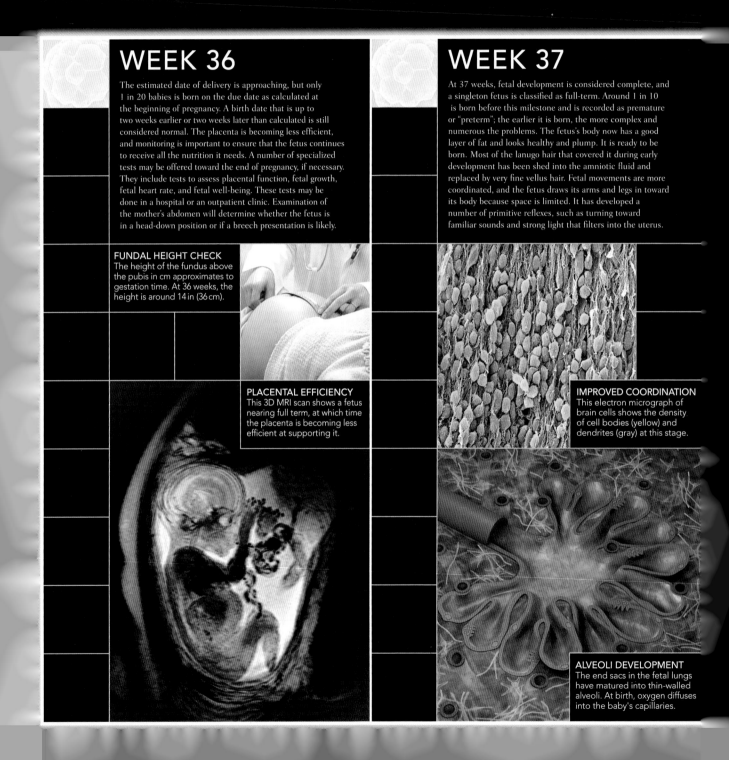

WEEK 36

The estimated date of delivery is approaching, but only 1 in 20 babies is born on the due date as calculated at the beginning of pregnancy. A birth date that is up to two weeks earlier or two weeks later than calculated is still considered normal. The placenta is becoming less efficient, and monitoring is important to ensure that the fetus continues to receive all the nutrition it needs. A number of specialized tests may be offered toward the end of pregnancy, if necessary. They include tests to assess placental function, fetal growth, fetal heart rate, and fetal well-being. These tests may be done in a hospital or an outpatient clinic. Examination of the mother's abdomen will determine whether the fetus is in a head-down position or if a breech presentation is likely.

FUNDAL HEIGHT CHECK
The height of the fundus above the pubis in cm approximates to gestation time. At 36 weeks, the height is around 14 in (36 cm).

PLACENTAL EFFICIENCY
This 3D MRI scan shows a fetus nearing full term, at which time the placenta is becoming less efficient at supporting it.

WEEK 37

At 37 weeks, fetal development is considered complete, and a singleton fetus is classified as full-term. Around 1 in 10 is born before this milestone and is recorded as premature or "preterm"; the earlier it is born, the more complex and numerous the problems. The fetus's body now has a good layer of fat and looks healthy and plump. It is ready to be born. Most of the lanugo hair that covered it during early development has been shed into the amniotic fluid and replaced by very fine vellus hair. Fetal movements are more coordinated, and the fetus draws its arms and legs in toward its body because space is limited. It has developed a number of primitive reflexes, such as turning toward familiar sounds and strong light that filters into the uterus.

IMPROVED COORDINATION
This electron micrograph of brain cells shows the density of cell bodies (yellow) and dendrites (gray) at this stage.

ALVEOLI DEVELOPMENT
The end sacs in the fetal lungs have matured into thin-walled alveoli. At birth, oxygen diffuses into the baby's capillaries.

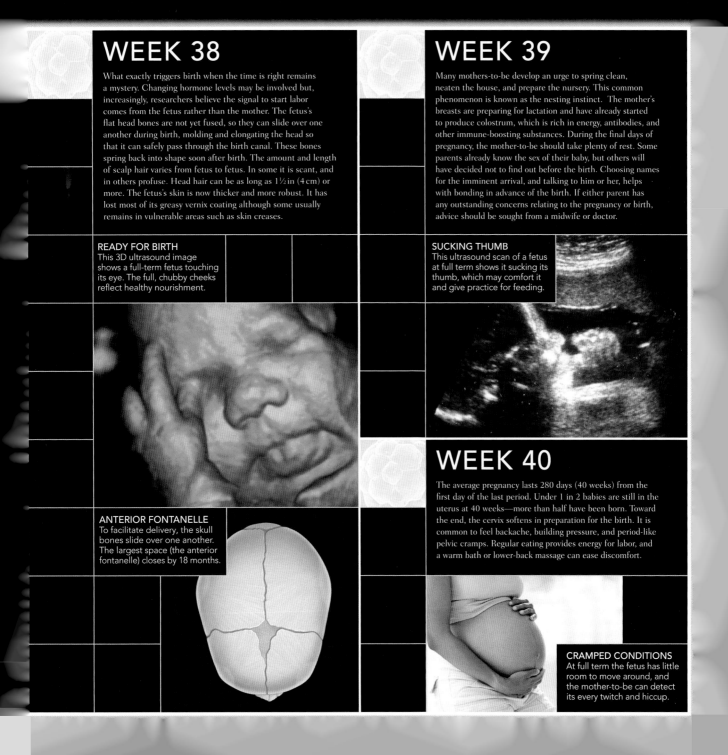

WEEK 38

What exactly triggers birth when the time is right remains a mystery. Changing hormone levels may be involved but, increasingly, researchers believe the signal to start labor comes from the fetus rather than the mother. The fetus's flat head bones are not yet fused, so they can slide over one another during birth, molding and elongating the head so that it can safely pass through the birth canal. These bones spring back into shape soon after birth. The amount and length of scalp hair varies from fetus to fetus. In some it is scant, and in others profuse. Head hair can be as long as 1½ in (4 cm) or more. The fetus's skin is now thicker and more robust. It has lost most of its greasy vernix coating although some usually remains in vulnerable areas such as skin creases.

READY FOR BIRTH
This 3D ultrasound image shows a full-term fetus touching its eye. The full, chubby cheeks reflect healthy nourishment.

ANTERIOR FONTANELLE
To facilitate delivery, the skull bones slide over one another. The largest space (the anterior fontanelle) closes by 18 months.

WEEK 39

Many mothers-to-be develop an urge to spring clean, neaten the house, and prepare the nursery. This common phenomenon is known as the nesting instinct. The mother's breasts are preparing for lactation and have already started to produce colostrum, which is rich in energy, antibodies, and other immune-boosting substances. During the final days of pregnancy, the mother-to-be should take plenty of rest. Some parents already know the sex of their baby, but others will have decided not to find out before the birth. Choosing names for the imminent arrival, and talking to him or her, helps with bonding in advance of the birth. If either parent has any outstanding concerns relating to the pregnancy or birth, advice should be sought from a midwife or doctor.

SUCKING THUMB
This ultrasound scan of a fetus at full term shows it sucking its thumb, which may comfort it and give practice for feeding.

WEEK 40

The average pregnancy lasts 280 days (40 weeks) from the first day of the last period. Under 1 in 2 babies are still in the uterus at 40 weeks—more than half have been born. Toward the end, the cervix softens in preparation for the birth. It is common to feel backache, building pressure, and period-like pelvic cramps. Regular eating provides energy for labor, and a warm bath or lower-back massage can ease discomfort.

CRAMPED CONDITIONS
At full term the fetus has little room to move around, and the mother-to-be can detect its every twitch and hiccup.

MONTH 9 | WEEKS 36–40
MOTHER AND FETUS

By 37 weeks, development is almost complete and the fetus is considered "full-term." It will still benefit from extra time in the uterus, however, and some babies are not born until 42 weeks. The fetus is gaining weight, and its skin is shedding the lanugo hair that covered it from 23 weeks. In its place fine, soft "vellus" hair is forming. The greasy vernix on its skin protects the fetus from the increasing amounts of concentrated urine now contained within the amniotic sac. Fingernails are growing fast, and they may need cutting soon after birth. The fetus's practice-breathing follows a regular rhythm, and it breathes quickly—around 40 times a minute. It may startle at loud sounds and will also recognize familiar voices. The mother's uterus rises farther up the abdomen and increases the pressure against the diaphragm, which can cause quicker, shallower breathing, fatigue, and indigestion.

MOTHER

- 75 beats per minute
- 108/68
- 9½ pints (5.5 l)

1,000
The number of times the capacity of the uterus can increase by, compared with a non-pregnant uterus.

25 oz (700 g)
The placenta now weighs around 25 oz (700 g) and has a diameter of 8–10 in (20–25 cm), and a thickness of ¾–1¼ in (2–3 cm).

MOTHER AT 40 WEEKS
The height of the uterus lowers during this month as the fetus's head "engages" or settles into the pelvis, in preparation for delivery.

Pressure eases on ribs
Engagement or "lightening" occurs during the ninth month and releases the pressure on the ribs, making breathing slightly easier.

Head presses on bladder
The mother may experience an increased urgency to urinate, as her bladder is compressed by the position of the fetus's head.

Pelvic joints loosen
The pubic symphysis joint loosens to increase flexibility, so that the baby can travel through the birth canal more easily.

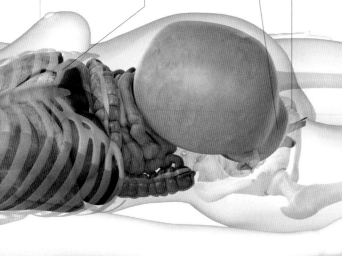

STATISTICS

FETUS

- 150 beats per minute
- 20 in (50 cm)
- 7¾ lb (3.5 kg)

Under 5%
The percentage of babies born on their due date. **30 percent** are born earlier than this date and **70 percent** are born later.

96%
The percentage of fetuses presenting in the upside-down position in the 40th week; **3 percent** of fetuses are in the breech position, and the remaining **1 percent** are in other positions.

| 1 | 2 | 3 | 4 | 5 | 6 | 7 | 8 | 9 | 10 | 11 | 12 | 13 | 14 | 15 | 16 | 17 | 18 | 19 | 20 | 21 | 22 | 23 | 24 | 25 | 26 | 27 | 28 | 29 | 30 | 31 | 32 | 33 | 34 | 35 | 36 | 37 | 38 | 39 | 40 |

Myometrium
This powerful muscular outer layer of the uterus is responsible for contractions during labor.

Endometrium

Chorionic villi

Perimetrium

Maternal artery

Maternal vein

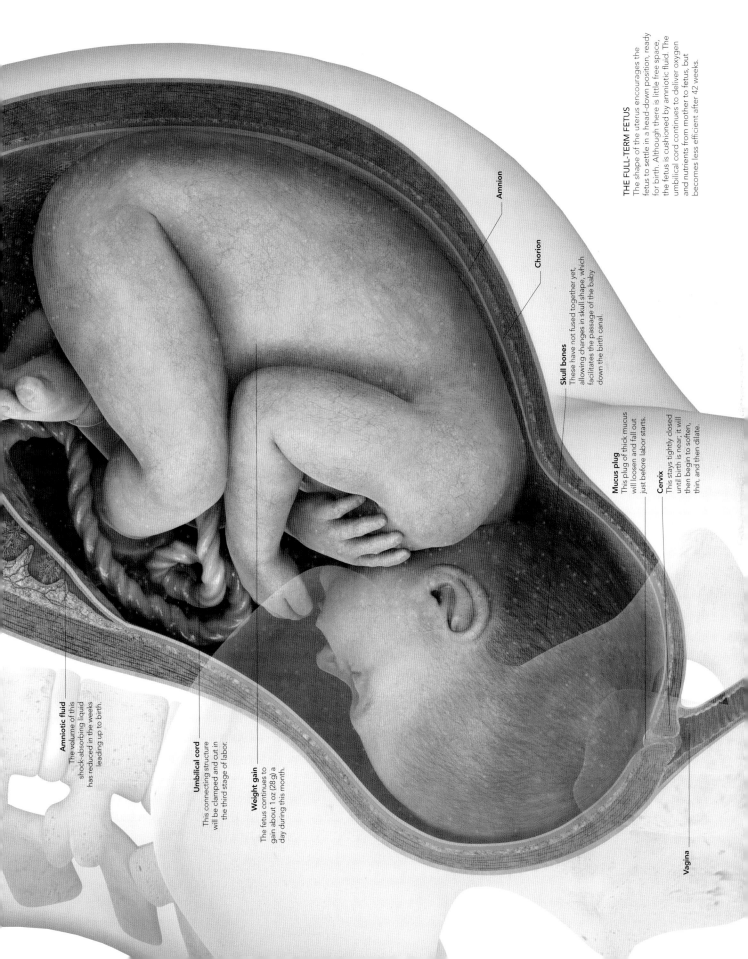

Amnion

Chorion

Skull bones
These have not fused together yet, allowing changes in skull shape, which facilitates the passage of the baby down the birth canal.

Mucus plug
This plug of thick mucus will loosen and fall out just before labor starts.

Cervix
This stays tightly closed until birth is near; it will then begin to soften, thin, and then dilate.

Vagina

THE FULL-TERM FETUS
The shape of the uterus encourages the fetus to settle in a head-down position, ready for birth. Although there is little free space, the fetus is cushioned by amniotic fluid. The umbilical cord continues to deliver oxygen and nutrients from mother to fetus, but becomes less efficient after 42 weeks.

Amniotic fluid
The volume of this shock-absorbing liquid has reduced in the weeks leading up to birth.

Umbilical cord
This connecting structure will be clamped and cut in the third stage of labor.

Weight gain
The fetus continues to gain about 1 oz (28 g) a day during this month.

MOTHER

THE PRODUCTION OF MILK

Toward the end of pregnancy, the breasts start to produce a rich, creamy pre-milk called colostrum. This can occasionally be discharged from the nipples involuntarily during the third trimester. After delivery and with the removal of the placenta, levels of estrogen, progesterone, and human placental lactogen (HPL) suddenly fall. However, prolactin levels remain high, and this is the hormone that stimulates full milk production. It is usually advised that babies be put to the breast as soon as possible after birth. The suckling helps stimulate milk production, and it is usual for milk to "come in" between the second and sixth day after delivery. Before this time, babies receive small amounts of colostrum, which provides energy, antibodies, and other immune-boosting substances. During the two to six days after birth, it is normal for babies to lose as much as 10 percent of their birth weight before full production of mature milk begins.

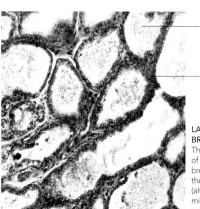

Milk
Produced by glands and secreted into saclike alveoli

Secretory lobule
Clusters of glands (lobules) group to form lobes

LACTATING BREAST TISSUE
This light micrograph of healthy lactating breast tissue shows the glandular spaces (alveoli) into which milk is secreted by specialized gland cells.

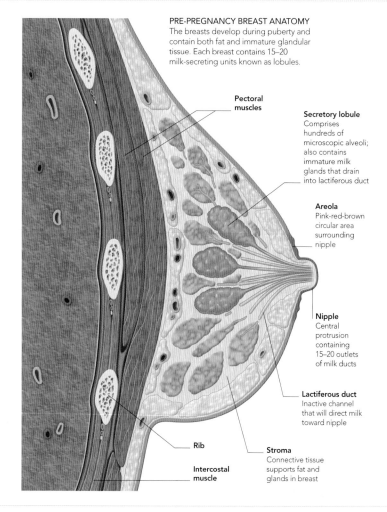

PRE-PREGNANCY BREAST ANATOMY
The breasts develop during puberty and contain both fat and immature glandular tissue. Each breast contains 15–20 milk-secreting units known as lobules.

Pectoral muscles

Secretory lobule
Comprises hundreds of microscopic alveoli; also contains immature milk glands that drain into lactiferous duct

Areola
Pink-red-brown circular area surrounding nipple

Nipple
Central protrusion containing 15–20 outlets of milk ducts

Lactiferous duct
Inactive channel that will direct milk toward nipple

Rib

Stroma
Connective tissue supports fat and glands in breast

Intercostal muscle

THE DELIVERY DATE

An estimated delivery date is calculated at the very beginning of pregnancy based on the first day of the last menstrual period. The age of the fetus is then assessed from measurements taken during early ultrasound scans. This can sometimes result in a new estimated date of delivery. A singleton fetus is considered "full term" and ready to leave the womb from 37 weeks onward, although three additional weeks of growth—bringing gestation time to 40 weeks—is usually beneficial. If a fetus is still in the womb at 42 weeks, delivery is usually induced because an aging placenta can no longer function at its best.

THE NESTING INSTINCT
Toward the end of pregnancy, it is common for women to have a strong urge to clean the house and prepare the nursery for the imminent arrival of the new family member.

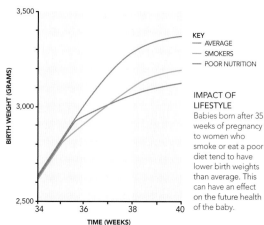

KEY
— AVERAGE
— SMOKERS
— POOR NUTRITION

BIRTH WEIGHT (GRAMS)

3,500

3,000

2,500

34   36   38   40

TIME (WEEKS)

IMPACT OF LIFESTYLE
Babies born after 35 weeks of pregnancy to women who smoke or eat a poor diet tend to have lower birth weights than average. This can have an effect on the future health of the baby.

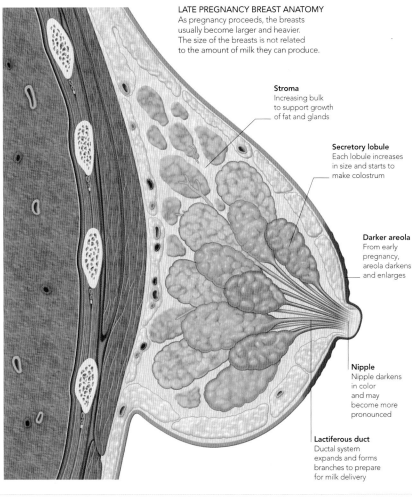

LATE PREGNANCY BREAST ANATOMY
As pregnancy proceeds, the breasts usually become larger and heavier. The size of the breasts is not related to the amount of milk they can produce.

Stroma
Increasing bulk to support growth of fat and glands

Secretory lobule
Each lobule increases in size and starts to make colostrum

Darker areola
From early pregnancy, areola darkens and enlarges

Nipple
Nipple darkens in color and may become more pronounced

Lactiferous duct
Ductal system expands and forms branches to prepare for milk delivery

THE HORMONES INVOLVED IN LACTATION

Like many aspects of pregnancy and childbirth, lactation occurs through a delicate interplay of hormonal activity. Some different hormones are secreted in addition to those already circulating in the pregnant body.

Progesterone	Progesterone is initially produced by the corpus luteum (the empty egg follicle after ovulation) and then by the placenta. High levels of progesterone stimulate the growth of alveoli and lobules within the breasts.
Estrogen	Before pregnancy, estrogen is involved in breast development at puberty. Increased estrogen levels during pregnancy are responsible for stimulating the growth and development of the milk duct system.
Prolactin	Produced in the pituitary gland, prolactin promotes milk production (lactation). Suckling the nipples causes release of prolactin so the breasts are constantly full. Oxytocin is usually secreted with prolactin.
Oxytocin	Oxytocin is secreted by the pituitary gland via an emotional trigger (baby crying) or stimulation of the nipples. Smooth muscle in the alveoli contracts and milk is ejected into the ducts—the "let-down" reflex.
Human placental lactogen (HPL)	Produced by the placenta from the second month of pregnancy, HPL mimics the action of both prolactin and growth hormone, causing the breasts, nipples, and areolae to increase in size.
Cortisol	Cortisol is present in relatively high amounts in colostrum during the first two days of breastfeeding. As it falls, the level of protective antibodies in milk (IgA) increases.
Thyroxine	Low amounts of thyroxine are present in breast milk. This hormone is thought to help prime the baby's digestive system.

AVERAGE DURATION OF PREGNANCY
Most pregnancies end about 280 days (40 weeks) after the first day of the woman's last menstrual period. This measure of pregnancy is known as gestational age.

KEY
PREMATURE
TERM
POSTMATURE

Quadruplets
The average duration of a pregnancy in which the end mother is carrying quadruplets is 32 weeks.

Full term
A fetus is considered full term at the end of the 37th week of pregnancy.

Within a week
Half of all babies are born within a week of their due date.

MONTHS

| 5 | 6 | 7 | 8 | 9 | 10 |

| 18 | 19 | 20 | 21 | 22 | 23 | 24 | 25 | 26 | 27 | 28 | 29 | 30 | 31 | 32 | 33 | 34 | 35 | 36 | 37 | 38 | 39 | 40 | 41 | 42 | 43 | 44 | 45 |

GESTATIONAL WEEKS

Youngest premature birth
The youngest premature baby that has gone on to lead a normal, healthy life was born at a mere 21 weeks and 5 days.

Viability
The threshold of viability is the point at which a premature baby has a 50 percent chance of survival outside the uterus.

Quintuplets
The average duration of a pregnancy in which the mother is carrying quintuplets is 30 weeks.

Triplets
The average duration of a pregnancy in which the mother is carrying triplets is 34 weeks.

Twins
The average duration of a pregnancy in which the mother is carrying twins is 38 weeks.

Within 2 weeks
The majority of babies (90%) are born within 2 weeks of their due date.

Induced labor
Labor is usually induced by 42 weeks; otherwise, placental deterioration will occur.

THE FORMATION OF THE BRAIN

Starting as a small thickening of the embryo's outer layer, the brain becomes a highly complex organ by the time the baby is born, containing 100 billion specialized cells known as neurons.

The first sign of the developing nervous system is a differentiation of cells to form the neural plate. This thickens and folds to form the neural tube, the precursor to the brain and spinal cord. The three main sections of the brain are evident within six weeks. The cerebellum starts to form at 13 weeks and is involved in regulating movements. The cerebrum is the largest part of the brain and comprises two different tissue types: gray and white matter. The former is the brain's processing center, while the latter carries information to different parts of the brain.

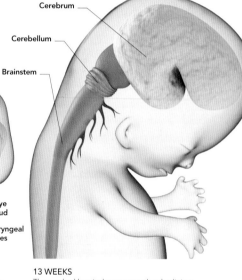

BRAIN COMPONENTS
In this color-enhanced electron micrograph, each fetal brain cell has a yellow cell body and is surrounded by many branching extensions, known as dendrites. These allow neurons to pass messages to neighboring brain cells.

Axon
Neuron cell body
Dendrite

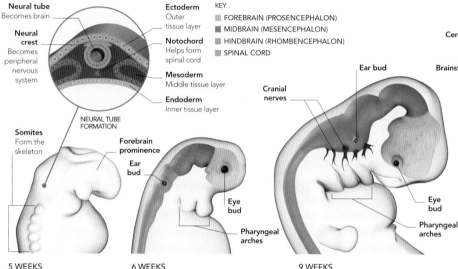

Neural tube Becomes brain

Neural crest Becomes peripheral nervous system

Ectoderm Outer tissue layer

Notochord Helps form spinal cord

Mesoderm Middle tissue layer

Endoderm Inner tissue layer

NEURAL TUBE FORMATION

KEY
- FOREBRAIN (PROSENCEPHALON)
- MIDBRAIN (MESENCEPHALON)
- HINDBRAIN (RHOMBENCEPHALON)
- SPINAL CORD

Somites Form the skeleton

Forebrain prominence

Ear bud

Eye bud

Pharyngeal arches

Cranial nerves

Ear bud

Eye bud

Pharyngeal arches

Cerebrum
Cerebellum
Brainstem

5 WEEKS
The neural tube forms in the fifth week from a groove that folds in on itself. The expanding neural tube at the head end forms the forebrain prominence.

6 WEEKS
The head end forms three hollow swellings that will develop into the forebrain, midbrain, and hindbrain. The main divisions of the central nervous system are now in place.

9 WEEKS
Swellings that will become the brainstem, cerebellum, and cerebrum grow at varying rates and start to fold in on one another. The cerebrum divides into hemispheres. Cranial and sensory nerves are forming.

13 WEEKS
The cerebral hemispheres expand and split into lobes. Connections start to form between brain cells. The hindbrain divides into the cerebellum and brainstem, the latter of which is involved in regulating basic functions such as breathing.

NEURAL NETWORKS

At birth, basic neural connections are in place, which help control vital functions such as breathing, heart beat, digestion, and reflexes. As more links form, and nerve cell axons become myelinated (insulated), higher mental functions develop, such as memory, increased attention span, language, intellect, and social skills. By early adulthood, the complex neural network allows for reasoning, judgment, and original thought.

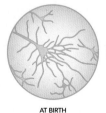

AT BIRTH | AGED SIX | AGED EIGHTEEN

FISSURES AND RIDGES

The cerebrum, the largest part of the brain, is divided into the right and left cerebral hemispheres. During development, each hemisphere enlarges forward to form a frontal lobe, upward and sideways to form a parietal lobe, and backward and underneath to form occipital and temporal lobes. As more neurons climb up into the outer layer of the brain (the cerebral cortex), the surface develops folds to accommodate them. This results in the formation of shallow grooves (sulci), deep grooves (fissures), and convolutions (gyri). Each lobe forms its own major sulci, gyri, and fissures that can be identified in most individuals. For example, the postcentral gyrus is the main area where sensations from the body are interpreted, and the precentral gyrus is where voluntary movement is controlled.

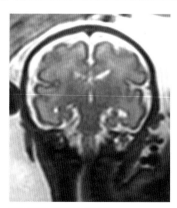

CORTICAL DEVELOPMENT
This MRI scan of a 25-week-old fetus reveals the complex folds in the developing brain. Fissures and gyri can be seen clearly in cross section.

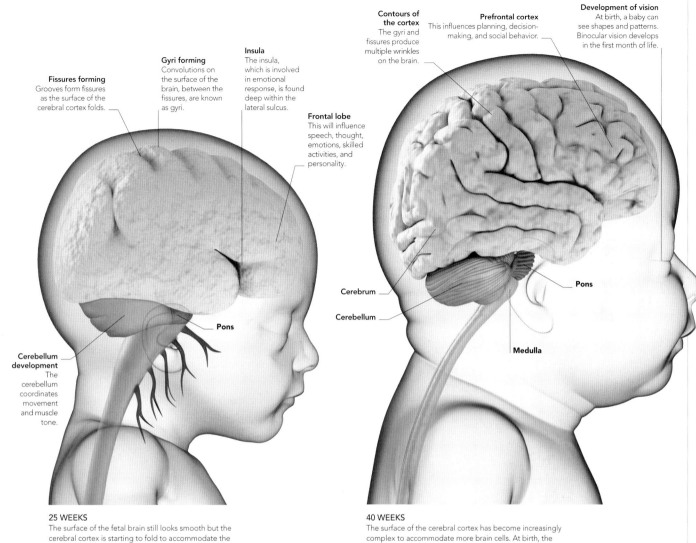

Fissures forming
Grooves form fissures as the surface of the cerebral cortex folds.

Gyri forming
Convolutions on the surface of the brain, between the fissures, are known as gyri.

Insula
The insula, which is involved in emotional response, is found deep within the lateral sulcus.

Frontal lobe
This will influence speech, thought, emotions, skilled activities, and personality.

Cerebellum development
The cerebellum coordinates movement and muscle tone.

Pons

Contours of the cortex
The gyri and fissures produce multiple wrinkles on the brain.

Prefrontal cortex
This influences planning, decision-making, and social behavior.

Development of vision
At birth, a baby can see shapes and patterns. Binocular vision develops in the first month of life.

Cerebrum

Cerebellum

Pons

Medulla

25 WEEKS
The surface of the fetal brain still looks smooth but the cerebral cortex is starting to fold to accommodate the rapidly increasing number of cells. From now until the first few months after birth, the developing brain rapidly increases in size. This is known as the brain growth spurt.

40 WEEKS
The surface of the cerebral cortex has become increasingly complex to accommodate more brain cells. At birth, the brain contains 100 billion brain cells, but their connections are not yet fully laid down. This part of the brain will not be fully mature until the person reaches the mid-twenties.

FORMATION OF GRAY MATTER

Support, or glial, cells in the developing brain act like scaffolding onto which newly divided brain cells (neurons) climb when emerging from the neural tube, in order to reach the outer part of the cerebral emispheres. Here, in the so-called gray matter, the cortex begins to develop six layers of cells. Neurons climbing up the glial cells are thought to follow chemical signals that indicate the right point at which to jump off and begin forming a layer. As the framework for one layer is completed, the next wave of neurons climbs higher, through the initial layers, to form a new layer on top. The way these layers form is vital for ordered thought processes in later life.

KEY
■ VENTRICULAR ZONE
■ WHITE MATTER
■ SUBPLATE
■ CORTICAL PLATE
■ LAYERS 1–6

THE SIX LAYERS OF GRAY MATTER
Layers of neurons develop until, by birth, there are six layers. Neurons here become specialized for different tasks, such as thinking, writing, and speaking.

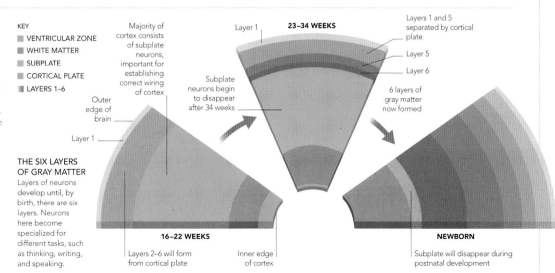

Majority of cortex consists of subplate neurons, important for establishing correct wiring of cortex

Outer edge of brain

Layer 1

16–22 WEEKS

Layers 2–6 will form from cortical plate

Inner edge of cortex

Layer 1

23–34 WEEKS

Subplate neurons begin to disappear after 34 weeks

Layers 1 and 5 separated by cortical plate

Layer 5

Layer 6

6 layers of gray matter now formed

NEWBORN

Subplate will disappear during postnatal development

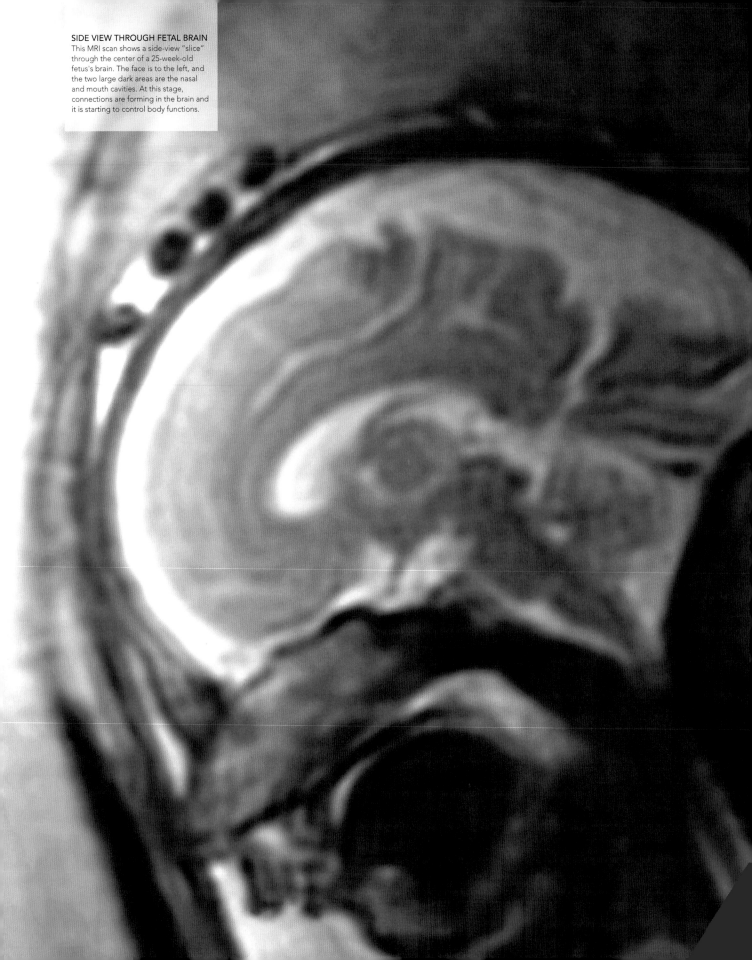

SIDE VIEW THROUGH FETAL BRAIN
This MRI scan shows a side-view "slice" through the center of a 25-week-old fetus's brain. The face is to the left, and the two large dark areas are the nasal and mouth cavities. At this stage, connections are forming in the brain and it is starting to control body functions.

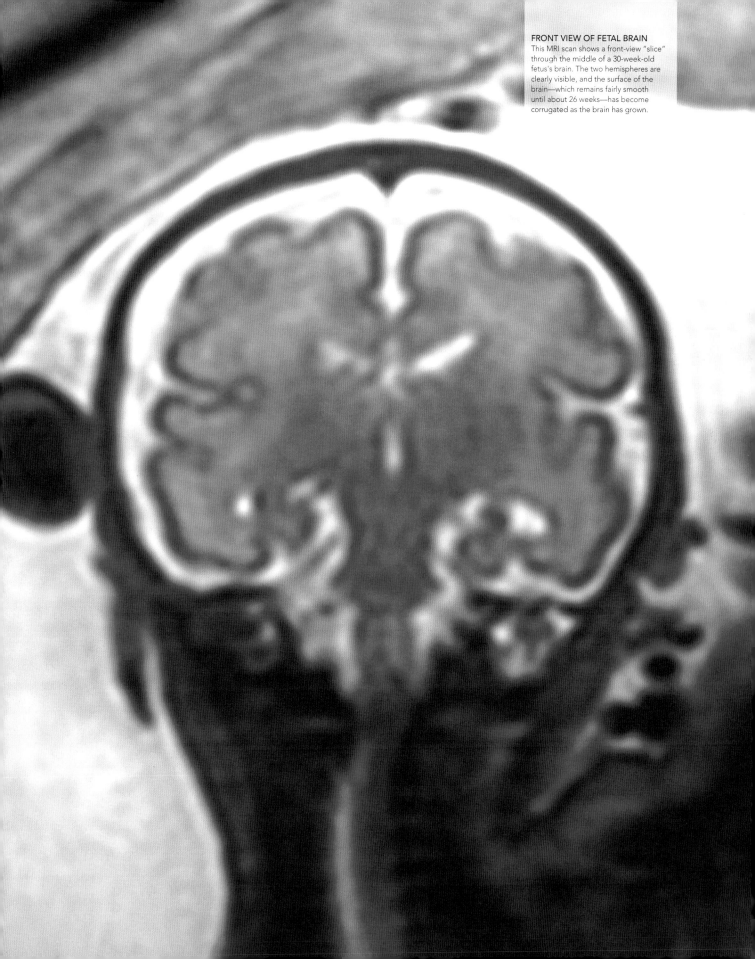

FRONT VIEW OF FETAL BRAIN
This MRI scan shows a front-view "slice" through the middle of a 30-week-old fetus's brain. The two hemispheres are clearly visible, and the surface of the brain—which remains fairly smooth until about 26 weeks—has become corrugated as the brain has grown.

FETUS

FETAL SKULL BONES

Rapid brain development results in the head of a full-term fetus being 2 percent larger than the birth canal. To overcome this problem, the fetal brain is protected by a series of flat, soft skull bones that have not fused and have the ability to slide over one another. This allows the fetal skull to contract sufficiently to avoid damage as it passes through the birth canal. Two large spaces occur where the fetal skull bones meet at the front and back of the head—the anterior and posterior fontanelles. There are four other fontanelles at the sides of the skull. Sutures are seams of connective tissue where the flat bones abut.

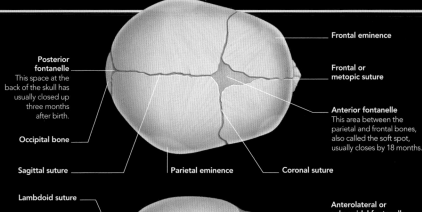

Frontal eminence

Posterior fontanelle
This space at the back of the skull has usually closed up three months after birth.

Frontal or metopic suture

Occipital bone

Anterior fontanelle
This area between the parietal and frontal bones, also called the soft spot, usually closes by 18 months.

Sagittal suture

Parietal eminence

Coronal suture

Lambdoid suture

Posterolateral or mastoid fontanelle
This space is behind the ear, between the parietal and temporal bones.

Anterolateral or sphenoidal fontanelle
This area forms between the frontal, temporal, and sphenoid bones.

Maxilla
The upper jaw (like the lower jaw) contains teeth buds that slowly start to erupt after birth.

Mandible
The lower jaw develops slowly to let the baby latch onto the breast and suckle.

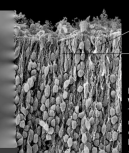

Posterior fontanelle

VISIBLE FONTANELLE
This 3D ultrasound scan shows the posterior fontanelle, which forms in the space between the occipital bone and the two parietal bones.

SKULL AT 9 MONTHS
The skull bones of a newborn baby are not fused. The fontanelles and sutures are protected by tough membranes that turn to bone (ossify) during the first two years of life.

INCREASED COORDINATION

Nerve cells (neurons) in the brain of a fetus multiply at an astonishing rate of 50,000–100,000 cells per second. The gray matter, or cortex, of the brain develops in successive layers. When the framework for one layer is completed, the next wave of neurons emerges to form a new layer on top. As the brain rapidly enlarges, these brain cells make more and more connections with other brain cells, and the coordination of fetal movements improves and becomes ever more complex.

Nerve cell body
Control center housing the nucleus

Dendrite
Communication fiber that disperses impulses

NEURONS INVOLVED WITH MOVEMENT
This color-enhanced electron micrograph shows fetal brain cells (green) in the part of the brain that controls posture and movement

SPECIALIZED MONITORING

As the fetus approaches full term, the mature placenta becomes less efficient at providing all the nutrients required for growth and sustenance. A range of different tests is used to assess whether the fetus is being deprived nutritionally. These tests help assess fetal growth and well-being and may check breathing, movement, and heart rate. They require specialized equipment and are usually carried out in a hospital or an outpatient clinic.

TESTS TO ASSESS FETAL HEALTH	
TEST PERFORMED	**DESCRIPTION**
Fetal growth	If the fetus is growing slowly, ultrasound scans are carried out regularly. The circumference of the fetal head and size of the liver are measured, as well as the thigh bone (femur) length. If the placenta is not working well, the fetal head will seem relatively large compared with the liver, because the baby's fat stores are used up (or have not been laid down).
Fetal well-being	A biophysical profile assesses fetal well-being by monitoring the heart rate on a cardiotocograph (or CTG) and by using ultrasound to record the amount of amniotic fluid, the fetus's movements, extensions of the limbs, and breathing. This profiling is carried out if the fetus is not growing as expected and if the blood flow in the umbilical arteries is poor.

ABDOMINAL EXAMINATION
A healthcare professional carries out an examination on the abdomen of a woman whose fetus is full term

FINAL DEVELOPMENTS

At nine months the fetus is fully formed and its head is in proportion with the rest of the body. Increasing amounts of fat have been laid down, and the face has lost most of its wrinkles, making it appear plump. The fetus is covered in protective vernix, which is especially thick in skin creases such as the armpits. Small amounts of body hair (lanugo) may still be present but these disappear soon after birth. The fingernails and toenails are almost fully grown and may extend to the ends of the digits. The fetus tends to lie with its arms and legs drawn up and can grip quite strongly with its fingers. Many—but not all—babies are now in the head-down position ready for birth.

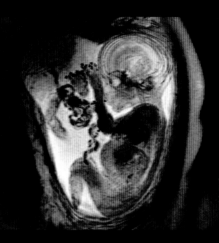

TIGHT FIT
Toward the end of pregnancy there is little room left in the fully stretched uterus. Although bunched up, the fetus can still move around within its protective sac of amniotic fluid.

BEFORE AND AFTER BIRTH
A comparison between a 3D ultrasound scan of the face of a fetus and the same infant after birth reveals the accuracy of prenatal imaging.

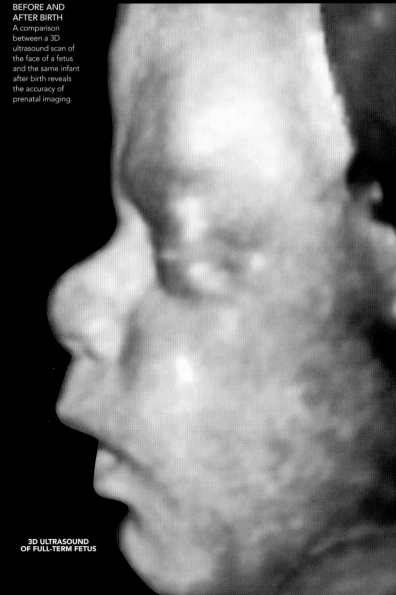

3D ULTRASOUND OF FULL-TERM FETUS

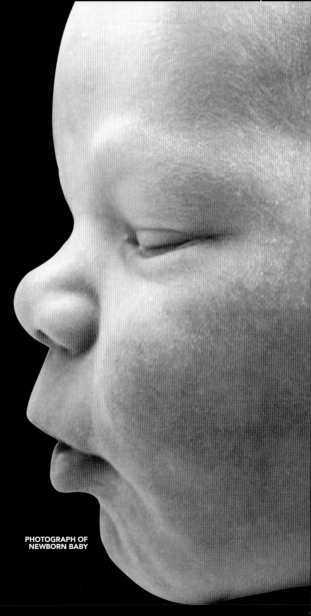

PHOTOGRAPH OF NEWBORN BABY

THE MOTHER'S CHANGING BODY

A woman's body undergoes profound changes during pregnancy. A number of these changes are beneficial, such as the development of stronger nails and a glowing complexion, but there are also some potential discomforts, such as back pain, breathlessness, and fatigue.

The mother's body must supply the developing fetus's rising demand for oxygen and nutrients, creating increased work for her own lungs, heart, and digestive system. In addition to carrying the baby, her body must support the growth of the placenta and production of amniotic fluid. As the pregnancy progresses, the uterus expands upward and outward to push against her intestines and diaphragm. Her breasts begin to enlarge in preparation for lactation, and her blood volume, body fluids, and fat stores increase. Altogether, these changes account for a normal weight gain of around 22–29 lb (10–13 kg).

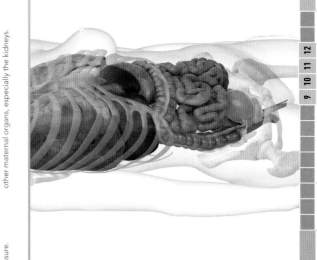

BLOOD VOLUME
Blood volume increases steadily in pregnancy (until around 32 weeks, when it tends to level off) to allow extra blood flow to the uterus and other maternal organs, especially the kidneys.

Rapid rise in blood volume occurs between months 6 and 8

BLOOD PRESSURE
Blood pressure tends to go down during the early stages of pregnancy and then increases during the last trimester. Changes in posture, such as lying flat on the back, can affect blood pressure.

Systolic (maximum pressure)

Diastolic (minimum pressure)

HEART RATE
Maternal heart rate increases during pregnancy in response to increased blood volume and the extra work performed by the heart as it pumps blood through the placenta.

Heart rate increases in steps and eventually levels out in month 9

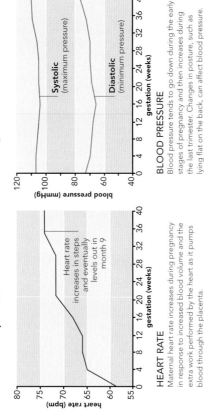

WEEKS 1 2 3 4

MONTH 1
During this month, the mother may not even be aware that she is pregnant. The first sign is usually a missed period. Some women notice changes in taste sensation, tingling of breasts, nausea, or unusual fatigue.

5 6 7 8

MONTH 2
The mother has usually missed a period by now and knows she is pregnant. Breast tenderness, areolae enlargement, increased urinary frequency, and food cravings may occur. Fatigue is also common.

9 10 11 12

MONTH 3
As the first trimester ends, the uterus grows to reach the top of the pelvic cavity. Vaginal discharge may increase. Blood volume has increased, and some women already have a healthy pregnancy glow.

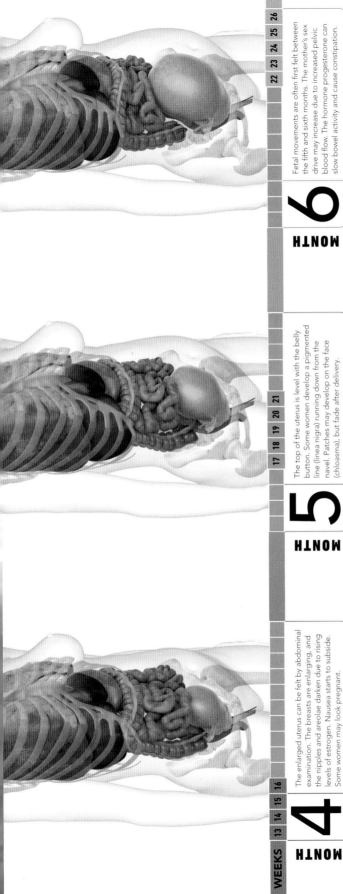

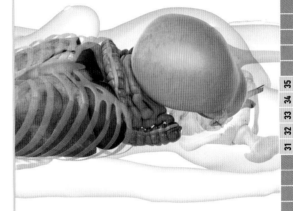

WEEKS | 13 | 14 | 15 | 16

4
MONTH

The enlarged uterus can be felt by abdominal examination. The breasts are enlarging, and the nipples and areolae darken due to rising levels of estrogen. Nausea starts to subside. Some women may look pregnant.

| 17 | 18 | 19 | 20 | 21

5
MONTH

The top of the uterus is level with the belly button. Some women develop a pigmented line (linea nigra) running down from the navel. Patches may develop on the face (chloasma), but fade after delivery.

| 22 | 23 | 24 | 25 | 26

6
MONTH

Fetal movements are often first felt between the fifth and sixth months. The mother's sex drive may increase due to increased pelvic blood flow. The hormone progesterone can slow bowel activity and cause constipation.

WEEKS | 27 | 28 | 29 | 30

7
MONTH

Rapid abdominal expansion and hormone changes can lead to stretch marks on the abdomen, thighs, buttocks, or breasts. As the uterus pushes up against the intestines, indigestion and heartburn can occur.

| 31 | 32 | 33 | 34 | 35

8
MONTH

Fat stores may be deposited in odd places: between the shoulders, on the upper back, around the knees. If the uterus compresses the diaphragm, deep breathing is difficult. Braxton-Hicks' contractions may occur.

| 36 | 37 | 38 | 39 | 40

9
MONTH

If the fetal head "engages," pressure may be felt in the pelvis. Increasing fatigue is normal. The breasts are making colostrum. As the cervix softens, loss of the cervical mucus plug may show that delivery is imminent.

THE FETUS'S CHANGING BODY

The 40 weeks of pregnancy encompass the remarkable transformation of a single-celled fertilized egg to a breathing baby. During this time, the 11 major systems of the body take shape, undergoing predictable periods of growth and development.

The organization of the baby's body is incredibly complex. Each of its trillions of cells communicates with its neighbors, following chemical and hormone signals that direct its movements and the kind of cell it will become. These interactions depend on the genes inherited from the parents. The basic blueprint for each of the body systems is laid down

during the first eight weeks of life—the embryonic stage—after which the embryo is known as a fetus. By the end of the second trimester, the fetal systems have developed to the point at which it has a chance of survival if born prematurely. The third trimester is a period of rapid growth, helping prepare the fetus for the world outside the uterus.

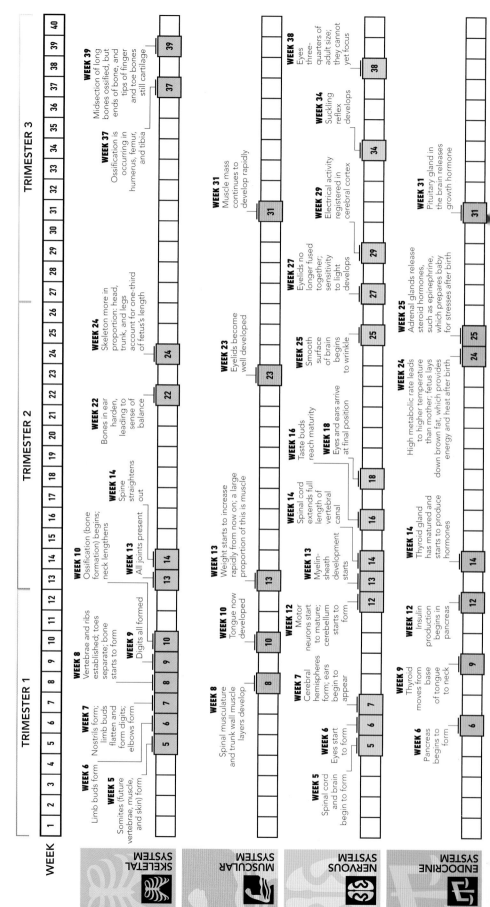

TIMELINE OF MAJOR EVENTS
Each of the 11 major body systems undergoes specific stages of growth, which occur in a predictable sequence. Most of the fetus's body systems are mature enough to function after 37–40 weeks of development, when it is considered "full-term."

WEEK 1 2 3 4 5 6 7 8 9 10 11 12 13 14 15 16 17 18 19 20 21 22 23 24 25 26 27 28 29 30 31 32 33 34 35 36 37 38 39 40

TRIMESTER 1   TRIMESTER 2   TRIMESTER 3

SKELETAL SYSTEM

WEEK 5 Somites (future vertebrae, muscle, and skin) form

WEEK 6 Limb buds form

WEEK 7 Nostrils form; limb buds flatten and form digits; elbows form

WEEK 8 Vertebrae and ribs established; toes separate; bone starts to form

WEEK 9 Digits all formed

WEEK 10 Ossification (bone formation) begins; neck lengthens

WEEK 13 All joints present

WEEK 14 Spine straightens out

WEEK 22 Bones in ear harden, leading to sense of balance

WEEK 24 Skeleton more in proportion: head, trunk, and legs account for one-third of fetus's length

WEEK 37 Ossification is occurring in humerus, femur, and tibia

WEEK 39 Midsection of long bones ossified, but ends of bone, and tips of finger and toe bones still cartilage

MUSCULAR SYSTEM

WEEK 8 Spinal musculature and trunk wall muscle layers develop

WEEK 10 Tongue now developed

WEEK 13 Weight starts to increase rapidly from now on; a large proportion of this is muscle

WEEK 23 Eyelids become well developed

WEEK 25 Smooth surface of brain begins to wrinkle

WEEK 31 Muscle mass continues to develop rapidly

NERVOUS SYSTEM

WEEK 5 Spinal cord and brain begin to form

WEEK 6 Eyes start to form

WEEK 7 Cerebral hemispheres form; ears begin to appear

WEEK 12 Motor neurons start to mature; cerebellum starts to form

WEEK 13 Myelin-sheath development starts

WEEK 14 Spinal cord extends full length of vertebral canal

WEEK 16 Taste buds reach maturity

WEEK 18 Eyes and ears arrive at final position

WEEK 27 Eyelids no longer fused together; sensitivity to light develops

WEEK 29 Electrical activity registered in cerebral cortex

WEEK 34 Suckling reflex develops

WEEK 38 Eyes three-quarters of adult size; they cannot yet focus

ENDOCRINE SYSTEM

WEEK 6 Pancreas begins to form

WEEK 9 Thyroid moves from base of tongue to neck

WEEK 12 Insulin production begins in pancreas

WEEK 14 Thyroid gland has matured and starts to produce hormones

WEEK 24 High metabolic rate leads to higher temperature than mother; fetus lays down brown fat, which provides energy and heat after birth

WEEK 25 Adrenal glands release steroid hormones, such as epinephrine, which prepares baby for stresses after birth

WEEK 31 Pituitary gland in the brain releases growth hormone

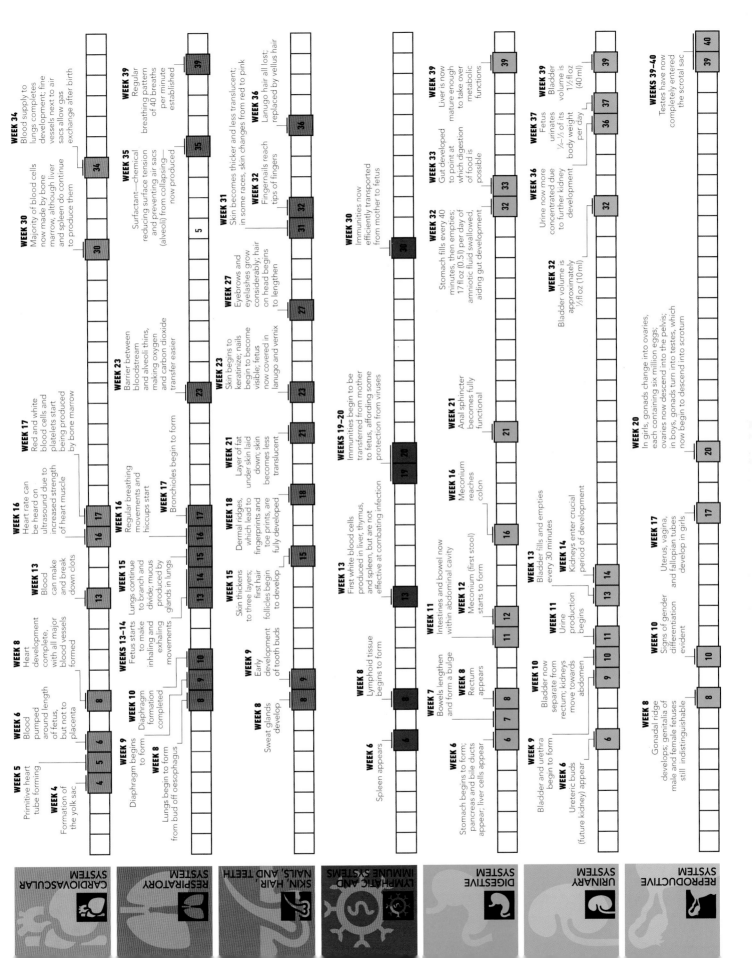

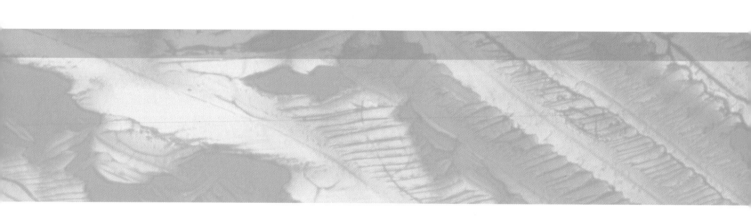

AN INCREDIBLE SERIES OF CHANGES TAKES PLACE IN THE
MOTHER AND FETUS OVER THE COURSE OF PREGNANCY,
CULMINATING IN THE MOST REMARKABLE OF EVENTS: BIRTH.
THE BEGINNING OF THIS PROCESS—LABOR—STARTS WHEN
THE MUSCULAR WALL OF THE UTERUS CONTRACTS WITH
INCREASING STRENGTH AND FREQUENCY, PUSHING THE
BABY DOWN AND OPENING THE CERVIX IN PREPARATION
FOR THE BABY'S PASSAGE THROUGH THE BIRTH CANAL.
THE BABY TWISTS AND TURNS AS IT DESCENDS, THE BONES
OF ITS SKULL SHIFTING SLIGHTLY TO ENABLE THE HEAD TO
PASS THROUGH. THE FIRST BREATH TRIGGERS IMMEDIATE
CHANGES IN THE BABY'S LUNGS AND HEART, HERALDING
THE START OF INDEPENDENT LIFE.

LABOR AND BIRTH

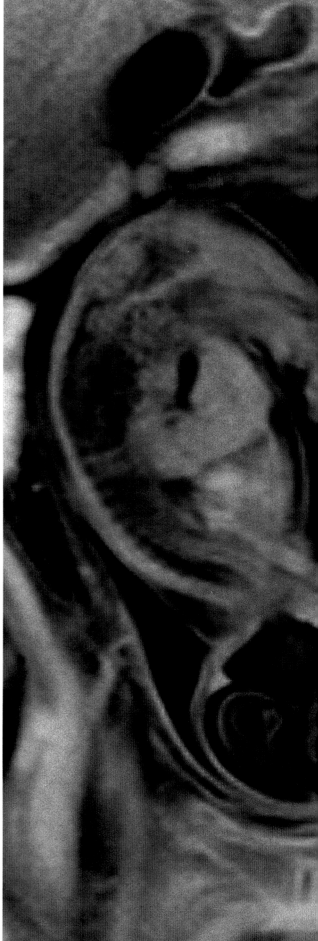

PREPARING FOR BIRTH

During the final weeks of pregnancy, hormonal changes in the mother and the pressure of the fetus as it moves down in the pelvis prepare the uterus for the imminent birth.

EARLY CONTRACTIONS

In the second trimester, very mild contractions begin to occur in the uterus, which gradually increase in intensity and frequency over the course of pregnancy. These painless tightenings, known as Braxton Hicks' contractions, tend to last about 30 seconds each. They cause an increase in blood flow to the placenta, thereby increasing the delivery of oxygen and nutrients to the fetus in its final stages of growth. Close to birth, Braxton Hicks' contractions can become uncomfortable, and some women, particularly with their first baby, mistake this "false" labor for the onset of true labor.

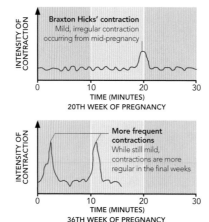

Braxton Hicks' contraction
Mild, irregular contraction occurring from mid-pregnancy

INTENSITY OF CONTRACTION

TIME (MINUTES)
20TH WEEK OF PREGNANCY

More frequent contractions
While still mild, contractions are more regular in the final weeks

INTENSITY OF CONTRACTION

TIME (MINUTES)
36TH WEEK OF PREGNANCY

CONTRACTIONS
Braxton Hicks' contractions become more frequent as a pregnancy progresses. Although distinctive in character, they are the forerunners of the strong, regular contractions that herald the onset of true and established labor.

THE LATENT PHASE

This very early part of labor is characterized by mild and irregular contractions. These contractions cause the changes in the cervix necessary for birth, making it softer, thinner, and much shorter than its original ¾ in (2 cm) length. The latent phase tends to last around eight hours, but can be shorter in women who have had several babies. The mild contractions may be felt as backache or menstrual period-type pain and do not usually cause distress. Some women are unaware that this phase is taking place. With the onset of established labor (see p.190), the opening (dilating) of the cervix begins, caused by the stronger and more frequent contractions that occur.

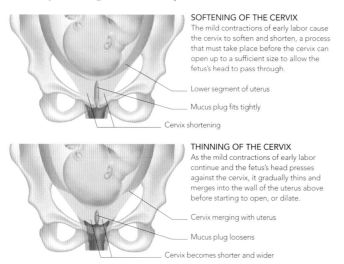

SOFTENING OF THE CERVIX
The mild contractions of early labor cause the cervix to soften and shorten, a process that must take place before the cervix can open up to a sufficient size to allow the fetus's head to pass through.

— Lower segment of uterus

— Mucus plug fits tightly

— Cervix shortening

THINNING OF THE CERVIX
As the mild contractions of early labor continue and the fetus's head presses against the cervix, it gradually thins and merges into the wall of the uterus above before starting to open, or dilate.

— Cervix merging with uterus

— Mucus plug loosens

— Cervix becomes shorter and wider

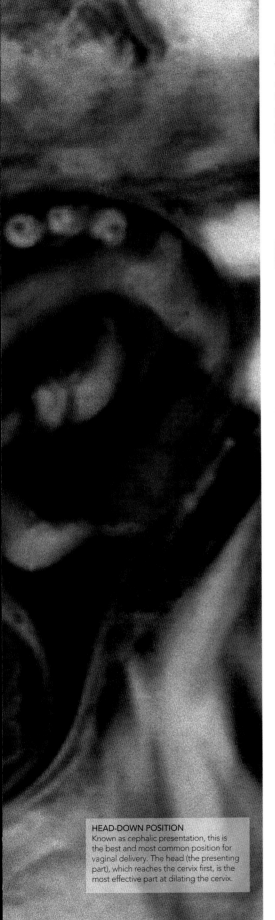

FETAL LIE

The fetus can lie in a vertical, horizontal, or diagonal position in the uterus. A vertical lie can be cephalic (head-down) or, less commonly, breech (bottom-down). With a horizontal or diagonal lie, there is no presenting part. By 35 weeks, most babies will be in the cephalic position. At term, 95 percent of babies are cephalic, 4 percent breech, and 1 percent transverse (horizontal) or oblique (diagonal).

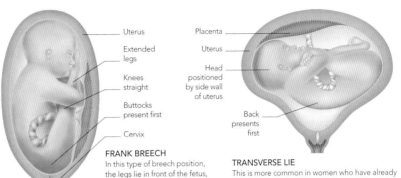

FRANK BREECH
In this type of breech position, the legs lie in front of the fetus, in contrast to complete breech, where the fetus sits cross-legged.

TRANSVERSE LIE
This is more common in women who have already had a baby. The looseness of the uterine muscles allows the fetus to lie horizontally across it.

ENGAGEMENT

The term "engagement" is used when three-fifths or more of the fetus's head has passed through the pelvic inlet. By feeling the abdomen, the physician or midwife can assess how much of the head lies above the pubic bone at the front of the pelvic inlet and determine whether the head is engaged. During labor, engagement is assessed by vaginal examinations.

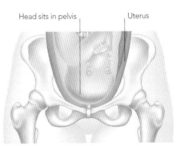

BEFORE THE HEAD ENGAGES
If three-fifths or more of a fetus's head lies above the pelvic inlet, it is not engaged. Most first babies engage at about 36 weeks. Some babies' heads do not engage until after the beginning of labor.

AFTER ENGAGEMENT
Once two-fifths or less of the head lies above the pelvic inlet—with most of it lying below—the head is engaged. The fetus is able to move down because the lower section of the uterus expands.

LATE-PREGNANCY HORMONE CHANGES

Estrogen levels rise in the last weeks of pregnancy, while progesterone levels stabilize. Estrogen triggers contractions of the uterus, while progesterone loosens joints to ease the passage of the fetus through the pelvis. Levels of hCG do not change significantly after the fourth month because this hormone's principal role of maintaining the corpus luteum in the ovaries has been fulfilled by then.

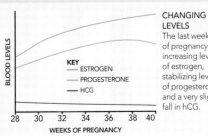

KEY
— ESTROGEN
— PROGESTERONE
— HCG

WEEKS OF PREGNANCY

CHANGING LEVELS
The last weeks of pregnancy see increasing levels of estrogen, stabilizing levels of progesterone, and a very slight fall in hCG.

THE FIRST STAGE OF LABOR

This period of labor is characterized by the onset of regular, painful contractions and is complete when the cervix is fully dilated to allow the baby to pass through. During this stage, contractions become stronger and closer together.

EARLY SIGNS OF LABOR

Before the first stage of labor becomes established, there are mild and irregular contractions (see p.188). These are then overtaken by strong, regular contractions. As labor approaches, the mucus plug that has been present in the cervix throughout pregnancy is dislodged (and is then known as the "show"). The waters usually break during labor or just before it begins. Occasionally, the waters break prematurely, before 37 weeks.

UTERINE CONTRACTIONS

In the early part of the first stage, contractions are very mild and produce only a small amount of cervical dilation. Later on, in established labor, the forceful contractions drive the baby down toward the cervix, which opens at a much faster rate. The muscles of the uterine wall have a rich supply of blood. With each contraction, the blood vessels that supply oxygen and nutrients to the muscles are squeezed, reducing the oxygen supply and causing pain. This pain becomes more severe as contractions become stronger and more prolonged.

LOWER ABDOMINAL PAIN
With the onset of strong contractions comes pain in the lower abdomen and often in the lower back. There are a number of options to help ease the discomfort.

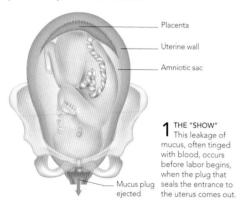

Placenta
Uterine wall
Amniotic sac

1 THE "SHOW"
This leakage of mucus, often tinged with blood, occurs before labor begins, when the plug that seals the entrance to the uterus comes out.

Mucus plug ejected

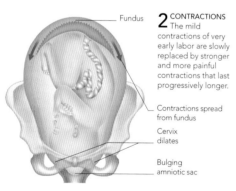

Fundus

2 CONTRACTIONS
The mild contractions of very early labor are slowly replaced by stronger and more painful contractions that last progressively longer.

Contractions spread from fundus

Cervix dilates

Bulging amniotic sac

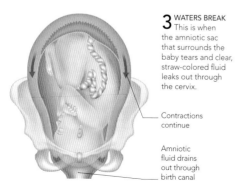

3 WATERS BREAK
This is when the amniotic sac that surrounds the baby tears and clear, straw-colored fluid leaks out through the cervix.

Contractions continue

Amniotic fluid drains out through birth canal

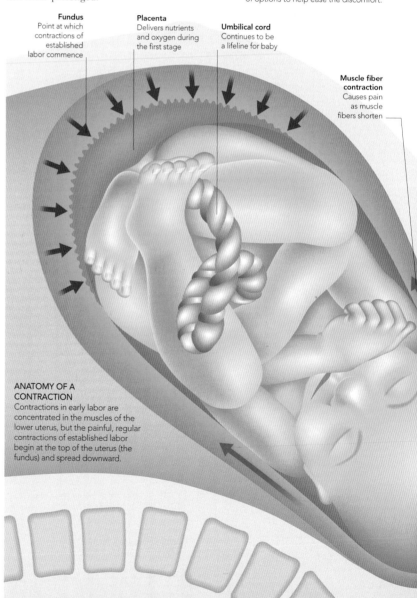

Fundus
Point at which contractions of established labor commence

Placenta
Delivers nutrients and oxygen during the first stage

Umbilical cord
Continues to be a lifeline for baby

Muscle fiber contraction
Causes pain as muscle fibers shorten

ANATOMY OF A CONTRACTION
Contractions in early labor are concentrated in the muscles of the lower uterus, but the painful, regular contractions of established labor begin at the top of the uterus (the fundus) and spread downward.

CERVICAL DILATION

During labor, the cervix opens to 4 in (10 cm) wide. Early on, the physician or midwife performs a vaginal examination to assess various aspects of the cervix, including degree of dilation, length, consistency, and position. How far the baby has descended into the pelvis is also recorded, as well as the baby's lie (see p.189). Throughout the first stage, the mother is assessed regularly, both with abdominal and vaginal examinations, to ensure that adequate progress is being made in terms of cervical dilation and descent of the baby into the pelvis.

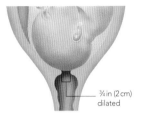

¾ in (2 cm) dilated

1 INITIAL DILATION
In the early stages of labor the cervix opens slowly, as the uterine contractions are still mild at this stage.

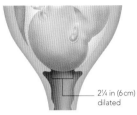

2¼ in (6 cm) dilated

2 CERVIX WIDENS
Once labor is established and the contractions are effective, the cervix dilates from 1½ in (4 cm) to 4 in (10 cm).

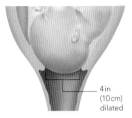

4 in (10 cm) dilated

3 FULLY DILATED
Once the opening is 4 in (10 cm) across (fully dilated), the mother can soon begin to push the baby out.

THE TRANSITION PHASE

For some women, there is a period of time between full dilation and the onset of the urge to push. Known as the transition phase, it may last a few minutes or as long as an hour. The contractions are very strong and frequent during this stage, so it can be difficult for the mother as she waits for the second stage of labor to begin.

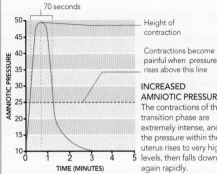

Height of contraction

Contractions become painful when pressure rises above this line

INCREASED AMNIOTIC PRESSURE
The contractions of the transition phase are extremely intense, and the pressure within the uterus rises to very high levels, then falls down again rapidly.

Fundus
Softens as muscles relax between contractions

BETWEEN CONTRACTIONS
Welcome moments of respite between contractions give the mother a chance to breathe more easily and try to relax before the next contraction. These periods become shorter as labor progresses.

Muscle fiber relaxing
This lengthens the muscles into the relaxed state

Cervix
Dilated by pressure from the baby's head

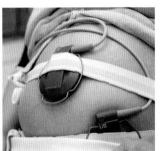

Pubic bone

Bladder
Becomes more compressed as the baby moves down

Vaginal rugae
Make up the corrugated lining that allows vagina to stretch

Skull bones
Can move, to allow head to change shape during delivery

Cervix
Softens and thins; dilates as head of baby presses against it

Rectum

FETAL MONITORING

The main indicator for fetal well-being during labor is the rate of the fetal heart and how it fluctuates in response to contractions. The simplest methods used to listen to the fetal heart are a Pinard stethoscope or a hand-held sonicaid machine, both of which are held against the mother's abdomen. Electronic fetal monitoring

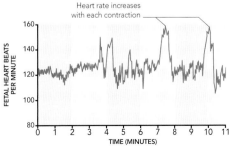

is used over longer periods, usually with two monitors strapped to the abdomen. Sometimes the heart rate is monitored via an electrode attached to the baby's head.

ELECTRONIC FETAL MONITOR
This measures the fetal heart rate and the intensity of contractions. Two sensors are linked to a cardiotocograph machine, which shows the results on a continuous trace.

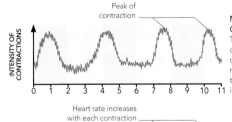

Peak of contraction

MATERNAL CONTRACTIONS
These regular contractions are typical of a normal labor. As revealed by the trace, they gradually increase in intensity.

Heart rate increases with each contraction

FETAL HEART BEAT
The heart rate is constantly fluctuating, and a certain degree of variability indicates that a baby is active and coping well with the labor. The rate increases when contractions occur.

THE BIRTH

The second stage of labor is the birth, culminating in the emergence of a new human being. Great effort from the mother, together with strong, frequent contractions, are needed to push the baby down the birth canal.

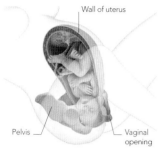

Wall of uterus

Pelvis

Vaginal opening

POSITION OF THE BABY IN THE PELVIS

The second stage begins once the cervix is fully dilated, contractions are strong and regular, and the woman has the desire to push. The baby rotates and the position of its head changes as it passes down the birth canal so that the widest part of its head is in line with the widest part of the mother's pelvis. Once the head has emerged, the baby turns again so that its shoulders can come out easily, one after the other. As soon as the baby emerges, the umbilical cord is checked to make sure it is not around the baby's neck, and mucus is cleared from the baby's nose and mouth to aid the baby's breathing. The birth typically lasts about one to two hours.

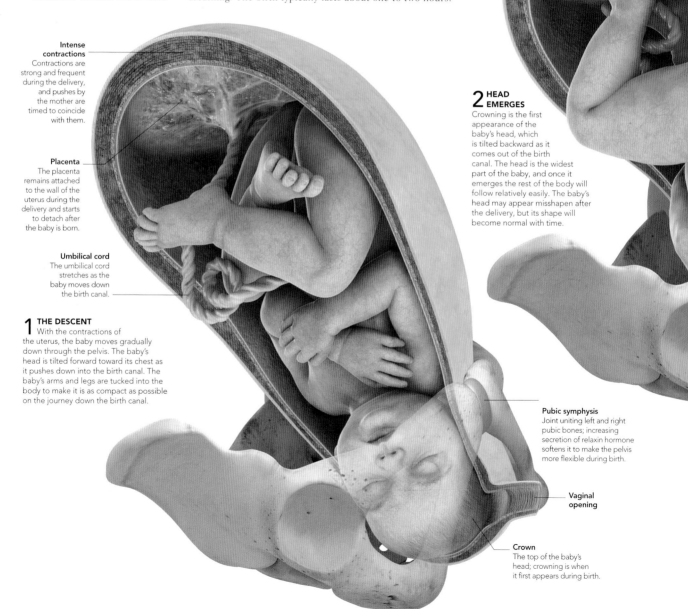

Shrinking uterus
The top of the uterus lowers as the baby moves down through the pelvis.

Intense contractions
Contractions are strong and frequent during the delivery, and pushes by the mother are timed to coincide with them.

Placenta
The placenta remains attached to the wall of the uterus during the delivery and starts to detach after the baby is born.

Umbilical cord
The umbilical cord stretches as the baby moves down the birth canal.

1 THE DESCENT
With the contractions of the uterus, the baby moves gradually down through the pelvis. The baby's head is tilted forward toward its chest as it pushes down into the birth canal. The baby's arms and legs are tucked into the body to make it is as compact as possible on the journey down the birth canal.

2 HEAD EMERGES
Crowning is the first appearance of the baby's head, which is tilted backward as it comes out of the birth canal. The head is the widest part of the baby, and once it emerges the rest of the body will follow relatively easily. The baby's head may appear misshapen after the delivery, but its shape will become normal with time.

Pubic symphysis
Joint uniting left and right pubic bones; increasing secretion of relaxin hormone softens it to make the pelvis more flexible during birth.

Vaginal opening

Crown
The top of the baby's head; crowning is when it first appears during birth.

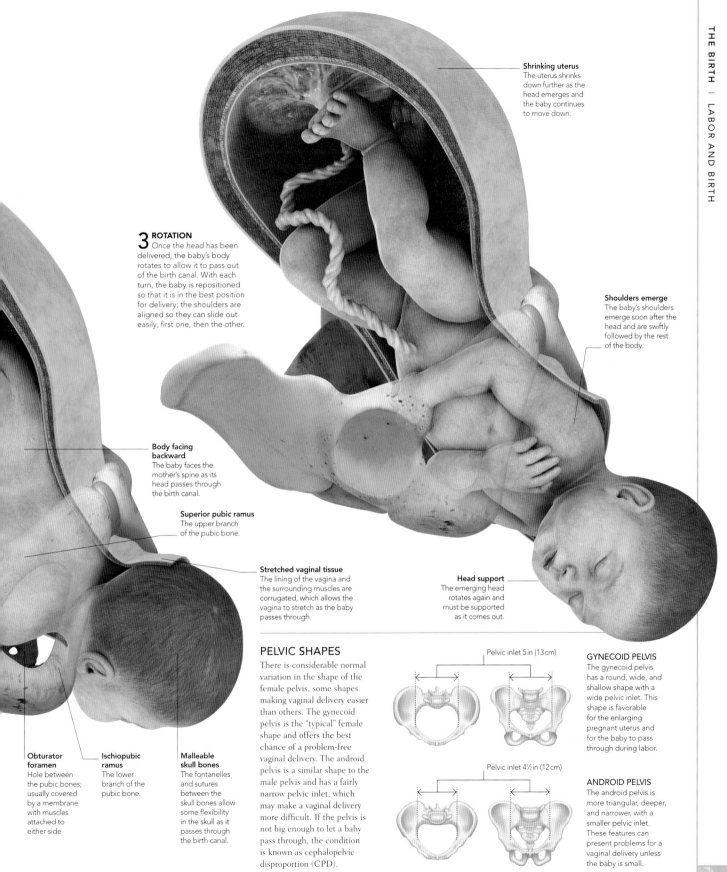

Shrinking uterus
The uterus shrinks down further as the head emerges and the baby continues to move down.

3 ROTATION
Once the head has been delivered, the baby's body rotates to allow it to pass out of the birth canal. With each turn, the baby is repositioned so that it is in the best position for delivery; the shoulders are aligned so they can slide out easily, first one, then the other.

Shoulders emerge
The baby's shoulders emerge soon after the head and are swiftly followed by the rest of the body.

Body facing backward
The baby faces the mother's spine as its head passes through the birth canal.

Superior pubic ramus
The upper branch of the pubic bone.

Stretched vaginal tissue
The lining of the vagina and the surrounding muscles are corrugated, which allows the vagina to stretch as the baby passes through.

Head support
The emerging head rotates again and must be supported as it comes out.

Obturator foramen
Hole between the pubic bones; usually covered by a membrane with muscles attached to either side

Ischiopubic ramus
The lower branch of the pubic bone.

Malleable skull bones
The fontanelles and sutures between the skull bones allow some flexibility in the skull as it passes through the birth canal.

PELVIC SHAPES
There is considerable normal variation in the shape of the female pelvis, some shapes making vaginal delivery easier than others. The gynecoid pelvis is the "typical" female shape and offers the best chance of a problem-free vaginal delivery. The android pelvis is a similar shape to the male pelvis and has a fairly narrow pelvic inlet, which may make a vaginal delivery more difficult. If the pelvis is not big enough to let a baby pass through, the condition is known as cephalopelvic disproportion (CPD).

Pelvic inlet 5in (13cm)

Pelvic inlet 4½in (12cm)

GYNECOID PELVIS
The gynecoid pelvis has a round, wide, and shallow shape with a wide pelvic inlet. This shape is favorable for the enlarging pregnant uterus and for the baby to pass through during labor.

ANDROID PELVIS
The android pelvis is more triangular, deeper, and narrower, with a smaller pelvic inlet. These features can present problems for a vaginal delivery unless the baby is small.

193

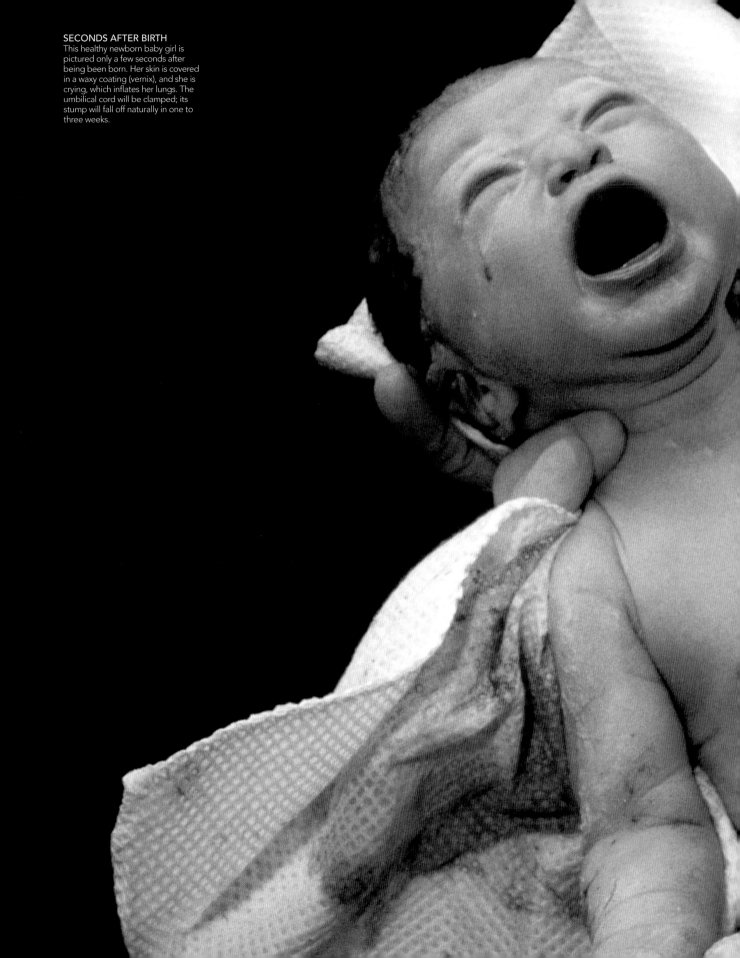

SECONDS AFTER BIRTH
This healthy newborn baby girl is pictured only a few seconds after being been born. Her skin is covered in a waxy coating (vernix), and she is crying, which inflates her lungs. The umbilical cord will be clamped; its stump will fall off naturally in one to three weeks.

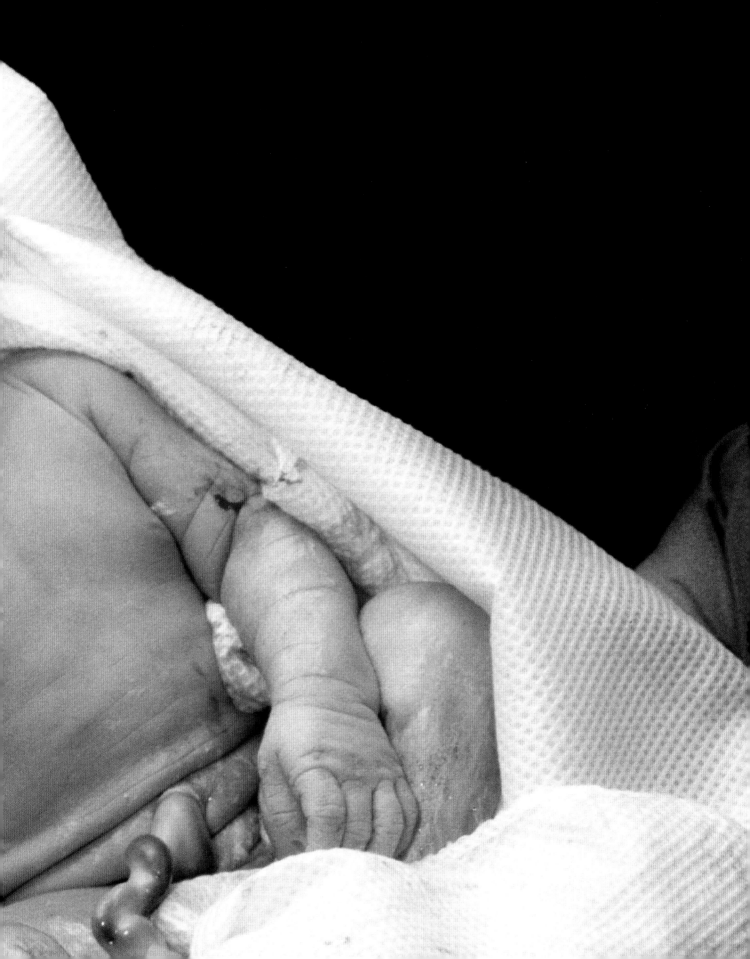

BIRTH POSITIONS

There are many possible options for the position in which to go through labor and give birth. Many women find it helpful to move around during the first stage of labor, and then to try one of a number of positions that are better for delivery than simply lying flat on the back. Some women feel more comfortable sitting on a bed with their back supported by pillows, whereas others prefer to kneel, squat, or use a birthing stool.

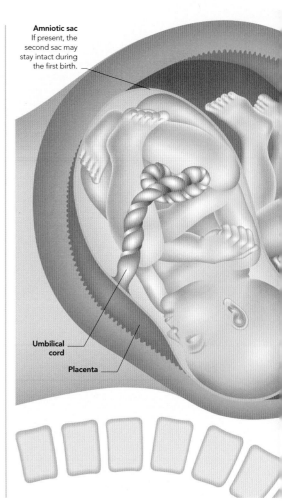

Amniotic sac
If present, the second sac may stay intact during the first birth.

Umbilical cord

Placenta

SITTING UPRIGHT
This position, supported behind by pillows, can be comfortable and good for pushing because the woman can pull against her thighs.

KNEELING
The woman can kneel upright with support or on all fours. Gravity can be helpful in upright positions in aiding the descent of the baby.

SQUATTING
When squatting, the pelvis is opened up, which, with the aid of gravity, makes it easier for the baby to be delivered.

BREECH BIRTH

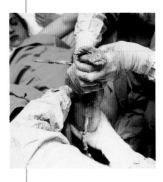

Many women with a breech presentation (see p.189)—in which the baby presents buttocks-first—have a cesarean section. However, a vaginal delivery may be considered in some cases, although it may not be possible to continue if problems develop, such as a cord prolapse (see p.232), in which the umbilical cord comes out first. If the cord is compressed, this can deprive the baby of oxygen and cause fetal distress or death (see p.232).

FEET FIRST
In a breech birth, the baby's buttocks and legs appear first, followed by the body. The widest part of the baby, its head, appears last.

PAIN RELIEF

There are a number of options available for pain relief in labor. Some have a generalized effect, relieving pain and having effects all around the body. These include opioid analgesics, the most commonly used being morphine as well as drugs derived from morphine. In contrast, the effects of regional analgesics are largely limited to one area of the body. There are also other, nondrug methods, which may help with relaxation as well as pain control.

GAS AND AIR

Commonly known as Entonox (a brand name), gas and air consists of a combination of oxygen and nitrous oxide and is often used to provide pain relief during labor. Entonox can be given through a mouthpiece or through a mask. When using gas and air, the woman should inhale and exhale with deep, regular breaths. It does not completely eliminate the pain but reduces it while also helping the woman to feel calmer. The effect begins to be felt after about 30 seconds, so the woman must start breathing it as soon as a contraction starts to feel the benefit at the right time. Gas and air can cause nausea and lightheadedness, but the effects wear off quickly. Gas and air is not used in some countries, including the USA.

DRUG INJECTIONS

The analgesic drugs used during labor are either administered by injection or through an IV tube. They act by relieving pain throughout the body and tend to be given early in labor. Morphine is the most common but stadol and nubaine are also used. All have potential side effects, but they are often chosen because they are easy to administer and relieve pain relatively quickly.

TYPE	HOW IT WORKS	SIDE EFFECTS
Morphine	Morphine may be given by injection into a muscle or through an IV tube inserted into the arm and attached to a pump that the woman controls herself (known as patient-controlled analgesia).	Nausea, vomiting, and sedation in the mother; sedation and depressed breathing in the baby.
Stadol	Like morphine, Stadol is given by IV injection.	Effects are similar to those of morphine, but it is effective for a shorter period of time.
Nubaine	Nubaine is also given by IV injection.	Effects are similar to those of morphine, but it is effective for a shorter period of time.

MULTIPLE BIRTHS

In the majority of cases, multiple births are delivered by cesarean section, although a vaginal delivery may be attempted, especially for twins. In such cases, the twins will be carefully monitored throughout labor by electronic fetal monitoring. Usually, the first twin is monitored via an electrode attached to its scalp while the second twin is monitored by sensors strapped to the mother's abdomen. bstetricians, midwives, pediatricians, and an anesthetist will be close by in case any problems develop. Also, an epidural anesthetic may be given so that the mother is ready if it does become necessary to perform a cesarean section.

Placenta

Umbilical cord

Pubic bone

Compressed bladder

DELIVERY OF TWINS
The most common position for twins is both head-down, so it may be possible to deliver first one twin then the other. The second twin continues to be carefully monitored while the first twin is delivered.

Emerging head
The first twin is seen here being born in the usual way, head-first.

Dilated cervix
The cervix is fully dilated to allow the first then the second twin to pass through.

IMMEDIATELY AFTER THE BIRTH

Within seconds of being born, the baby will take its first breath, inflating its lungs and crying for the first time. In addition to assessing the baby's condition and physical appearance, the midwife will weigh the baby and measure the head circumference. The baby is dried and wrapped to keep it from losing too much body heat. A vitamin K supplement, which helps with blood clotting, will be offered for the baby.

APGAR SCORE

The Apgar score is a method of rapidly assessing the condition of a baby after birth to see if emergency care is required. It is done at both one and five minutes after birth. In dark-skinned babies, "color" refers to the mouth, palms of the hands, and soles of the feet.

SIGN	SCORE: 0	SCORE: 1	SCORE: 2
Heart rate	None	Below 100 beats per minute	Above 100 beats per minute
Breathing rate	None	Irregular; weak cry	Regular; strong cry
Muscle tone	Limp	Moderate bending of limbs	Active movements
Reflex response	None	Moderate reaction or grimace	Crying or intense grimace
Color	Pale or blue	Pink, with blue hands and feet	Pink

EPIDURAL AND SPINAL BLOCKS

In these forms of anesthesia, a local anesthetic is injected around the spinal cord in the lower back, which blocks feeling below the level of the injection. However, as well as numbing pain in the abdomen, they may also make it difficult to move the legs. An epidural takes 20–30 minutes to work, whereas a spinal anesthetic starts to work almost immediately after it has been given.

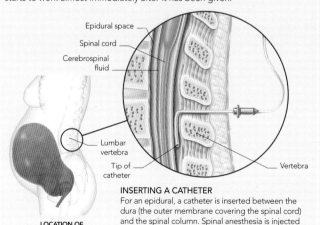

Epidural space

Spinal cord

Cerebrospinal fluid

Lumbar vertebra

Tip of catheter

Vertebra

LOCATION OF INSERTION POINT

INSERTING A CATHETER
For an epidural, a catheter is inserted between the dura (the outer membrane covering the spinal cord) and the spinal column. Spinal anesthesia is injected through the dura into the fluid around the cord.

NONPHARMACOLOGICAL RELIEF

Nondrug options for pain relief include breathing techniques (see below), reflexology, acupuncture, hypnotherapy, relaxation techniques, water immersion (see p.198), and massage. Transcutaneous electrical nerve stimulation (TENS) uses tiny electric currents to stimulate the release of endorphins, the body's own natural painkillers.

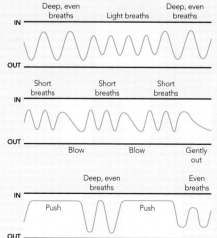

LATE FIRST STAGE
This stage involves taking deep, even breaths at the start and end of a contraction, and light breaths during its peak.

Deep, even breaths — Light breaths — Deep, even breaths

IN
OUT

TRANSITION STAGE
To avoid pushing too early, the mother should alternate between taking short breaths and blowing out, and exhale gently when the contraction ends.

Short breaths — Short breaths — Short breaths

IN
OUT
Blow — Blow — Gently out

SECOND STAGE
The mother should take and hold a deep breath while pushing down smoothly. After a push, deep, even breaths should be taken.

Deep, even breaths — Even breaths

IN
Push — Push
OUT

BIRTH SETTINGS

Women have a number of options for delivery, including where and how to have the baby. Personal preference, well-being, and the baby's safety are key factors involved in the decision.

WATER BIRTHS

A variety of settings, including hospital, a midwife-led birthing center, or home, are suitable for water births. Giving birth in water can provide pain relief, as well as aiding relaxation. The buoyancy of the water also makes a woman feel lighter and more mobile. Water births may be less traumatic for babies because they leave the fluid of the uterus to enter the waters of the pool. Birthing pools may be available in the hospital prenatal unit, although most have only one, or can be rented for use at home.

BIRTHING POOL
Many hospitals now have birthing pools. They can be used during the first stage of labor to help ease the contractions. The woman is then usually taken to a delivery room but can give birth in the pool.

HOME BIRTHS

This option is suitable for women who have had previous normal pregnancies and deliveries and who have no medical problems. It is generally recommended that first deliveries take place in a hospital. The prenatal care for a woman hoping for a home birth is provided by community midwives, who also perform the delivery. A hospital prenatal unit should be easily accessible in case of unforeseen complications during labor.

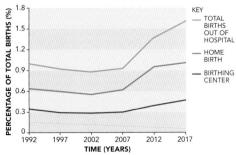

OUT-OF-HOSPITAL BIRTHS
Home and birthing-center births make up a small percentage of total births. A birthing center is a midwife-led maternity unit offering a homey approach. This graph, constructed from US data, shows increasing numbers of out-of-hospital births since 2007.

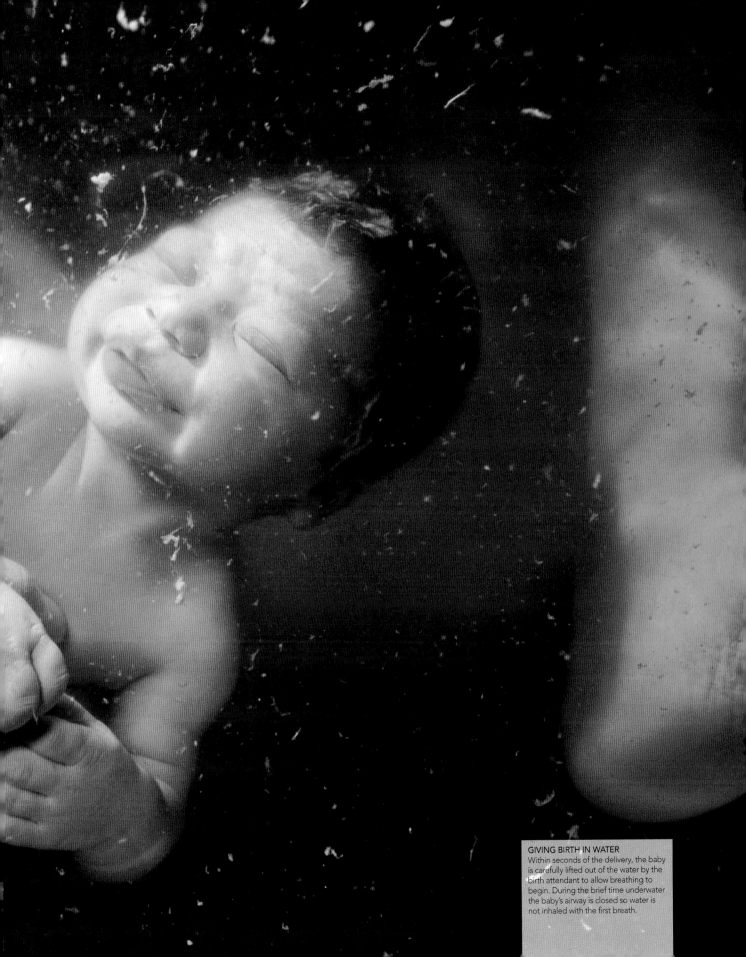

Within seconds of the delivery, the baby is carefully lifted out of the water by the birth attendant to allow breathing to begin. During the brief time underwater the baby's airway is closed so water is not inhaled with the first breath.

AFTER THE BIRTH

Within seconds of delivery, a series of events occurs, beginning with the baby taking its first breath. The umbilical cord is clamped and cut soon afterward, and the baby can then begin to feed without being directly connected to the mother.

DELIVERY OF THE PLACENTA

Soon after the baby has been delivered and the cord cut, the placenta must be removed. This is known as the third stage of labor. After the delivery, and once the uterus has contracted, the midwife or physician may gently pull on the cord and ease the placenta out with one hand while placing the other hand on the lower abdomen to keep the uterus in place. An injection of oxytocine may be given to the mother after the baby's head is delivered to help the uterus contract rapidly. The placenta must be carefully checked because any pieces of retained tissue may cause prolonged bleeding and prevent the uterus from contracting fully.

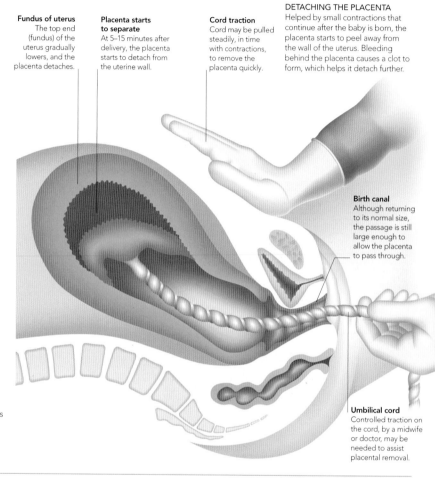

Fundus of uterus
The top end (fundus) of the uterus gradually lowers, and the placenta detaches.

Placenta starts to separate
At 5–15 minutes after delivery, the placenta starts to detach from the uterine wall.

Cord traction
Cord may be pulled steadily, in time with contractions, to remove the placenta quickly.

DETACHING THE PLACENTA
Helped by small contractions that continue after the baby is born, the placenta starts to peel away from the wall of the uterus. Bleeding behind the placenta causes a clot to form, which helps it detach further.

Birth canal
Although returning to its normal size, the passage is still large enough to allow the placenta to pass through.

Umbilical cord
Controlled traction on the cord, by a midwife or doctor, may be needed to assist placental removal.

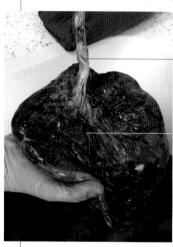

Umbilical cord
An uncut umbilical cord will pulsate for up to 3 minutes

Network of vessels
Multiple tiny blood vessels radiate from the umbilical cord

A HEALTHY PLACENTA
The placenta usually weighs about 1 lb (500 g) and is 8–10 in (20–25 cm) in diameter. In addition to the placenta, the membranes need to be removed from the uterus to avoid the risk of serious bleeding and infection.

CUTTING THE CORD

The umbilical cord has been the baby's lifeline throughout the 40 weeks of pregnancy. The baby has been dependent on this collection of blood vessels for intake of oxygen and nutrients and removal of wastes. Soon after delivery, the cord is cut because the baby can now live independently of the mother. It may be advantageous for the cord to remain connected momentarily so blood in the placenta can pass into the baby's circulation to boost the blood volume. This takes up to 3 minutes and allows the baby to be placed on the mother's abdomen—with the cord intact—for a short time without any problems.

CLAMPING AND CUTTING
Two clamps are placed around the cord about 1½ in (4 cm) apart, and the cut is made in the middle of them. This prevents leakage of blood from either the baby or the placenta.

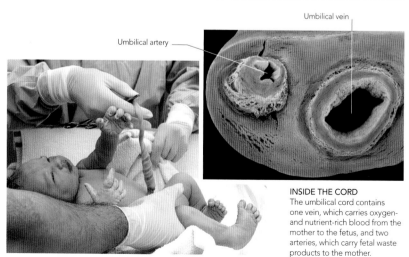

Umbilical artery

Umbilical vein

INSIDE THE CORD
The umbilical cord contains one vein, which carries oxygen- and nutrient-rich blood from the mother to the fetus, and two arteries, which carry fetal waste products to the mother.

CIRCULATION IN THE FETUS

The fetus cannot use its lungs until birth; before then they are deflated. In the uterus, it receives oxygen from maternal blood, which is transferred into the fetal blood in the placenta. Most of the fetal blood is directed from one side of the heart to the other, via a small opening called the foramen ovale. A vessel called the ductus arteriosus allows blood to enter the aorta without having to pass through the lungs. The blood leaves the heart via the aorta to supply the fetal body.

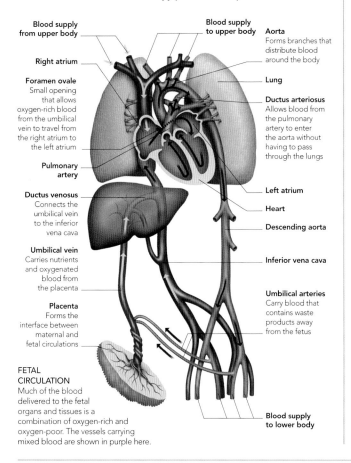

FETAL CIRCULATION
Much of the blood delivered to the fetal organs and tissues is a combination of oxygen-rich and oxygen-poor. The vessels carrying mixed blood are shown in purple here.

CIRCULATION AT BIRTH

From the baby's first breath, the circulatory set-up changes so that blood travels from the right side of the heart to the lungs for oxygen and then back to the left side of the heart, from where it passes into the aorta. The ductus arteriosus, ductus venosus, and the umbilical vessels close and become ligaments. The foramen ovale is also forced shut by the pressure of the blood returning to the left atrium (after collecting oxygen from the lungs).

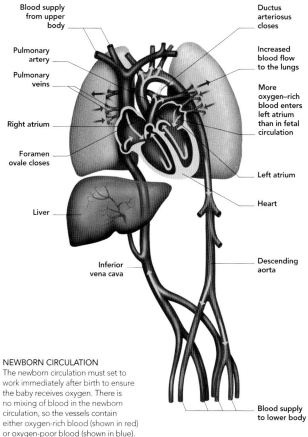

NEWBORN CIRCULATION
The newborn circulation must set to work immediately after birth to ensure the baby receives oxygen. There is no mixing of blood in the newborn circulation, so the vessels contain either oxygen-rich blood (shown in red) or oxygen-poor blood (shown in blue).

SUCKLING REFLEX

This is a primitive reflex that is present from birth and is closely linked to the rooting reflex (see p.210). Gently touching the roof of a baby's mouth triggers the suckling reflex. For this to happen, the baby needs to take a nipple (or bottle teat) into the mouth. Many newborn babies are put to the breast soon after birth and can feed right away. However, for others it will take time and patience to encourage a baby to suckle effectively. Suckling the nipple stimulates the production of oxytocin and prolactin, the hormones that are needed for the production and release of milk.

FIRST FEEDING
A rich, creamy substance called colostrum, packed with antibodies, is released for the first few days. True breast milk then starts to flow.

HORMONE CHANGES AFTER BIRTH

The levels of estrogen, progesterone, and other hormones fall dramatically following the birth of a baby. Effects of the drop include shrinkage of the uterus and increased tone in the pelvic floor muscles. The mother's circulating blood volume—raised to meet the demands of the fetus—returns to normal.

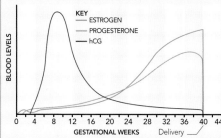

PLUMMETING LEVELS
The rapid fall in the levels of estrogen and progesterone is thought to play a role in the baby blues. It is not known why some women are more susceptible to the sudden drop.

201

ASSISTED BIRTH

Situations in which help may be needed to deliver a baby include being overdue, slow progress during labor, fetal distress, or an abnormal lie. Assisted deliveries may be planned or are required urgently if problems arise before or during labor.

INDUCING LABOR

Induction of labor may be recommended if a pregnancy goes beyond 42 weeks, if labor fails to start after the water has broken, and with certain medical conditions, such as preeclampsia. Sweeping the membranes, in which the membranes are gently pulled away from the cervix, may be performed during a vaginal examination. Another method is to insert prostaglandin into the vagina. If these methods fail, pitocin (synthetic oxytocin) in a drip may help increase contractions.

OXYTOCIN CRYSTALS
This light micrograph shows the structure of oxytocin, a hormone that is released by the pituitary gland. One of its main functions is to instigate labor, but it is not known what triggers its release.

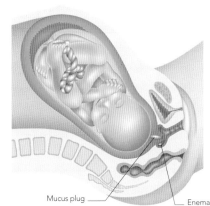

Amniotic fluid

Mucus plug — Enema

INSERTION OF PROSTAGLANDIN
Prostaglandin is used to induce labor and is inserted into the vagina close to the cervix as an enema or on a thin strip of gauze. The hormonelike substance helps ripen the cervix and stimulate contractions.

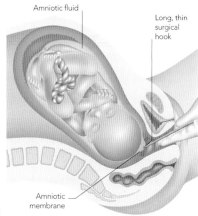

Long, thin surgical hook

Amniotic membrane

RUPTURE OF MEMBRANES
A hook is passed through the vagina to tear the amniotic membranes, allowing the fluid to drain out. This method tends to be used if labor is progressing slowly rather than if it has not started at all.

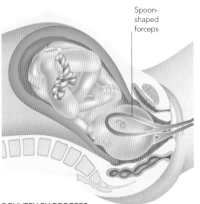

Spoon-shaped forceps

DELIVERY BY FORCEPS
The two blades of the forceps are placed around the baby's head and locked together. The physician pulls the forceps while the mother pushes with each contraction.

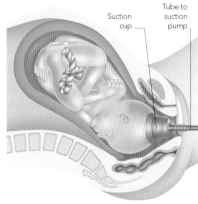

Suction cup

Tube to suction pump

DELIVERY BY VACUUM EXTRACTION
The cup is placed on the baby's head, and suction is then applied to fix it securely. The device is pulled gently to help the baby out.

DELIVERY BY FORCEPS AND VACUUM

Forceps or vacuum deliveries are used in about 5–15 percent of births, for a number of reasons, but most commonly, for fetal distress (usually indicated by the fetal heart rate) and maternal exhaustion after a labor of many hours. Either one of these methods may be used to help the delivery of a baby when it is low in the pelvis, but the cervix must be fully dilated so that the baby can pass through. Forceps are similar to large salad servers, which come apart in two pieces but lock to avoid crushing the baby's head during delivery. The ends are curved to cradle the baby's head. The vacuum (also known as the silastic) extractor has a suction cup, which is attached to the baby's head. An episiotomy is necessary with a forceps delivery, but may not be needed for a vacuum extraction.

EPISIOTOMY

An episiotomy is a cut made in the tissues between the vagina and anus to create a bigger opening and prevent tissue damage. It is carried out to prevent a bad tear or if there is fetal distress and the baby needs to be delivered quickly. The procedure is done under local, epidural, or spinal anesthesia. The cut is sewn up afterward.

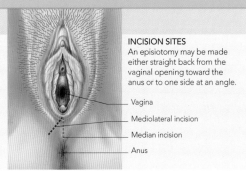

INCISION SITES
An episiotomy may be made either straight back from the vaginal opening toward the anus or to one side at an angle.

Vagina

Mediolateral incision

Median incision

Anus

Temporary ring from suction cup

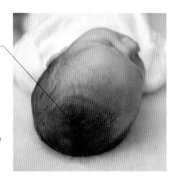

SILASTIC MARKS
The suction received through the cup on the top of the baby's head can leave a red circular bruise, known as a chignon. Although it looks alarming, the mark only lasts a week or so.

CESAREAN SECTION

In a cesarean section, the baby is removed from the uterus through an incision in the abdominal wall. There are a number of reasons why the vaginal route becomes impossible or undesirable. A cesarean may be planned, due to a nonurgent reason, for example if the mother is carrying twins, or it may be unplanned, due to an urgent reason, such as the development of fetal distress, or a less urgent one, such as no progress in labor. Before the operation, the abdomen is numbed, either by a regional anesthetic (epidural or spinal), which leaves the mother aware, or by general anesthesia, with the mother unconscious.

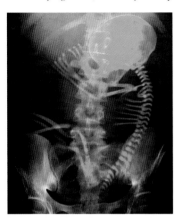

BREECH PRESENTATION
This X-ray shows a fetus in a breech lie (where the head is not the presenting part). If the baby cannot be manipulated, before labor, into a head-first position, a cesarean is the safest delivery option.

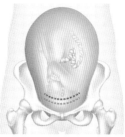

LOW TRANSVERSE INCISION

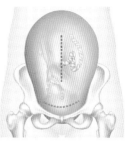

CLASSICAL INCISION

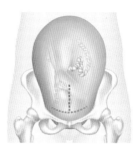

LOW VERTICAL INCISION

TYPES OF INCISION
The most common type of incision into the uterus is the low transverse incision. In some cases, a larger vertical cut (classical incision) may be made, for example if the baby lies across the abdomen. A low vertical incision may be used for other types of abnormal lie. The initial incision in the abdominal wall is usually the same in each case.

KEY
---- INCISION IN ABDOMINAL WALL
---- INCISION IN UTERUS

HOW THE PROCEDURE IS PERFORMED
A cut is made through the skin of the abdomen and the layers of tissue and muscle beneath are parted to reveal the uterus. The uterus is opened up, following one of the incision lines (top right), and the baby is lifted out.

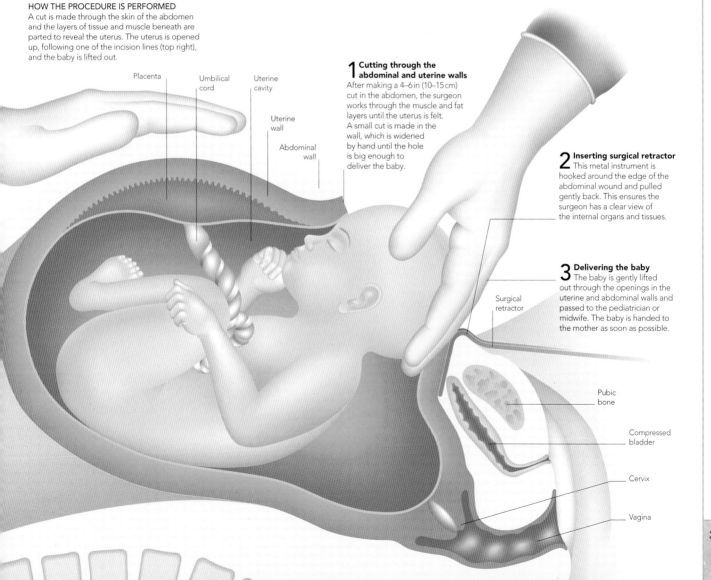

Placenta

Umbilical cord

Uterine cavity

Uterine wall

Abdominal wall

1 Cutting through the abdominal and uterine walls
After making a 4–6 in (10–15 cm) cut in the abdomen, the surgeon works through the muscle and fat layers until the uterus is felt. A small cut is made in the wall, which is widened by hand until the hole is big enough to deliver the baby.

2 Inserting surgical retractor
This metal instrument is hooked around the edge of the abdominal wound and pulled gently back. This ensures the surgeon has a clear view of the internal organs and tissues.

3 Delivering the baby
The baby is gently lifted out through the openings in the uterine and abdominal walls and passed to the pediatrician or midwife. The baby is handed to the mother as soon as possible.

Surgical retractor

Pubic bone

Compressed bladder

Cervix

Vagina

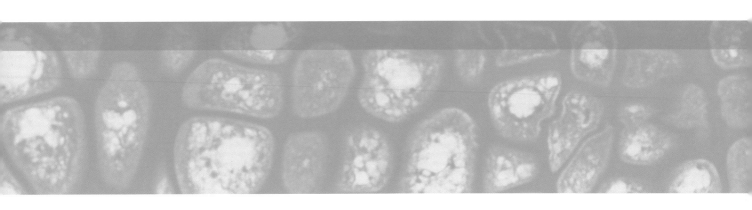

THE SPECIAL FEATURES A FETUS DEVELOPS TO COEXIST
WITH THE MOTHER DURING PREGNANCY CHANGE SOON
AFTER BIRTH TO ENABLE THE BABY TO LIVE INDEPENDENTLY.
A NEWBORN BABY ACQUIRES SKILLS VERY QUICKLY, IN
RESPONSE TO THE MANY STIMULI SURROUNDING IT. THESE
ACCOMPLISHMENTS EVOLVE IN A RECOGNIZED PATTERN,
WITH THE FIRST DEVELOPMENTAL BUILDING BLOCKS LAID
DOWN WITHIN DAYS OF BIRTH. THE KEY SKILLS A BABY
ACQUIRES ARE KNOWN AS THE DEVELOPMENTAL
MILESTONES. THESE, ALONG WITH OTHER FACTORS SUCH
AS WEIGHT AND HEAD CIRCUMFERENCE, ARE CAREFULLY
NOTED BY HEALTHCARE PROFESSIONALS AND SEEN AS
A MEASURE OF GOOD HEALTH AND WELL-BEING.

POSTNATAL DEVELOPMENT

RECOVERY AND FEEDING

The weeks after birth bring great emotional and physical changes for the mother, not least the onset of feeding. Hormonal changes, the responsibilities of becoming a parent, and a severe lack of sleep all add to the impact.

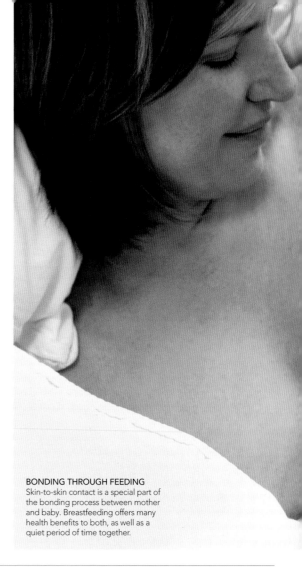

THE RECOVERING MOTHER

The few first weeks with a new baby are wonderful but exhausting, especially as the mother undergoes various physical changes. The enlarged uterus and loose abdominal muscles can make the abdomen continue to appear pregnant, and there may be cramping, similar to contractions, as the uterus shrinks. Bleeding occurs for the first two to six weeks; the discharge is initially bright red, turning to pink and then brown. An episiotomy scar (see p.202) will be sore initially, and urinating may be uncomfortable. Constipation is also a common problem. During the early days of breastfeeding, the breasts can be sore and engorged; any nipple tenderness will improve if the baby latches on well (see below). All these problems should be resolved with time.

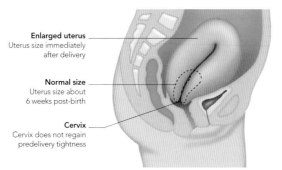

Enlarged uterus
Uterus size immediately after delivery

Normal size
Uterus size about 6 weeks post-birth

Cervix
Cervix does not regain predelivery tightness

SHRINKING UTERUS
By six weeks after the birth, the uterus has almost returned to its size before pregnancy. Breastfeeding can help with this process due to the production of oxytocin (see opposite), which stimulates muscle contractions.

BONDING THROUGH FEEDING
Skin-to-skin contact is a special part of the bonding process between mother and baby. Breastfeeding offers many health benefits to both, as well as a quiet period of time together.

KEGEL EXERCISES

Strengthening the pelvic floor—the slinglike muscles that support the bladder, bowel, and uterus (see p.91) —is as important after birth as it is during pregnancy. It can help with bladder control and make urine leakage less likely to occur. The muscles can be located by imagining trying to stop the flow when urinating. The muscles can be squeezed repeatedly, several times a day, either pulsing or holding for several seconds. These exercises should be built up over time.

EXERCISING WITH YOUR BABY
Kegel exercises can be incorporated into a daily routine, perhaps when the baby is sleeping. A few minutes spent strengthening these muscles will pay dividends later.

EMOTIONS

Most women find they experience a wide range of emotions in the days following the birth of a baby—from absolute elation to feeling down and tearful. The ups and downs experienced are understandable given the great hormonal changes that have taken place and the sleep deprivation that is almost universal when parenting a newborn baby. Feelings of exhilaration and achievement are common following the delivery, but these may soon give way to sadness. The so-called baby blues are common and usually resolve with time. However, if the feelings of sadness and not coping persist, postpartum depression may be the cause, and this requires specialist help (see p.243).

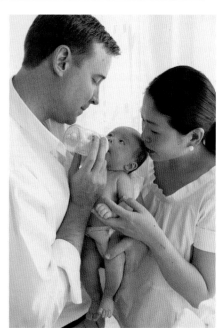

BONDING
Getting the father involved early is important, not just to ease physical and emotional pressure on the mother, but also so he can develop his own bond with the baby.

BREASTFEEDING

Breast milk is often considered to be the ideal food for a baby because it contains all the nutrients needed for early growth and development, and also provides antibodies that help fight many diseases, such as gastroenteritis and pneumonia. This lowers the risk of illness during the first year of life. Production and release of breast milk rely on two hormones that are produced by the pituitary gland in the brain: prolactin stimulates milk production; and oxytocin initiates milk ejection or "let down." Initially, the breasts produce a thick substance called colostrum (see below), which is then superseded by the mature milk after a couple of days. At each feeding, both breasts produce thirst-quenching "foremilk" followed by nutrient-rich "hindmilk."

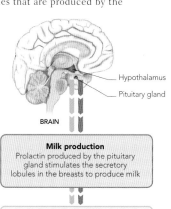

Hypothalamus

Pituitary gland

BRAIN

Milk production
Prolactin produced by the pituitary gland stimulates the secretory lobules in the breasts to produce milk

Milk release
The pituitary gland also releases oxytocin, which contracts the smooth muscle of the secretory lobules, forcing milk into the lactiferous ducts to the nipple

KEY

▮ PROLACTIN RELEASE

▮ OXYTOCIN RELEASE

THE LETDOWN REFLEX
The squeezing of milk out of the breast is stimulated by oxytocin. There may be temporary pain or tingling. At first, let down is triggered by suckling, but once breastfeeding is established other triggers, such as hearing the baby cry, can cause hormone release.

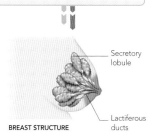

Secretory lobule

BREAST STRUCTURE

Lactiferous ducts

LATCHING ON

Latching on occurs when a baby correctly positions its mouth on the breast and can suckle effectively. This does not always happen naturally and can be painful if done incorrectly. The breast should be well inside the mouth with the nipple near the back and most of the areola (the dark area around the nipple) also in the mouth. The baby moves its jaw up and down, and tongue movements cause release of milk. This position avoids the nipple being pulled or pinched and becoming sore, and maximizes the amount of milk taken in.

Lactose	Energy	Sodium
5.3g	55 kcal	48mg
Fat		Calcium
2.9g		28mg
Protein		Vitamins
2.0g		189mmg

COLOSTRUM (100ML)

Lactose	Energy	Sodium
7.0g	67 kcal	15mg
Fat		Calcium
4.2g		30mg
Protein		Vitamins
1.1g		134mmg

BREAST MILK (100ML)

THE CHANGING COMPOSITION OF BREAST MILK
Colostrum and breast milk differ in their composition. Also known as "first milk," colostrum is rich in the antibodies that help the baby's immature immune system fight infection. Colostrum is also very rich in vitamins.

1 STIMULATING THE REFLEX
Brushing the baby's lip against the nipple encourages the mouth to open and accept the nipple. The baby's head can be cradled in the hand and guided into position.

2 CORRECT POSITION
Once the mouth opens to its widest, insert the nipple and areola, positioning them deep in the mouth, while continuing to support the baby's head.

BOTTLE FEEDING

Not all mothers want to breastfeed and some are unable to for health or other reasons. Formula is intended to replicate breast milk as closely as possible: it is made from cows' milk that is fortified with extra minerals and vitamins. It is important that a mother is not made to feel guilty if she feeds her baby formula. Bottle feeding still gives the opportunity for bonding while providing the key nutrients a new baby needs. It also allows the father to spend extra time with the baby and to give feedings during the night, allowing the mother more opportunity for sleep without the need to express milk.

THE NEWBORN BABY

A healthy newborn baby has the same complement of organs and tissues as an adult, but these change and mature as it develops. Over the first six weeks of life, the baby's appearance will begin to change.

ANATOMY

A new baby weighs on average 7¾ lb (3.5 kg). Although well prepared for the world outside, a baby's organs and tissues will continue to change and develop until adulthood. Some are relatively large in the newborn, reflecting the crucial roles they play during pregnancy and early childhood. For example, the large thymus gland in the chest is vital for developing early immunity; later in childhood, when it is no longer needed, it starts to shrink. Changes in the circulation take place at birth, triggered by the baby's first breath, which causes the lungs to start working and allows the baby to breathe independently (see p.201). Some features of a newborn's appearance, such as a conical-shaped head, reflect what has happened during the birth and will resolve with time.

EYES
A newborn's eyelids tend to look puffy as a result of pressure in the birth canal. Early vision is poor, and the eyes can appear crossed due to underdeveloped muscles.

SKULL AND BRAIN
The skull is made up of bony plates, which meet at seams (sutures), and two soft spots called fontanelles. These allow the bones to slide over each other so that the skull can change shape as it moves through the birth canal. This accounts for the temporary cone-shaped appearance of some newborn heads. Later, the fontanelles will close: the back (posterior) one by about six weeks and the front (anterior) one by 18 months.

NEURAL DEVELOPMENT
This CT scan of a newborn baby's brain shows large areas of developing neural networks (green). Multiple connections between the nerve cells of the brain are laid down from the moment of birth.

EARLY BODY STRUCTURE
The relative size and composition of a baby's anatomy change with time. Bodily structures are monitored during the early weeks to check that the baby is developing normally.

Wrist
The carpal bones of the wrist are largely made up of cartilage.

Heart
Blood pumps from the heart to the lungs for the first time.

Jaw
Fully formed teeth are present within the jawbone.

Lungs
The first breath draws air into the lungs and enables them to function.

Eye socket

Frontal bone

Trachea

Anterior fontanelle

Parietal bone

Posterior fontanelle

Occipital bone

Ear

Neck
Undeveloped muscles cannot support the large, heavy head in the first few weeks.

Thymus gland
Oversized at birth, this gland plays a key role in developing a functioning immune system.

Ribcage

Fingernails
A newborn's nails grow quickly and can be sharp.

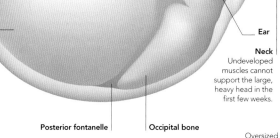

Cerebral hemispheres

Developing neural networks

Fluid-filled brain ventricles

TEMPERATURE CONTROL

Newborn babies do not have a fully developed ability to regulate body temperature. Because of the large surface area of the skin relative to body weight, a new baby loses heat easily and cannot shiver to generate body temperature. Babies can cool down by sweating and through the dilation of blood vessels in the skin. Their ability to lose heat is not as effective as an adult's so a baby must not be allowed to overheat.

SWADDLING
This is a technique of wrapping up a baby to create a sense of security while making sure to avoid overheating.

POSTNATAL CHECK-UPS

Every baby is checked soon after birth and again about six weeks later. This consists of a check of the external anatomy, such as the hands and feet, as well as the heart, lungs, hips, and other internal structures. The physician checks in the mouth for a cleft palate and shines a light in the eyes, as well as listens to the chest. The back is examined to look for spinal problems, and the legs are moved to assess the stability of the hips. The scrotum is checked in boys to make sure both testes are present. The doctor also looks for any birthmarks. As well as a physical examination, newborn babies are given a heel-prick blood test to check for various metabolic disorders (see p.237).

LISTENING TO THE HEART
Soon after birth, the physician checks the baby for any unusual heart sounds, known as murmurs. These may be normal or may indicate a problem.

Liver
Relatively large at birth, the liver is the site of new blood-cell formation in the fetus.

Stomach **Small intestine** **Large intestine** **Rectum**

GENITALS AND BREAST TISSUE
In both boys and girls, the genitals may appear enlarged, swollen, and dark in color due to the high levels of female hormones in the mother before delivery. These hormones pass from mother to baby across the placenta. One or both breasts may be enlarged, and a small amount of fluid may leak from the nipple soon after birth. Girls may also have a vaginal discharge that sometimes contains a small amount of blood.

Genitals

Pelvis

Feet
Newborn babies often lie with their feet turned outward.

Gallbladder **Appendix**

UMBILICAL CORD
A plastic clip is left on the cut cord until it dries and seals, stemming all blood flow (see p.200). The stump blackens and falls off within 10 days.

Hip
If the femur does not sit securely in the pelvic socket, the hips may be unstable.

Bones
Some bones will fuse as they mature.

Cartilage
Cartilage at the ends of long bones enables bones to lengthen, before gradually ossifying.

FLAKING SKIN
A newborn baby's skin can appear flaky, and this may persist for a few days or even weeks. Postmature babies may also have slightly dry and wrinkly skin.

209

EARLY RESPONSES AND PROGRESS

On average, a baby spends more than half the day asleep. Despite this apparent inactivity, the first weeks of life are extremely eventful—growth is rapid, and skills are acquired on a daily basis.

GROWTH

A baby grows at an incredible rate during the early weeks and months, while at the same time, organs are developing and maturing. This rapid growth needs to be fueled by frequent feedings and it also relies on periods of inactivity, when the baby is sleeping. Growth and weight are carefully monitored because they are key indicators of health and development. Percentile charts are widely used to record changes in size over time. If measurements are plotted regularly they will show whether a child falls within the average range and if growth is occurring at a steady rate. Growth that tails off may indicate that there is an underlying health problem.

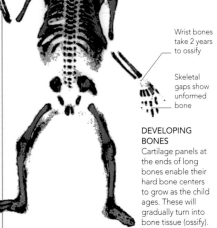

Wrist bones take 2 years to ossify

Skeletal gaps show unformed bone

KEY
— 99.6TH CENTILE
— 75TH CENTILE
— 50TH CENTILE
— 25TH CENTILE
— 0.4TH CENTILE
▨ FULL CENTILE RANGE

DEVELOPING BONES
Cartilage panels at the ends of long bones enable their hard bone centers to grow as the child ages. These will gradually turn into bone tissue (ossify).

GROWTH CHARTS
Measurements that fall between the top and lowest centiles are considered average. These charts show rates of growth for girls—separate charts are used for boys, who grow at different speeds.

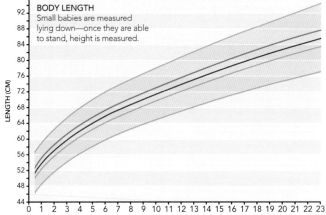

HEAD CIRCUMFERENCE
Measurements are taken from the widest part of the skull.

Top centile

Middle centile
Corresponds to average head size (or length or weight)

Lowest centile

HEAD CIRCUMFERENCE (CM) / AGE (MONTHS)

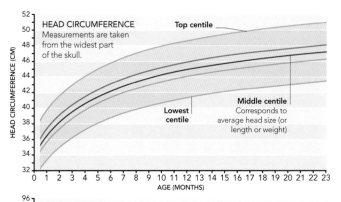

BODY LENGTH
Small babies are measured lying down—once they are able to stand, height is measured.

LENGTH (CM) / AGE (MONTHS)

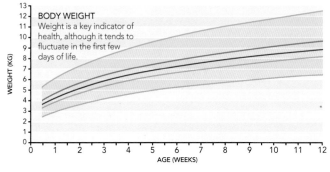

BODY WEIGHT
Weight is a key indicator of health, although it tends to fluctuate in the first few days of life.

WEIGHT (KG) / AGE (WEEKS)

PRIMITIVE REFLEXES

Various reflex reactions in response to specific stimuli are expected to appear and then disappear at particular stages in infant development. Their presence is an indicator that the neurological system is functioning and developing well. Physicians look for these reflexes at routine early checkups, and often they can be observed in a baby's daily activities. The rooting reflex is used to latch a baby onto the nipple when feeding (see p.205).

STARTLE REFLEX
A baby will fling out its arms in shock if its head suddenly falls back. Present for three months, this is also called the Moro reflex.

STEPPING REFLEX
If held upright on a firm surface, a baby will take steps as if walking. This is present for the first six weeks.

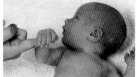

GRASP REFLEX
For about the first three months, a baby will close its hand into a fist if an object is placed in the palm.

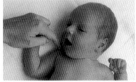

ROOTING REFLEX
If a baby is touched near the mouth, its head will turn to the stimulus. This usually disappears by four months.

SLEEPING AND WAKING

A baby's day is punctuated by frequent naps—on average six or seven a day in the newborn—and periods of wakefulness, when he or she becomes increasingly responsive. Because of their small stomachs and an almost constant need for food, babies usually wake up every two to four hours. It can be one or two years—or for a small number, even longer—before a baby sleeps consistently through the night. However, by about six weeks a baby's 24-hour clock is established and longer periods of sleep can take place during the night.

SLEEP DEVELOPMENT
Within a few weeks, a baby may sleep for up to five hours at a time, reflecting the gradual increase in stomach size.

THE EFFECTS OF MELATONIN

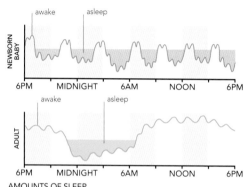

This hormone, secreted by the pineal gland in the brain, regulates other hormones and helps maintain the body's sleep–wake rhythm. High levels of melatonin are associated with an increased need to sleep. Maternal melatonin passes to the fetus via the placenta, and to a newborn baby via breast milk. Raised levels of melatonin are believed to help a baby sleep.

Pineal gland

LOCATION OF PINEAL GLAND

🕐 25 minutes

"ACTIVE SLEEP" (REM)
This stage of sleep involves high brain activity, which is believed to aid nerve-tissue development. Newborns spend 50 percent of their sleeping hours in REM sleep, double the amount in adults. During active sleep, the baby's eyes move rapidly back and forth, and the baby is restless and easily awakened.

🕐 25 minutes

"QUIET SLEEP" (NON-REM)
Quiet sleep has two key stages: light and deep sleep. Babies pass from light to deep sleep and back again, before moving into REM.

Light sleep
Brain activity slows as the baby falls asleep. The baby may be twitchy and respond to light and noise.

THE SLEEP–WAKE CYCLE
A newborn has a roughly 50-minute sleep cycle, made up of quiet and active sleep. This latter stage is when significant nerve development is thought to occur.

AWAKENS
During the transition from deep sleep back to light sleep, the baby is most liable to wake up and break the cycle.

Deep sleep
Brain activity is at its lowest. The baby is quiet, motionless, and at the most difficult stage to rouse.

AMOUNTS OF SLEEP
A newborn baby spends on average 16 hours asleep each day (this can range from 12 to 20 hours). The average adult requires half this amount of sleep.

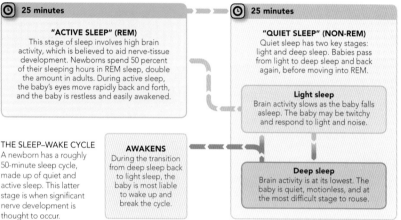

EXPRESSING FEELINGS
It is possible to recognize a baby's needs by the nature of its crying. A different cry is used to denote pain from hunger, and this forms a language to which parents become attuned.

FIRST SMILES
The timing of the first genuine smile can vary, but most babies are thought to smile for the first time after about four to six weeks. This is usually in response to the sight of their parents' faces or the sound of their voices. Before this, babies can pull facial expressions that resemble smiles but are often in response to wind or fatigue.

EARLY COMMUNICATION

Babies communicate with other people from birth—indeed their very survival depends on their ability to express their needs—and they do this in a variety of ways, primarily through crying. Babies instinctively cry to show hunger, distress, discomfort, pain, and also loneliness, and mothers become finely tuned to the sound of their own baby's cry and what it means. A combination of other sounds develop after about two weeks: first squealing; then gurgling and cooing. Parents quickly come to have an understanding of their babies' feelings without a word ever being spoken.

GENUINE RESPONSE
A baby's first true smile is an amazing event, involving a reflex response from both the eyes as well as the mouth.

THE SENSES

Babies are highly responsive to sound from birth, as demonstrated by the way a new baby is startled by loud noises and within a few weeks starts to turn toward voices. Parents are offered a screening audio test for their baby within a few weeks of birth. Vision, however, is thought to be relatively poorly developed at birth, newborn babies seeing best at a distance of about 8–10 in (20–25 cm).

CONTRASTING PATTERNS
Poor vision means that young babies respond most to primary contrasting colors, or black and white, and geometric shapes.

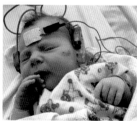

AUDIO TESTING
If the basic audio test detects a problem, this more complex test gauges a baby's response to clicks through headphones.

THE FIRST TWO YEARS

The early part of a child's life is a time of remarkable physical and developmental change. The complex nerve networks in the brain enable great achievements, such as sitting, standing, first steps, and first words. Even at this early stage, a child is clearly an individual, able to communicate needs and wishes.

PHYSICAL CHANGES

In addition to the reduction in head size relative to the rest of the body (see right), the first two years see a child's appearance change in other ways—the limbs and trunk lose some of their baby fat, reflecting increased movement and growth, the hair thickens and grows, and the face takes on a more mature appearance. This is a result of the eruption of many of the milk teeth and the loss of some of the subcutaneous fat around the cheeks and chin.

CHANGING PROPORTIONS
At birth, the head is as wide as the shoulders and makes up about one-quarter of body length; by two years, head size has reduced relative to body size.

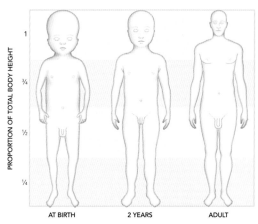

AT BIRTH   2 YEARS   ADULT

PROPORTION OF TOTAL BODY HEIGHT

TEETHING

Milk teeth usually first emerge at six to eight months and continue erupting until almost three years. Adult teeth begin to appear at about six years. Opinions vary as to whether teething can cause symptoms, such as fever; many experts believe they just happen to occur together. However, teething may cause swollen gums, drooling, and sleep problems.

MILK TEETH
The milk teeth erupt through the gums in a recognized sequence, with the two lower central incisors usually appearing first and then the upper central incisors.

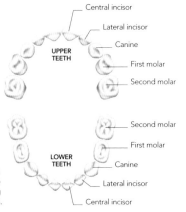

UPPER TEETH
- Central incisor
- Lateral incisor
- Canine
- First molar
- Second molar

LOWER TEETH
- Second molar
- First molar
- Canine
- Lateral incisor
- Central incisor

APPEARANCE OF UPPER TEETH	
TOOTH	**TIME OF ERUPTION**
Central incisor	8–12 months
Lateral incisor	9–13 months
Canine	16–22 months
First molar	13–19 months
Second molar	25–33 months

APPEARANCE OF LOWER TEETH	
TOOTH	**TIME OF ERUPTION**
Central incisor	6–10 months
Lateral incisor	10–16 months
Canine	17–23 months
First molar	14–18 months
Second molar	23–31 months

WEANING

Introducing solids into a baby's diet while reducing milk intake is called weaning. Its timing varies, but the general advice is that solids can be started from six months—before this time, the digestive system is still developing. Many parents give babies puréed or mashed foods for a few weeks before introducing finger food—small pieces of food that a baby will be able to pick up and eat. Breastmilk or formula usually remains the main source of nutrients for the first year.

FIRST SOLIDS
Simple purées of vegetables and fruit are often the first foods. Finger foods encourage a baby to feed independently.

DEVELOPING BRAIN FUNCTION

The newborn brain is made up of billions of nerve cells (neurons) that send and receive messages along nerve fibres. Almost a complete set of neurons is present but they have limited links. In the early years, multiple new connections are formed, as the senses encounter new stimuli and the body responds. Brain development occurs at its fastest rate in the first six years of life, during which time the brain almost reaches its full size.

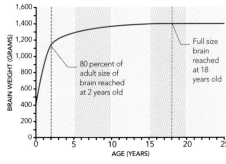

80 percent of adult size of brain reached at 2 years old

Full size brain reached at 18 years old

BRAIN WEIGHT (GRAMS)

AGE (YEARS)

INCREASING BRAIN SIZE
The brain's rapid development, as measured by its weight, can be seen on this graph. At birth, the brain weighs about 14 oz (400 g); by 2 years, it has reached 80 percent of its final adult weight of 3 lbs (1,400 g).

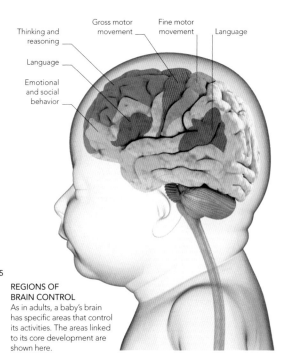

- Thinking and reasoning
- Gross motor movement
- Fine motor movement
- Language
- Language
- Emotional and social behavior

REGIONS OF BRAIN CONTROL
As in adults, a baby's brain has specific areas that control its activities. The areas linked to its core development are shown here.

MOVEMENT AND COORDINATION

Holding up the head or controlling its side-to-side and back-to-front movements is not possible for a newborn baby, so the head needs support at all times. This need lessens after a few weeks, and gradually complete head control is achieved. This fundamental skill, along with the control of body posture, form the basis for all movement skills. There is a specific sequence: a baby learns weight bearing and then balance.

Many attempts are needed before independent walking is possible—it rarely starts before 10 months. Movements become more complex as a range of actions are coordinated at the same time.

INDEPENDENT MOVEMENT
Crawling starts at around the age of seven months followed by walking with support. Some babies shuffle on their bottoms as a first means of getting around.

LANGUAGE AND COMMUNICATION

Babies use both verbal and nonverbal means to communicate how they feel and what they need. Crying is an instinctive way of communicating (see p.211), but the early gurglings of the first few weeks gradually give way to articulated sounds as the baby gains awareness and experiments with its voice. Hand gestures, such as pushing something unwanted away, also become a key means of communication. At around six months, babbled conversations begin, and by one year the baby should say recognizable words, such as "mama" and "dada," and enjoy repeating familiar sounds.

BABY SIGNING
From as young as six months, babies can learn a simple sign language to convey what they want. In this image, a mother teaches her baby to sign for "more."

DEVELOPMENTAL MILESTONES

The core skills that are achieved throughout early childhood are known as developmental milestones. These can broadly be divided into three categories: physical achievements; reasoning and communication skills; and emotional development combined with social skills. The milestones usually occur in a specific order and, for most children, within a certain age range. However, some children achieve particular milestones earlier or later, and some skills are skipped altogether. Developmental milestones form the basis for acquiring more complex abilities later. By the age of two years, children have gained an impressive degree of independence, and the ability to walk allows them to express the innate desire to explore the surrounding world.

AGE (MONTHS)

| 0 | 2 | 4 | 6 | 8 | 10 | 12 | 14 | 16 | 18 | 20 | 22 | 24 |

PHYSICAL ABILITIES
Control of posture, balance, and movement are the vital early motor skills. A baby first learns head control, and eventually can sit. Once neural connections for these skills are laid down, crawling, standing, and walking are possible.

- Lifts head and chest
- Brings hand to mouth
- Grasps objects with hands

- Crawls
- Walks holding furniture
- Bangs objects together
- Eats finger foods unaided

- Walks unaided
- Carries or pulls toys
- Starts to run
- Can throw and kick a ball
- Walks up stairs unaided
- Can hold and use a pencil
- Gains control of bowels

- Reaches for objects
- Rolls over
- Supports own weight on feet
- Sits unsupported
- Stands by hoisting up own weight

- Crawls up stairs
- Squats to pick up objects
- Jumps with both feet
- Starts to drink from a cup

THINKING AND LANGUAGE SKILLS
Successful communication relies on an understanding of language. Imitating sounds made by parents is the first step toward gaining language and higher skills such as thinking, reasoning, and logic.

- Smiles at parents' voices
- Starts to imitate sounds

- Recognizes own name
- Responds to simple commands
- Uses first words
- Imitates behavior

- Points to named objects
- Sorts shapes and colors
- Says simple phrases
- Follows simple instructions
- Engages in fantasy play

- Begins to babble
- Investigates with hands and mouth
- Reaches for out-of-reach objects
- Understands "no," "up" and "down"

- Says "dada" and "mama" to parents
- Can put two words together

SOCIAL AND EMOTIONAL DEVELOPMENT
Social interaction begins with watching people and smiling. Play helps build social skills, so by one year most babies interact happily with others; gaining independence and an understanding of social behavior are also key.

- Makes eye contact
- Recognizes familiar people
- Cries when needing attention
- Smiles at mother, then socially
- Watches faces intently
- Recognizes parents' voices

- Cries when parent leaves
- Shows preferences for people and objects
- Repeats sounds and gestures

- Responds to own name
- Plays peekaboo

- Imitates others' behavior
- Enjoys company of other children
- Demonstrates defiant behavior
- Stays dry during the day

A WIDE RANGE OF CONDITIONS CAN AFFECT THE HUMAN
REPRODUCTIVE SYSTEM. SOME MAY AFFECT FERTILITY, WHILE
OTHERS ARE EXCLUSIVE TO PREGNANCY OR BIRTH. BABIES
MAY ALSO HAVE A VARIETY OF MEDICAL CONDITIONS,
SOME ARISING FROM PROBLEMS IN DEVELOPMENT DURING
EARLY PREGNANCY, OTHERS RELATING TO EVENTS LATER IN
PREGNANCY OR DURING BIRTH. BECAUSE OF THERAPEUTIC
ADVANCES AND AN INCREASED UNDERSTANDING OF THE
WAY IN WHICH CONDITIONS DEVELOP, MANY CAN BE
TREATED SUCCESSFULLY, PRODUCING A HEALTHY OUTCOME
FOR MOTHER AND BABY. FERTILITY TREATMENTS HAVE
PERHAPS SEEN THE GREATEST IMPROVEMENTS, OFFERING
HOPE TO THOUSANDS OF CHILDLESS COUPLES.

DISORDERS

FERTILITY DISORDERS

INFERTILITY IS A COMMON PROBLEM, AFFECTING AS MANY AS 1 IN 10 COUPLES WHO WISH TO CONCEIVE.
THE PROBLEM MAY LIE WITH EITHER THE MALE OR THE FEMALE PARTNER, OR THERE MAY BE A COMBINATION
OF FACTORS INVOLVED. ASSISTED CONCEPTION TECHNIQUES NOW OFFER HOPE TO MANY INFERTILE COUPLES.

FEMALE FERTILITY DISORDERS

In around half of all instances of couples experiencing fertility problems, the problem lies with the woman. The underlying causes of infertility can be broadly divided into problems relating to egg production, egg transportation toward the uterus, the egg meeting the sperm, and conditions that prevent the fertilized egg either from implanting or growing in the uterus. Age is also a key factor in women because fertility reaches its peak by the age of about 27 and then falls, gradually at first, then more rapidly from the age of 35.

DAMAGED FALLOPIAN TUBES

WHEN A FALLOPIAN TUBE IS DAMAGED, OFTEN AS A RESULT OF AN INFECTION, THE MONTHLY JOURNEY OF AN EGG TO THE UTERUS MAY BE PREVENTED.

One or both tubes may be damaged due to an infection of the pelvic organs (see p.218). Endometriosis (see p.218) may also affect the fallopian tubes. Tubes may be assessed by keyhole surgery (laparoscopy) or hysterosalpingography, in which dye is injected through the cervix and its progress through the uterus and tubes is tracked on X-rays. Microsurgery may be suitable for some tubal damage, or drug treatment in cases of endometriosis. Otherwise, assisted conception may be considered.

BLOCKAGE IN A FALLOPIAN TUBE
This hysterosalpingogram reveals that the right tube (on the left, as shown) is blocked next to the uterus, and the left tube is abnormal and enlarged.

ABNORMALITIES OF THE UTERUS

PROBLEMS WITHIN THE UTERUS CAN PREVENT A FERTILIZED EGG FROM EITHER IMPLANTING OR DEVELOPING NORMALLY.

The lining of the uterus may be damaged by an infection, or hormonal factors may result in a failure to prepare the lining for pregnancy during the menstrual cycle. Fibroids (see p.219) or an abnormally shaped uterus (see p.221) may prevent normal fetal growth. Hysteroscopy (in which a viewing instrument is passed into the uterus via the cervix) or an ultrasound scan may be used to check the uterus. Causes are treated if possible; for example, large fibroids may be removed.

OVULATION PROBLEMS

FAILURE OF THE OVARIES TO RELEASE A MATURE EGG EVERY MONTH IS A COMMON CAUSE OF INFERTILITY. IT HAS VARIOUS POSSIBLE CAUSES.

The release of eggs from the ovaries is controlled by a complex system of hormones from the hypothalamus, pituitary gland, and ovaries, which work in harmony to maintain the system. Problems can arise if the system is interrupted. Polycystic ovarian syndrome (see p.219) is a common cause. Others include noncancerous pituitary gland tumors and thyroid problems (thyroid hormones are also important for fertility). Excessive exercise, obesity, being extremely underweight, and stress may result in a hormone imbalance. Early menopause may also cause a failure to ovulate. Blood tests can check hormone levels, and an ultrasound scan can check the ovaries. Causes are treated when possible, but sometimes no cause is found. Drugs may be given to stimulate ovulation or, in some cases, assisted conception is considered.

EXCESSIVE EXERCISE
The delicate balance of hormones that leads to the monthly release of a mature egg from the ovaries can be disrupted by frequent strenuous exercise.

CERVICAL PROBLEMS

VARIOUS FACTORS CAN AFFECT CERVICAL-MUCUS PRODUCTION AND PREVENT A SPERM'S NORMAL PASSAGE INTO THE UTERUS VIA THE CERVIX.

To meet a mature egg, a sperm must first pass through the cervix. The mucus produced by the cervix acts as a temporary repository and transport medium for the sperm. For a variety of reasons (see panel, below), mucus can become hostile to normal sperm, or the amount produced or its consistency may change. If cervical mucus antibodies are the suspected cause, a sample of mucus is analyzed shortly after intercourse. If antibodies are present, corticosteroid drugs may be given to suppress their production or intrauterine insemination may be used to introduce the sperm directly into the uterus. Other underlying causes, such as medications, may be dealt with as appropriate.

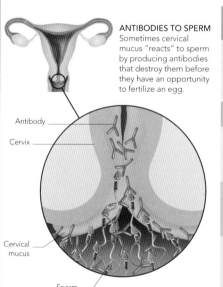

ANTIBODIES TO SPERM
Sometimes cervical mucus "reacts" to sperm by producing antibodies that destroy them before they have an opportunity to fertilize an egg.

Antibody

Cervix

Cervical mucus

Sperm

FACTORS THAT AFFECT CERVICAL MUCUS

There are various conditions that can affect cervical mucus, making it hostile toward sperm in some way, reducing the amount produced, or having a damaging effect on its quality (see p.41).

DRUGS THAT AFFECT MUCUS	HEALTH CONDITIONS THAT AFFECT MUCUS
Clomifene citrate, used in the treatment of infertility, is a common cause of hostile mucus.	Polycystic ovarian syndrome (see p.219) can be associated with poor cervical-mucus production.
Antihistamines reduce the production of mucus (the sperm's transport medium through the cervix).	Infections, such as yeast or vaginosis (see p.220), can affect mucus production by the cervix.
Dicyclomine, used to treat irritable bowel syndrome, can also reduce cervical-mucus production.	Damage to the cervix, perhaps as a result of a biopsy, can affect its ability to produce mucus.

MALE FERTILITY DISORDERS

In about one-third of couples with infertility, the problem lies with the man. Male fertility problems can be divided into two main groups—those that affect sperm production and those in which sperm delivery is a problem. Difficulties delivering sperm can occur at any point within the complex system of tubes that carry sperm from the testes to the penis or may relate to ejaculation itself.

PROBLEMS WITH SPERM PRODUCTION

A SPERM COUNT MAY BE LOW, OR THE SPERM PRODUCED ARE ABNORMAL AND UNABLE TO FERTILIZE AN EGG. VARIOUS CAUSES ARE POSSIBLE BUT, OFTEN, NO CAUSE IS FOUND.

Factors that elevate the temperature within the scrotum, such as a varicocele (see p.222), can affect sperm production, as can long-term illnesses, damage to the testes, smoking, alcohol, and certain drugs. Testosterone-related problems may be the cause, for which, rarely, there is an underlying chromosomal abnormality. Such causes are considered during an examination and investigated with blood tests. Underlying causes are treated where possible; otherwise, assisted conception may be considered (see panel below).

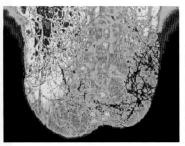

THERMOGRAM OF VARICOCELE
This thermogram shows the elevated temperature (red area) within a varicocele (enlarged vein in the testis) compared with the rest of the testes.

DIFFICULT PASSAGE OF SPERM

THE PASSAGE OF SPERM THROUGH THE COMPLEX SYSTEM OF TUBES WITHIN THE MALE REPRODUCTIVE SYSTEM MAY BE COMPROMISED FOR SEVERAL REASONS.

Damage to the sperm-transporting tubes (vasa deferentia and epididymides), perhaps due to a sexually transmitted infection, may affect the passage of sperm. It may be possible to treat the damage with microsurgery. Sometimes, following prostate surgery, the valves that prevent backflow of semen into the bladder at ejaculation (retrograde ejaculation) do not close properly. In this case, artificial insemination may be used to try to achieve conception.

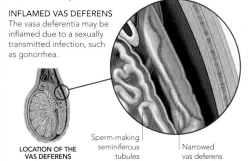

INFLAMED VAS DEFERENS
The vasa deferentia may be inflamed due to a sexually transmitted infection, such as gonorrhea.

LOCATION OF THE VAS DEFERENS

Sperm-making seminiferous tubules

Narrowed vas deferens

EJACULATION PROBLEMS

SOMETIMES, HEALTHY SPERM CANNOT BE DELIVERED TO THE TOP OF THE VAGINA, COMMONLY DUE TO ERECTILE DYSFUNCTION.

Erectile dysfunction (ED), or the inability to achieve or maintain an erection, is a common cause of male infertility; it may reflect emotional problems, such as anxiety or depression or, less commonly, medical conditions, such as a long-standing vascular disease that affects the blood supply to the penis or diabetes mellitus, if the nerve supply to the penis is impaired. Prescribed drugs, including some of those for treating high blood pressure, may also be involved in causing ED. In addition, heavy drinking of alcohol and smoking can be associated with it. Treatment aims to identify and resolve the underlying cause of the problem through psychological or medical treatments as appropriate, but if this is not possible, artificial insemination may provide the solution.

ASSISTING CONCEPTION

Since the birth of the first IVF baby in 1978, major advances have been made. Simple forms of treatment involve the use of fertility drugs, while more complex techniques can introduce sperm directly into the uterus at the time of ovulation (intrauterine insemination) or inject a single sperm into an egg. The use of donor eggs and sperm is now common, and surrogacy is also considered as an available option.

IVF
In vitro fertilization (IVF) can be used in many situations, including when fallopian tubes are damaged or no cause has been found for fertility problems. Fertility drugs are usually given to stimulate egg production and the resulting eggs are removed using a needle passed through the vaginal wall. The eggs are combined with the sperm in a laboratory, and the fertilized eggs may then be tested for genetic abnormalities. One or two healthy embryos are inserted into the uterus using a catheter (a thin tube) passed through the cervix. If the cycle is successful, one or both embryos will implant in the uterine wall.

ICSI
Intracytoplasmic sperm injection (ICSI) is used to help couples when the problem lies with the man and can be used as part

of an IVF program. Only one sperm is needed, which is injected straight into an egg taken from the ovary of the woman. Sperm can be taken either from a semen sample or directly from the epididymis or testes. As with IVF, fertility drugs can be given first and the best embryos are transplanted directly into the uterus via the cervix.

GIFT
Gamete intrafallopian transfer (GIFT) is similar to IVF except that eggs and sperm are transferred directly to the fallopian tubes, where fertilization takes place. Less commonly used, zygote intrafallopian transfer (ZIFT) involves the transfer of a zygote (newly fertilized egg) into the tubes. These methods are used when the sperm count is low, sperm motility is poor, or infertility is unexplained.

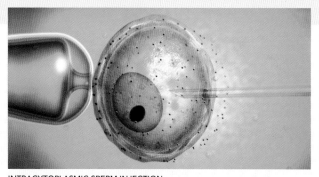

INTRACYTOPLASMIC SPERM INJECTION
In this microscopic technique, an egg is fertilized by using a needle (seen to the right in this image) to inject a single sperm cell directly into the egg.

IVF SUCCESS RATES
The success rates of IVF are related to a woman's age; these are greatest in women under the age of 35, then decrease gradually with increasing age.

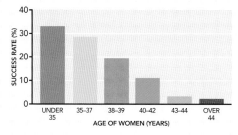

SUCCESS RATE (%)

AGE OF WOMEN (YEARS)
UNDER 35 · 35–37 · 38–39 · 40–42 · 43–44 · OVER 44

FEMALE REPRODUCTIVE DISORDERS

MANY CONDITIONS CAN IMPACT UPON THE COMPLEX SYSTEM OF FEMALE REPRODUCTIVE ORGANS, AFFECTING THE REPRODUCTIVE PROCESS AT ONE OF ITS DIFFERENT STAGES. FOR EXAMPLE, THE PRODUCTION OF EGGS MAY BE IMPAIRED, THE PASSAGE OF THE EGG ALONG THE FALLOPIAN TUBE MAY BE BLOCKED, OR A CONDITION AFFECTING THE UTERUS MAY PREVENT THE NORMAL IMPLANTATION OF A FERTILIZED EGG. MANY OF THESE CONDITIONS CAN BE TREATED, OR THE PROBLEM MAY BE BYPASSED BY ONE OF SEVERAL METHODS OF FERTILITY TREATMENTS NOW AVAILABLE.

ENDOMETRIOSIS

IN THIS COMMON DISORDER, PIECES OF TISSUE THAT USUALLY LINE THE UTERUS ARE FOUND ELSEWHERE IN THE PELVIS AND ABDOMEN, WHICH CAN CAUSE INFERTILITY.

The lining of the uterus (endometrium) thickens every month in preparation for pregnancy, then is shed if fertilization does not occur. Fragments of endometrium may become attached to other tissues and organs in the abdomen and pelvis, where they continue to respond to the hormonal changes of the menstrual cycle, bleeding and causing pain when menstruation occurs. Scar tissue eventually forms at these bleeding sites, and ovarian cysts may develop. The cause of endometriosis is not fully understood, but endometrial fragments may pass along the fallopian tubes and into the abdomen during menstruation.

This condition may reduce fertility in several ways; one is thought to be blockage of the fallopian tubes by scar tissue. Symptoms, if present, may include pain, heavy or irregular periods, pain on urinating, and painful intercourse. Endometriosis may be diagnosed by laparoscopy, in which the internal organs are examined using a viewing instrument passed through the abdominal wall. Treatment options include drugs, such as the combined oral contraceptive pill or other hormones that temporarily stop menstruation, or laser treatment of the lesions. Hysterectomy and removal of the ovaries and other affected tissue may be recommended for women who no longer wish to have children.

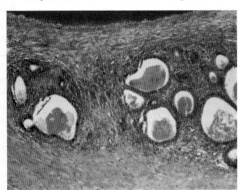

ENDOMETRIAL TISSUE IN THE VAGINA
This microscopic view shows the multiple "chocolate" cysts (named because of the color) of abnormal tissue typical of endometriosis. These cysts bleed during menstruation.

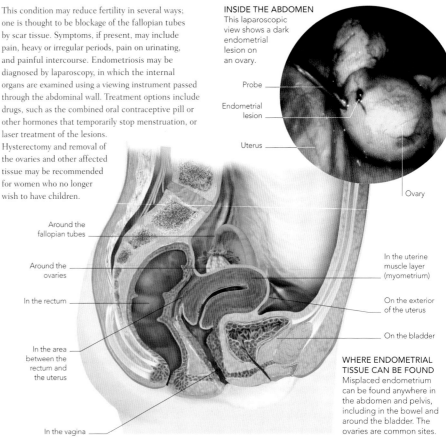

INSIDE THE ABDOMEN
This laparoscopic view shows a dark endometrial lesion on an ovary.

Probe

Endometrial lesion

Uterus

Ovary

Around the fallopian tubes

Around the ovaries

In the rectum

In the area between the rectum and the uterus

In the vagina

In the uterine muscle layer (myometrium)

On the exterior of the uterus

On the bladder

WHERE ENDOMETRIAL TISSUE CAN BE FOUND
Misplaced endometrium can be found anywhere in the abdomen and pelvis, including in the bowel and around the bladder. The ovaries are common sites.

PELVIC INFLAMMATORY DISEASE

INFLAMMATION OF PELVIC ORGANS, IN PARTICULAR THE FALLOPIAN TUBES, CAN BLOCK THE PASSAGE OF EGGS AND SPERM. SEXUALLY TRANSMITTED DISEASES, SUCH AS CHLAMYDIA, ARE COMMON CAUSES.

Pelvic inflammatory disease (PID) may be symptomless, discovered only when a woman is investigated for fertility problems. The infection starts in the vagina, and passes up into the uterus, fallopian tubes, and, sometimes, the ovaries. Having a coil (intrauterine device) may increase the risk of PID. Women with PID may have abnormal vaginal discharge, fever, pain during intercourse, and heavy or prolonged periods. Urgent treatment may be required if the condition comes on suddenly and is associated with severe pain and a high temperature. In addition

to increasing the risk of infertility, PID may make an ectopic pregnancy more likely. Investigations may include taking swabs from the cervix to look for infections, an ultrasound scan to check for swelling in the fallopian tubes, and a laparoscopy to look for inflammation. PID is treated with antibiotics.

X-RAY VIEW OF A PELVIS
A fallopian tube may become filled with pus in PID, making it swell up, as seen in this contrast X-ray taken following the introduction into the vagina of special dye via a catheter.

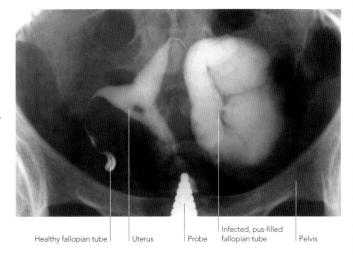

Healthy fallopian tube | Uterus | Probe | Infected, pus-filled fallopian tube | Pelvis

UTERINE FIBROIDS AND POLYPS

THESE ARE NONCANCEROUS GROWTHS WITHIN THE UTERUS AND CERVIX. FIBROIDS DEVELOP WITHIN THE MUSCULAR WALL, WHILE POLYPS PROTRUDE FROM THE INNER LINING. LARGE FIBROIDS CAN CAUSE FERTILITY PROBLEMS.

Fibroids are common and consist of muscle and fibrous tissue. The cause of their development is unknown, but may relate to the female sex hormone estrogen. As fibroids enlarge, they may begin to produce symptoms, such as painful, prolonged, and heavy periods. If large, they can distort the cavity within the uterus, causing recurrent miscarriages and affecting fertility if implantation becomes a problem. Fibroids can also cause a fetus to lie in an abnormal position. Polyps can cause bloodstained discharge, as well as bleeding after intercourse and between periods. They may be seen when looking at the cervix through a speculum (the instrument used to hold the vaginal walls apart) and may be removed at this time. Both polyps and fibroids may be diagnosed by ultrasound or hysteroscopy (in which a viewing instrument is passed up the vagina and through the cervix to look inside the uterus). Small fibroids and uterine polyps may be removed during hysteroscopy; larger fibroids can be removed through an abdominal incision. Hysterectomy may be considered for women who no longer wish to become pregnant.

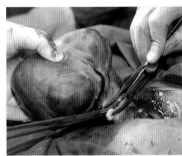

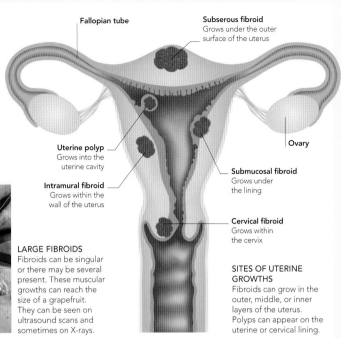

Fallopian tube

Subserous fibroid
Grows under the outer surface of the uterus

Uterine polyp
Grows into the uterine cavity

Intramural fibroid
Grows within the wall of the uterus

Ovary

Submucosal fibroid
Grows under the lining

Cervical fibroid
Grows within the cervix

LARGE FIBROIDS
Fibroids can be singular or there may be several present. These muscular growths can reach the size of a grapefruit. They can be seen on ultrasound scans and sometimes on X-rays.

SITES OF UTERINE GROWTHS
Fibroids can grow in the outer, middle, or inner layers of the uterus. Polyps can appear on the uterine or cervical lining.

OVARIAN CYSTS

THESE FLUID-FILLED GROWTHS CAN OCCUR SINGLY OR THERE MAY BE MANY. CYSTS DO NOT AFFECT FERTILITY, UNLESS THEY OCCUR AS A RESULT OF POLYCYSTIC OVARIAN SYNDROME.

There are many types of ovarian cyst. Some develop from the follicles in which eggs mature in the ovaries. Others form from the corpus luteum, which develops from the follicle after ovulation. Dermoid cysts contain tissues found in other parts of the body, such as skin. Ovarian cysts may occur singly or in multiples, as in polycystic ovarian syndrome (see right). They tend to be symptomless, but if symptoms do develop, they may include irregular periods, abdominal discomfort, and painful intercourse. Occasionally, cysts may cause an emergency situation if they burst or become twisted. Some types may become huge and fill most of the abdomen. Cysts may be diagnosed by ultrasound scan or by laparoscopy. They may resolve without treatment or may be removed surgically. Those that are removed are checked for cancerous cells because cysts are very occasionally malignant.

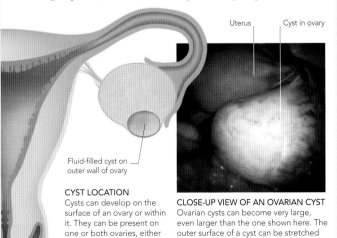

Uterus

Cyst in ovary

Fluid-filled cyst on outer wall of ovary

CYST LOCATION
Cysts can develop on the surface of an ovary or within it. They can be present on one or both ovaries, either singly or as multiples.

CLOSE-UP VIEW OF AN OVARIAN CYST
Ovarian cysts can become very large, even larger than the one shown here. The outer surface of a cyst can be stretched by the large amount of fluid inside.

POLYCYSTIC OVARIAN SYNDROME

THIS COMMON CONDITION, WHICH IS OFTEN ASSOCIATED WITH FERTILITY PROBLEMS RELATED TO AN IMBALANCE IN THE LEVELS OF THE SEX HORMONES, IS CHARACTERIZED BY MULTIPLE SMALL, FLUID-FILLED CYSTS PRESENT ON THE OVARIES.

In polycystic ovarian syndrome (PCOS), hormone levels are disrupted; often, the levels of testosterone and LH (luteinizing hormone), produced by the pituitary gland, are higher than normal, resulting, in some cases, in a failure to ovulate, associated with absent or irregular periods. Other features of PCOS include obesity, acne, and hirsutism. Affected women are more likely to develop resistance to the sugar-regulating hormone insulin and may develop diabetes mellitus. Diagnosis involves blood tests to check hormone levels and ultrasound scans to look for ovarian cysts. Drugs, in particular clomifene, may help restore fertility, and the combined oral contraceptive pill can be used if the aim is to restore regular periods.

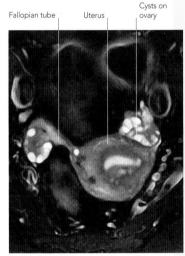

Fallopian tube   **Uterus**   **Cysts on ovary**

OVARIES WITH MULTIPLE CYSTS
This MRI scan of a uterus, fallopian tubes, and ovaries clearly shows multiple cysts (seen in white) on both ovaries, particularly the left.

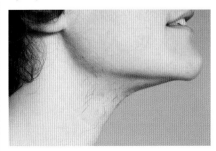

CLINICAL FEATURES
The hormonal imbalance of PCOS has some unwanted effects, including excessive body and facial hair. Acne may also be a problem.

VULVOVAGINITIS

THIS CONDITION IS AN INFLAMMATION OF THE VULVA AND VAGINA, WHICH MAY CAUSE DISCOMFORT, ITCHING, AND DISCHARGE. THE CAUSE IS USUALLY AN INFECTION.

Likely infective causes include *Candida albicans* (yeast), *Trichomonas vaginalis*, or an excessive quantity of the bacteria that normally inhabit the vagina (see bacterial vaginosis, below). Another cause may be irritants, such as substances found in laundry detergents. Swabs are taken and, if a bacterial cause is found, antibiotics are prescribed. Very rarely, cancerous cells are present, so in some cases, a tissue sample is taken to exclude cancer. Any possible irritants should be avoided. The condition usually clears with treatment, but may recur.

TRICHOMONAS VAGINALIS
This highly magnified image shows a parasitic microorganism that can cause vulvovaginitis.

Inflamed labia

AFFECTED GENITALIA
The inner surfaces of the labia, as well as the vaginal wall, become red and inflamed.

BACTERIAL VAGINOSIS

THIS CONDITION RESULTS FROM AN OVERGROWTH OF BACTERIA THAT ARE NORMALLY PRESENT IN THE VAGINA. ANTIBIOTICS MAY BE NEEDED TO TREAT THE CAUSE.

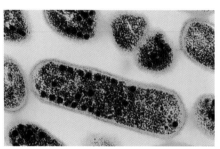

GARDNERELLA VAGINALIS
The bacterium seen in this electron micrograph can cause a watery vaginal discharge with a fishy odor.

There is a delicate balance of bacteria in the healthy vagina, the main ones being *Gardnerella vaginalis* and *Mycoplasma hominis*. If the balance is disrupted, symptoms such as discharge and itching around the vulva or vagina may develop, but bacterial vaginosis can be symptomless. The cause is unknown, but the presence of a sexually transmitted disease (STD) may disrupt the balance. Bacterial vaginosis may affect fertility by causing pelvic inflammatory disease (see p.218). Swabs are taken to look for an infection that might cause the condition and antibiotics are prescribed as appropriate. The woman's partner should be tested and treated if an STD is found.

BARTHOLINITIS

INFLAMMATION OF ONE OR BOTH OF THE SMALL GLANDS THAT OPEN INTO THE VULVA TO RELEASE A LUBRICANT DURING INTERCOURSE IS KNOWN AS BARTHOLINITIS.

The pea-sized Bartholin's glands open into either side of the vulval area, each via a tiny duct. These glands may become inflamed due to a bacterial infection. Poor hygiene or a sexually transmitted disease, such as gonorrhea, may be potential causes. The duct leading from a gland to the vulva may become blocked, causing a fluid-filled cyst (known as a Bartholin's cyst), or an abscess (a pus-filled swelling) may develop. Abscesses are caused by infection, most commonly due to the bacteria *Staphylococcus* or *E. coli*, and are very painful and require prompt treatment. Antibiotics are prescribed for bartholinitis, which should clear the condition. In the meantime, analgesics may be needed to relieve discomfort. Cysts are usually left alone, unless they become very large and cause problems. The pus may be drained from an abscess through a small cut made in its wall. The incision may be sutured open so that it heals in this position, preventing further abscesses from forming. Bartholinitis may recur.

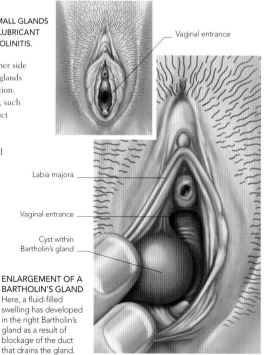

Vaginal entrance

Labia majora

Vaginal entrance

Cyst within Bartholin's gland

ENLARGEMENT OF A BARTHOLIN'S GLAND
Here, a fluid-filled swelling has developed in the right Bartholin's gland as a result of blockage of the duct that drains the gland.

MENSTRUAL PROBLEMS

The menstrual cycle and bleeding may be disrupted in a number of ways, some of which may cause problems when trying to conceive. Periods may be heavy, irregular, absent, or painful. Treatments will be available in many cases, either to relieve the symptoms or to treat the underlying cause. When conception is a problem, fertility treatments may be considered.

MENORRHAGIA

THIS TERM DESCRIBES HEAVY PERIODS THAT CANNOT BE CONTROLLED EFFECTIVELY BY SANITARY NAPKINS OR TAMPONS OR INVOLVE PASSING LARGE BLOOD CLOTS.

Excessively heavy periods can be prolonged and painful, and anemia may become a problem. Possible causes include fibroids, a uterine polyp (see p.219), and using a coil, although often no cause is found. Rarely, cancer is the cause. Blood will be tested for anemia. Other tests include ultrasound scanning of the uterus and hysteroscopy, in which a viewing instrument is passed into the uterus via the cervix. A sample of the lining may also be taken. Underlying causes are treated and, otherwise, drugs may be given to reduce bleeding.

THE ROLE OF PROGESTERONE
This highly magnified image shows crystals of progesterone. A drop in the levels of circulating progesterone is the trigger for menstrual bleeding.

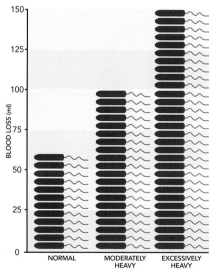

BLOOD LOSS DURING MENSTRUATION
This chart represents blood loss during one menstrual period. Normal blood loss is up to 60ml; moderately heavy is 60–100ml; and excessive loss is more than 100ml.

METRORRHAGIA

IRREGULAR PERIODS, WITH THE TIME BETWEEN EACH PERIOD VARYING, IS KNOWN AS METRORRHAGIA.

The most common cause of irregular periods is a disruption in the normal balance of the hormones that control the menstrual cycle. Such a disruption occurs quite naturally following pregnancy and childbirth. But hormonal imbalance may also occur during a long-term illness, and in times of stress and anxiety. Excessive exercise or dramatic weight loss may also affect hormonal balance. Irregular periods may be a feature of polycystic ovarian syndrome (see p.219). The onset of menopause is often signalled by irregular periods. However, for many affected women, there is no obvious cause. Often, irregular periods will settle back into a proper rhythm, but tests may be arranged to look for a cause, including blood tests to check hormone levels and an ultrasound scan of the uterus and ovaries. Underlying causes are treated as appropriate and, otherwise, drugs may be given to regulate periods, such as the combined oral contraceptive pill.

AMENORRHEA

WHEN THERE IS NO MENSTRUATION WHATSOEVER, THE CONDITION IS KNOWN AS AMENORRHEA.

Primary amenorrhea is the failure to start menstruation by the age of 16. It may be part of delayed puberty, and specialized investigations will look for a cause. Secondary amenorrhea is when periods stop for three months or more in a woman who has previously been menstruating and the cessation cannot be explained by the usual causes (for example, she is not breastfeeding, has not just had a baby, has not just stopped taking the combined oral contraceptive pill, or has not reached menopause). Disruption of the normal balance of the female sex hormones is usually the cause, often due to stress, excessive exercise, or weight loss; as part of polycystic ovarian syndrome (see p.219); or occasionally, due to a pituitary disorder, such as a tumor. Tests include blood tests to check hormone levels, an ultrasound scan of the uterus and ovaries, and a CT scan of the pituitary gland. Underlying causes will be treated and, if this is not possible, hormone treatments may be prescribed to trigger menstruation.

RIGOROUS ROUTINES
Frequent strenuous exercise can disrupt hormones, resulting in absent periods—a condition that notoriously affects ballerinas.

START OF PERIODS LINKED TO WEIGHT
Girls of normal weight tend to get their first period (menarche) around the age of 13 years. This age may vary for girls who are obese or overweight, and those who are underweight.

KEY
■ AVERAGE AGE OF MENARCHE
■ AGE OF MENARCHE FOR 50 PERCENT OF GIRLS
▨ FULL AGE RANGE FOR MENARCHE

(Chart: AGE OF MENARCHE (YEARS) on vertical axis, 9 to 15; horizontal axis categories: OBESE/OVERWEIGHT, NORMAL WEIGHT, UNDER-WEIGHT)

ABNORMALITIES OF THE UTERUS

An abnormally shaped uterus is present at birth, because it has not developed properly, but is often picked up only when a woman becomes pregnant or if she undergoes investigations for fertility problems. The shape can be abnormal in a number of ways (see diagrams, below). The abnormality can be identified by an ultrasound scan, which may show that there is a partial uterus or that the cavity is divided into two sections. Recurrent miscarriages or preterm labor may occur as a result of an abnormally shaped uterus.

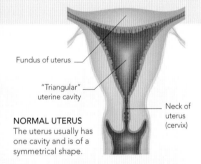

Fundus of uterus

"Triangular" uterine cavity

Neck of uterus (cervix)

NORMAL UTERUS
The uterus usually has one cavity and is of a symmetrical shape.

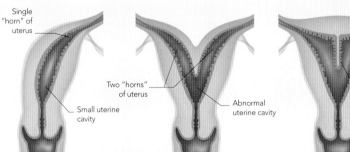

Single "horn" of uterus

Small uterine cavity

Two "horns" of uterus

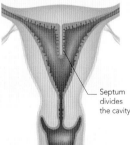

Abnormal uterine cavity

Septum divides the cavity

UNICORNUATE UTERUS
In this abnormality, only one side of the uterus is present, so that the uterine cavity is small and very narrow.

BICORNUATE UTERUS
Here, the uterus has two horns, so that both sides of the uterus are narrow, with a deep division down the center.

SEPTATE UTERUS
The septate uterus has a long central division that almost divides the cavity into two halves, thereby limiting the space for growth of a fetus.

DYSMENORRHEA

PAIN IN THE LOWER ABDOMEN JUST BEFORE AND DURING MENSTRUATION IS A COMMON PROBLEM THAT AFFLICTS UP TO 75 PERCENT OF WOMEN AT SOME TIME.

Dysmenorrhea can be either primary (no identified cause) or secondary (due to a problem in the pelvic organs). The former tends to be present in the teens and resolves with time. The latter is characterized by severe pain in women who have had little pain before, for which pelvic inflammatory disease (see p.218) and endometriosis (see p.218) are possible causes. Swabs to detect infection and pelvic ultrasound scanning are used in the diagnosis. Primary dysmenorrhea may be improved by nonsteroidal anti-inflammatory drugs or the combined oral contraceptive pill. In secondary dysmenorrhea, the underlying cause is treated.

PROSTAGLANDINS—MEDIATORS OF PAIN
Levels of prostaglandins rise shortly after ovulation, triggering contractions in the uterine muscles that affect the blood supply to the uterus, causing the pain of primary dysmenorrhea.

MALE REPRODUCTIVE DISORDERS

THE ORGANS OF THE MALE REPRODUCTIVE SYSTEM MAY BE AFFECTED BY A VARIETY OF DISORDERS, INCLUDING PROBLEMS CAUSED BY INFECTIONS AND THOSE RELATED TO ABNORMAL GROWTHS. SOME OF THE CONDITIONS MAY AFFECT NORMAL FUNCTION DURING SEXUAL INTERCOURSE WHILE OTHERS, SUCH AS INFLAMMATION OF THE TESTIS AND THE EPIDIDYMIS DUE TO MUMPS, MAY IMPACT ON A MAN'S FERTILITY. DISORDERS OF THE MALE REPRODUCTIVE SYSTEM THAT IMPAIR THE ABILITY TO PRODUCE HEALTHY SPERM OR TO DELIVER SPERM TO MEET AN EGG WILL OBVIOUSLY AFFECT FERTILITY.

EPIDIDYMAL CYSTS

THESE PAINLESS SWELLINGS CONTAINING CLEAR FLUID (ALSO KNOWN AS SPERMATOCELES) OFTEN DEVELOP IN THE EPIDIDYMIS, THE COILED TUBE THAT STORES SPERM AND TRANSPORTS THEM AWAY FROM THE TESTES.

Why these cysts develop is unknown. They tend to grow slowly, are usually symptomless, and do not become cancerous. In many cases, there are many cysts, often on both sides; but it is possible to have just one cyst on one side. It is very important to have a medical check if a swelling is discovered in the scrotum to rule out testicular cancer. Cysts can be diagnosed during a clinical examination by shining a light underneath so that the swelling lights up (transillumination); an ultrasound scan can confirm the diagnosis. Treatment is seldom needed because the cysts tend to be small. Occasionally, a large one may compress surrounding tissues, causing discomfort, so that removal is recommended. Surgical treatment does not affect fertility.

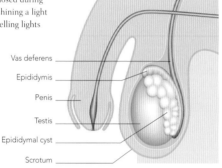

MULTIPLE CYSTS
Epididymal cysts are smooth and spherical in shape. They may occur alone, but often there are many cysts on both sides. Occasionally, they become infected, in which case they may be painful.

Vas deferens
Epididymis
Penis
Testis
Epididymal cyst
Scrotum

EPIDIDYMO-ORCHITIS

WHEN ONE OF THE TESTES AND ITS ADJACENT EPIDIDYMIS BECOME INFLAMED, IT OFTEN CAUSES SEVERE PAIN AND SWELLING ON THE AFFECTED SIDE.

The inflammation is usually caused by a bacterial infection, either from the prostate gland (see prostatitis, opposite) or the urinary tract or, in younger men, a sexually transmitted disease (see pp.224–25). Before mumps vaccinations were included in the routine childhood immunizations, mumps was a common cause of epididymo-orchitis in boys and young men. In some cases, it may affect fertility. The symptoms include pain, redness, and swelling on the affected side, often with a high temperature. Swabs may be taken from the urethra and samples of urine collected to discover the cause of the inflammation. Sometimes, an ultrasound scan is arranged to rule out testicular torsion (see opposite). Antibiotics are prescribed for a bacterial infection as well as analgesics. Ice packs can help relieve the discomfort too. The pain should subside within 48 hours, but the swelling may persist for a few weeks.

CAUSATIVE ORGANISM
Chlamydia bacteria, shown in pink on this color-enhanced electron micrograph, can cause epididymo-orchitis.

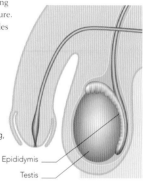

INFLAMED AREA
Both the testis and the epididymis are inflamed, causing tenderness, swelling, and redness. In severe cases there may be extreme pain and fever.

Epididymis
Testis

VARICOCELE

THIS KNOT OF DISTENDED VEINS IN THE SCROTUM MAY CAUSE DISCOMFORT FOR SOME MEN AND CAN RESULT IN A REDUCED SPERM COUNT. FOR SOME UNKNOWN REASON, THE LEFT-HAND SIDE IS MOST COMMONLY AFFECTED.

Varicoceles are varicose veins in the scrotum, which result from leaky valves within veins that take blood away from the testes; there is a backflow of blood into the scrotum and a buildup of blood, distending the veins and causing them to look like a bag of worms. Symptoms may include discomfort, a dragging sensation, and scrotal swelling. It is usually possible to confirm the diagnosis on clinical examination. In the majority of cases, varicoceles are small and do not require any treatment—they either cause no problems or resolve. Close- fitting underwear that provides support can help relieve the discomfort and the aching, dragging sensation. If the pain is problematic or fertility is affected, treatment may be recommended, which involves tying off the distended veins.

CONTRAST X-RAY OF A VARICOCELE
A varicocele is highlighted by a special dye that has been injected into the bloodstream before X-rays are taken.

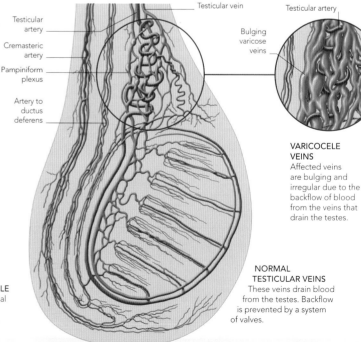

Testicular artery
Cremasteric artery
Pampiniform plexus
Artery to ductus deferens
Testicular vein
Testicular artery
Bulging varicose veins

VARICOCELE VEINS
Affected veins are bulging and irregular due to the backflow of blood from the veins that drain the testes.

NORMAL TESTICULAR VEINS
These veins drain blood from the testes. Backflow is prevented by a system of valves.

HYDROCELE

THIS SWELLING RESULTS FROM AN ABNORMAL ACCUMULATION OF FLUID BETWEEN LAYERS OF THE SCROTAL SAC THAT SURROUND EACH TESTIS. HYDROCELES ARE RARELY PAINFUL BUT CAN BE UNCOMFORTABLE IF LARGE.

When a hydrocele is present, there is an abnormally large volume of fluid present within layers of the scrotal sac (see p.29). Infections and testicular injuries are possible causes. Hydroceles tend to occur in young boys and elderly men. Diagnosis is made by a clinical examination—the swelling will light up when a flashlight is held against it—and by ultrasound scanning. If the symptoms of a hydrocele become troublesome, treatment options include drainage of the fluid with a needle or a small operation. Any infections are treated with a course of antibiotics.

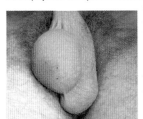

TESTICULAR SWELLING
A hydrocele is characterized by painless swelling on one side of the scrotum only; in this image the man's right testis is swollen, but the left looks normal.

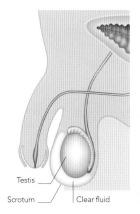

SWELLING WITHIN THE SCROTUM
A hydrocele is one possible cause of testicular swelling; fluid accumulates around a testis and, if large enough, can distort the shape of the scrotum.

Testis

Scrotum | Clear fluid

BALANITIS

INFLAMMATION AT THE END OF THE PENIS, OR BALANITIS, CAN BE SORE AND UNCOMFORTABLE. FORTUNATELY, MOST CAUSES ARE EASILY TREATED.

In this condition, the end of the penis (the glans) and the foreskin become inflamed, making them sore, itchy, and reddened. In addition, there may be some discharge from the urethra. Possible causes include bacterial infections, yeast (*Candida albicans*), and sexually transmitted diseases (see pp.224–25). In some cases, a tight foreskin may make it difficult to clean the end of the penis properly. After a physical examination of the penis, swabs may be taken from the end of the urethra and tested for possible infectious organisms, which are treated accordingly. A circumcision (removal of the foreskin) may be recommended in cases of a tight foreskin. Some cases of balanitis are due to an allergic reaction. If possible, the irritant is identified and then avoided. The end of the penis should be kept clean and dry.

TESTICULAR TORSION

THIS PAINFUL CONDITION NEEDS URGENT TREATMENT BECAUSE THE TESTIS MAY BE IRREPARABLY DAMAGED IF SURGERY IS NOT PERFORMED WITHIN ABOUT 24 HOURS.

For reasons that remain unclear, the spermatic cord, which contains the vas deferens and the blood vessels to the testis, becomes twisted, compromising the blood supply to the testis, which can cause permanent damage if not reversed quickly. The onset of symptoms is rapid, including pain in the scrotum, lower abdomen, and groin, as well as redness on one side of the scrotum. The condition is diagnosed by ultrasound scanning. After diagnosis, surgery to untwist the cord and then fix both testes in place is usually undertaken quickly. If the testis cannot be saved in time, it is removed—an implant may be put in its place for cosmetic reasons. If one testis is damaged during a torsion event, the unaffected testis can usually make enough sperm so that fertility is not significantly affected.

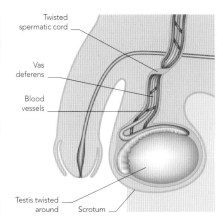

Twisted spermatic cord

Vas deferens

Blood vessels

Testis twisted around | Scrotum

TORSION OF THE TESTIS
In addition to the twisting of the spermatic cord, the testis lies in a different position in the scrotum. The usual shape of the scrotum may be distorted.

PROSTATITIS

THIS COMMON CONDITION MAY OCCUR IN TWO FORMS: A PAINFUL, ACUTE FORM AND AN OFTEN SYMPTOMLESS CHRONIC ONE. BOTH REQUIRE TREATMENT OF THE UNDERLYING CAUSE IF ONE CAN BE IDENTIFIED. THE PROBLEM MAY RECUR.

This condition particularly affects sexually active men. Often a cause cannot be found, but prostatitis may be due to a sexually transmitted disease (see pp.224–25) or a bacterial infection of the urinary tract. Acute prostatitis produces severe and rapidly developing symptoms including a high temperature, pain in the root of the penis, and pain in the lower back. Chronic prostatitis may be symptomless or produce only mild symptoms, which may include pain in the root of the penis, in the testes, and in the lower back, pain when ejaculating, and blood in the semen. In both types, there may be frequent and sometimes painful urination. A physician assesses the prostate gland during a rectal examination. Samples of urine, prostatic secretions, and swabs taken from the end of the urethra are tested for infections. Ultrasound or CT scanning may be used to look for an abscess in the prostate, a possible complication. The underlying cause, such as infection, is treated, but it may take months for the disorder to subside.

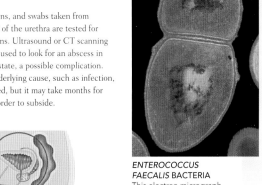

***ENTEROCOCCUS FAECALIS* BACTERIA**
This electron micrograph shows bacteria that are normally present in the gut, but may cause prostatitis and urinary-tract infections.

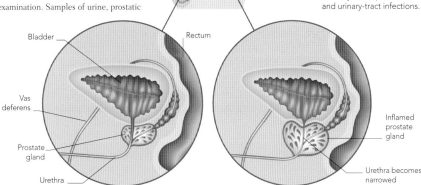

Bladder

Vas deferens

Prostate gland

Urethra

Rectum

NORMAL PROSTATE GLAND
Usually the prostate gland (the size of a walnut) sits just below the neck of the bladder and surrounds the urethra. Urine from the bladder flows freely into the urethra, which carries it out through the penis during urination.

Inflamed prostate gland

Urethra becomes narrowed

ENLARGED PROSTATE GLAND
In prostatitis, the gland is inflamed and can swell. The swollen prostate can compress the urethra so that urine cannot pass freely from the bladder as it should. This means that urine is passed frequently and in small amounts.

SEXUALLY TRANSMITTED DISEASES

MOST TYPES OF SEXUALLY TRANSMITTED DISEASES (STD) ARE PASSED FROM PERSON TO PERSON DURING SEXUAL INTERCOURSE. SOME, SUCH AS HIV AND SYPHILIS, CAN ALSO CROSS THE PLACENTA AND AFFECT THE FETUS, WHILE OTHERS, SUCH AS GONORRHEA AND CHLAMYDIA, CAN AFFECT FERTILITY. SOME CAN ALSO BE TRANSFERRED FROM MOTHER TO CHILD DURING DELIVERY AS THE BABY EMERGES THROUGH THE BIRTH CANAL.

HIV/AIDS

INFECTION WITH THE HUMAN IMMUNODEFICIENCY VIRUS (HIV), IF LEFT UNTREATED, LEADS TO ACQUIRED IMMUNODEFICIENCY SYNDROME (AIDS) AND SEVERELY IMPAIRED IMMUNITY. HIV CAN BE PASSED TO A FETUS IN UTERO AND TO A BABY VIA BREAST MILK.

HIV can be transferred by vaginal, anal, and oral intercourse, contaminated blood and blood products, and contaminated needles. It can also be passed to a fetus during pregnancy (HIV particles can cross the placenta), during childbirth, or after delivery in breast milk. The virus infects a type of white blood cell that has the CD4 receptor on its surface and replicates rapidly, killing the cells in the process. The body can cope for some time, but eventually the CD4 white cell count falls below a critical level. Most people who become infected with HIV have no symptoms initially. Some have the general symptoms typical of a viral illness, including fever, aching muscles and joints, swollen glands, and a sore throat. There usually follows a symptom-free interval, which may last for many years. Some people, however, may have further relatively mild symptoms, including thrush in the mouth, gum disease, and weight loss. Eventually, when the CD4 count falls below a certain level or certain conditions develop, such as particular infective illnesses and certain cancers, the person is said to have AIDS. Treatment of HIV and AIDS involves combinations of antiretroviral drugs and antibiotics. Condoms reduce the risk of HIV transmission.

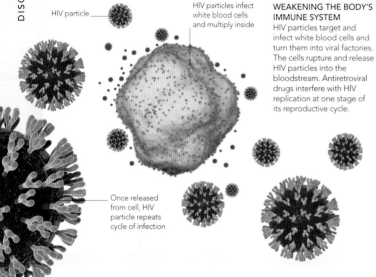

HIV particle

HIV particles infect white blood cells and multiply inside

Once released from cell, HIV particle repeats cycle of infection

WEAKENING THE BODY'S IMMUNE SYSTEM
HIV particles target and infect white blood cells and turn them into viral factories. The cells rupture and release HIV particles into the bloodstream. Antiretroviral drugs interfere with HIV replication at one stage of its reproductive cycle.

W & C EUROPE, NORTH AMERICA, AUSTRALIA, AND NEW ZEALAND

EASTERN EUROPE AND CENTRAL ASIA

LATIN AMERICA

CARIBBEAN

SUB-SAHARAN AFRICA

ASIA

NORTH AFRICA AND THE MIDDLE EAST

0   20   40   60   80   100
INFECTED PREGNANT WOMEN RECEIVING ANTIRETROVIRAL THERAPY (%)

PREVENTING HIV TRANSMISSION
Antiretroviral drugs are given in pregnancy. The proportion of infected women being treated is highest in developed countries and lowest in developing countries. It is crucial to treat pregnant women to improve their outlook and reduce the likelihood of transfer of infection to the baby.

SYPHILIS

THIS BACTERIAL ILLNESS BEGINS IN THE GENITALS, BUT LATER CAN AFFECT OTHER BODY TISSUES. A BABY CAN BE INFECTED IN UTERO OR DURING DELIVERY.

The cause of syphilis is the bacterium *Treponema pallidum*, which is transmitted during intercourse. There are three main stages; the first two being infectious for up to two years and the final stage being noninfectious. The primary stage progresses to the secondary stage if treatment is not given. A latent stage follows, succeeded by the third stage (tertiary syphilis), although this is now rare due to the availability of antibiotics. The diagnosis of syphilis is usually made through a blood test, and treatment consists of antibiotic injections, which can be given during pregnancy. Condoms should be used to avoid transmission of the causative bacteria. The incidence of syphilis has decreased since the introduction of penicillin.

SYMPTOM STAGES
Left untreated, syphilis infection progresses through a series of well-defined stages (primary, secondary, latent, and tertiary) on a relatively defined timescale.

Primary syphilis
A firm, painless sore, known as a chancre, appears usually in the genital region. Develops on average 21 days after exposure; lasts two to three weeks. Progresses to secondary syphilis without treatment.

Secondary syphilis
Generalized features appear and can include fever, sore throat, swollen glands, painful joints, rashes, and mouth and genital ulcers. Starts four to ten weeks after initial chancre appears. Progresses to latent stage without treatment.

Latent syphilis
Symptoms disappear, but blood tests show the infection is still present. Symptoms can recur within two years, or tertiary syphilis develops later.

Tertiary syphilis
Characteristic lesions, called gumma, develop mainly in the skin and in the bones, including those of the skull, leg, and collarbone. The cardiovascular and nervous systems may also be affected.

GENITAL HERPES

CAUSED BY THE HERPES SIMPLEX VIRUS, THIS INFECTION RESULTS IN PAINFUL ULCERS IN THE GENITAL AREA.

There are two types of the herpes simplex virus (HSV): HSV-1 usually causes cold sores, while HSV-2 results in genital herpes. HSV is highly contagious and can be passed from person to person via sexual contact. HSVs can cause problems in a newborn if transferred during delivery. The disease usually recurs, with the first attack being the worst. Blisters develop on and around the genitals, along with tingling and soreness. Other symptoms include painful urination, vaginal discharge, and fever. The symptoms can last up to three weeks. The diagnosis can usually be made from examination of the lesions. Treatment cannot cure the condition but may reduce its severity.

GENITAL WARTS

GROWTHS IN THE GENITAL AREA CAUSED BY THE HUMAN PAPILLOMAVIRUS (HPV) ARE PASSED ON VIA SKIN CONTACT.

Genital warts can take up to 20 months to appear after infection. They are painless and grow rapidly; they can also develop in the mouth as a result of oral sex. Various treatments are available including antiviral lotions. Infection with HPV in women increases their risk of cancer of the cervix. Condoms cannot offer complete protection, so HPV transmission may still occur. A baby can become infected with HPV during childbirth.

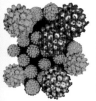

HUMAN PAPILLOMAVIRUS
This highly magnified image shows human papillomaviruses, the infective organisms responsible for genital warts.

GONORRHEA

THIS COMMON SEXUALLY TRANSMITTED BACTERIAL INFECTION CAUSES INFLAMMATION IN THE GENITAL AREA AND DISCHARGE IN MEN AND WOMEN, OFTEN WITHOUT SYMPTOMS.

The bacterial cause of gonorrhea is *Neisseria gonorrhoea*, which is transmitted by vaginal, oral, and anal sex. Usually, symptoms appear within two weeks of infection but may not appear for months, in which case infection may have spread around the body. Left untreated, the infection may spread to the fallopian tubes, causing damage that may affect fertility. The diagnosis is made by testing swabs of the affected areas; antibiotic treatment may be given intravenously if the infection has already spread. As with other STDs, both partners are tested. An infected woman can pass it on to her baby during delivery, resulting in an eye infection, which may cause blindness.

Eye infection with pain, swelling, and discharge

Symptomless infection in throat

Lower abdominal pain or tenderness

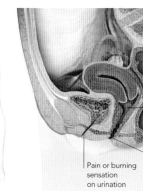

SYMPTOMS IN MEN AND WOMEN
The main symptoms are similar in both sexes. However, in up to 50 percent of women and 10 percent of men, there are no symptoms whatsoever.

Irregular vaginal bleeding

Inflammation in rectum, with pain, discomfort, or discharge

Pain or burning sensation on urination

Green or yellow vaginal discharge

CHLAMYDIA INFECTION

THIS BACTERIAL INFECTION OFTEN HAS NO SYMPTOMS AND IS A MAJOR CAUSE OF INFERTILITY IN WOMEN. HALF OF INFECTED MEN AND 80 PERCENT OF INFECTED WOMEN SHOW NO SYMPTOMS, MEANING IT MAY GO UNNOTICED.

It is estimated that 5 percent or more of sexually active women in the US are infected with *Chlamydia trachomatis*, the causative bacterium. If symptoms do occur, they include painful urination and discharge from the urethra in men, and in women, vaginal discharge, bleeding between periods and after sex, and pain in the lower abdomen. The infection can pass up to the fallopian tubes and may then cause infertility. *Chlamydia trachomatis* bacteria may be passed to a baby during delivery, causing conjunctivitis and pneumonia. Urine samples or urethral swabs are taken in men, and swabs of the cervix are tested in women. Treatment is with antibiotics, some of which cannot be taken during pregnancy. Condoms offer protection against transmission of this infection.

CHLAMYDIA BACTERIA IN VAGINA CELLS
This highly magnified view of a cervical smear shows *Chlamydia trachomatis* bacteria within the cells of the lining (epithelial cells). This infection is very common.

Spherical (dark pink) *Chlamydia* bacterium inside a vaginal epithelial cell (blue)

NONGONOCOCCAL URETHRITIS

THIS INFLAMMATION OF THE URETHRA IN MEN IS CAUSED BY AN INFECTION OTHER THAN GONORRHEA. A COMMON STD, IT PRODUCES CHARACTERISTIC EFFECTS, BUT IN AROUND 15 PERCENT OF CASES THERE ARE NO SYMPTOMS.

There are a variety of causes, including *Chlamydia trachomatis*, *Trichomonas vaginalis*, herpes simplex virus, and *Candida albicans*. Nearly half of the cases of non-gonococcal urethritis (NGU) identified are caused by *Chlamydia trachomatis*, which is the cause of chlamydia infections in women (see above). In one-quarter of cases, no cause can be found. The condition can take up to five weeks to develop after infection, but on average it takes two to three weeks. Discharge and pain on urination may be accompanied by soreness and redness around the urethral opening, which is at the end of the penis. The infection may spread to the epididymides, testes, and prostate gland. In addition, certain infections may travel in the bloodstream to cause inflammation and pain in the joints. Urine samples and swabs taken from the urethra are tested to look for gonorrhea and other possible infective causes. Using condoms can reduce the transmission of infection.

Urethra
Inflammation causes pain on urination

Epididymis
Can become inflamed if infection spreads

Penis
May be painful and itchy inside

Testis
May swell if infection spreads

NONGONOCOCCAL URETHRITIS SYMPTOMS
These features are typical of NGU, although there may be no symptoms. Consequently, an infected man can pass on the infection without being aware that he has it.

COMPLICATIONS IN PREGNANCY

IN THE VAST MAJORITY OF CASES, PREGNANCY PROCEEDS WITHOUT ANY MAJOR PROBLEMS. HOWEVER, SOMETIMES A PROBLEM DOES DEVELOP, AFFECTING THE MOTHER, THE FETUS, OR BOTH. FOR EXAMPLE, AN EMBRYO MAY FAIL TO IMPLANT OR DEVELOP PROPERLY, OR A PROBLEM MAY OCCUR LATER, WHEN THE FETUS SEEMS TO BE DEVELOPING NORMALLY. PROBLEMS IN PREGNANCY MAY BE DUE TO FETAL FACTORS, SUCH AS A GENETIC OR CHROMOSOME ABNORMALITY, OR TO MATERNAL FACTORS, SUCH AS AN INFECTION, OR A HORMONAL OR ANATOMICAL PROBLEM.

MISCARRIAGE

THIS IS THE SPONTANEOUS ENDING OF A PREGNANCY BEFORE 24 WEEKS. MOST OCCUR DURING THE FIRST 14 WEEKS.

Early miscarriages tend to result from a genetic or chromosomal abnormality in the fetus. Later miscarriages may be caused by a problem in the uterus. Other causes include cervical incompetence (see below) and maternal infections. Various factors increase the risk of miscarriage, including smoking, drinking, and drug abuse during pregnancy. The risk of having a miscarriage in early adult life is about 1 in 5 pregnancies (about 20 per cent) but the risk increases with age, especially over the age of 40. There are three main types of miscarriage. In a threatened miscarriage, there is vaginal bleeding but the fetus is alive and the cervix is closed. In an inevitable miscarriage, the cervix is open and the fetus is usually dead. In a missed miscarriage, the fetus has died but there is no bleeding. With a threatened miscarriage, the pregnancy may proceed to term. An inevitable miscarriage may be complete or incomplete, meaning that some tissue remains in the uterus. An incomplete or missed miscarriage may require surgery to empty the uterus.

THREATENED MISCARRIAGE
If the cervix remains closed and the fetus is still alive, the pregnancy can often continue to term. If a miscarriage becomes inevitable, the cervix opens so that tissue can pass through.

Amniotic fluid · Placenta · 12-week-old fetus · Umbilical cord · Blood clot · Blood traveling from uterus through cervical canal · Bleeding evident when via the vagina

CAUSES OF MISCARRIAGE

Miscarriage can occur as a result of various underlying problems, which may primarily be either maternal or fetal in origin. These can be classified into five main categories: inherited, hormonal, immunological, infective, and anatomical. However, it is not always possible to identify the cause.

CAUSE	POSSIBLE EXAMPLES
Inherited	Fetal genetic or chromosomal abnormalities are possible causes, such as the presence of too many or too few chromosomes.
Hormonal	Overactivity or underactivity of the thyroid gland, diabetes mellitus, and abnormally low levels of progesterone are possible causes.
Immunological	Miscarriage can be caused by rare immune disorders, such as antiphospholipid syndrome (placental clots reducing blood supply to the fetus).
Infective	Several infections affecting the mother can cause miscarriage, including rubella and toxoplasmosis (a protozoal infection).
Anatomical	Miscarriage can sometimes occur if the uterus is abnormally shaped or has large fibroids; cervical incompetence is another possible cause.

STILLBIRTH

A STILLBIRTH IS WHEN A FETUS DIES IN THE UTERUS AND IS BORN AFTER 20 WEEKS OF PREGNANCY.

A stillbirth may occur as a result of a wide variety of causes, from a genetic or structural disorder in the fetus to problems with the placenta or maternal infection or illness. However, in many cases, no cause can be identified. If the fetus dies late in pregnancy, labor may need to be induced, particularly if the mother's health is at risk, or it may be possible to wait for labor to start naturally. In either case, a stillbirth can be very distressing, and the mother will be offered various support options. In developed countries, the risk of stillbirth is small – about 1 in 200 pregnancies. To help minimize the risk, the mother should attend all antenatal appointments and tell the midwife or doctor if the fetus is moving less than normal, or if she develops abdominal pain, vaginal bleeding, or itching. Other measures include stopping smoking, going to sleep on one side, keeping a healthy weight, and avoiding alcohol and drugs. Certain foods should also be avoided, including soft cheeses, unpasteurized milk products, pâté, undercooked or raw meat, and uncooked shellfish.

CERVICAL INCOMPETENCE

IF THE CERVIX IS WEAK (INCOMPETENT), PRESSURE FROM THE GROWING FETUS AND AMNIOTIC FLUID MAY CAUSE IT TO OPEN EARLY, RESULTING IN A MISCARRIAGE.

Weakness of the cervix may follow surgery to the cervix or a number of procedures that require the cervix to be opened (including termination of pregnancy). Cervical incompetence tends to cause miscarriages after 14 weeks' pregnancy, and often there are no symptoms before miscarriage occurs. If a woman has had a late miscarriage, an ultrasound may be arranged to check the cervix. If the ultrasound confirms cervical incompetence, a stitch may be inserted in the cervix at 12–16 weeks in the next pregnancy (and any subsequent ones) and then removed at 37 weeks ready for the start of labor. If labor begins early, the stitch is removed straight away.

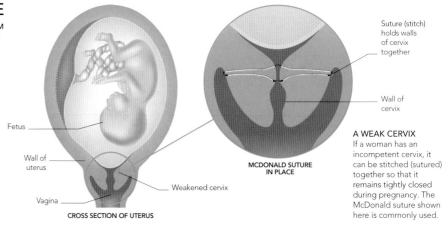

Fetus · Wall of uterus · Vagina · Weakened cervix
CROSS SECTION OF UTERUS

MCDONALD SUTURE IN PLACE

Suture (stitch) holds walls of cervix together · Wall of cervix

A WEAK CERVIX
If a woman has an incompetent cervix, it can be stitched (sutured) together so that it remains tightly closed during pregnancy. The McDonald suture shown here is commonly used.

ECTOPIC PREGNANCY

IN AN ECTOPIC PREGNANCY, THE FERTILIZED EGG IMPLANTS OUTSIDE THE UTERUS SO THAT THE EMBRYO CANNOT DEVELOP PROPERLY. THE CONDITION CAN BE LIFE-THREATENING FOR THE MOTHER.

In most ectopic pregnancies, the fertilized egg implants in the fallopian tube, although rarely it may implant elsewhere, such as in the cervix, ovary, or abdominal cavity.

Possible underlying causes include previous damage to the fallopian tube, perhaps due to surgery or an infection such as pelvic inflammatory disease (see p.218). Using a coil, or intrauterine contraceptive device (IUD), also increases the risk. The symptoms are vaginal bleeding and lower abdominal pain, usually on one side. To diagnose the condition, a pregnancy test may be arranged, followed by an ultrasound scan if the test is positive. A physician may also perform a laparoscopy (in which a

viewing instrument is passed through the abdominal wall). If an ectopic pregnancy is found, it will be removed during the laparoscopy. If an ectopic pregnancy leads to rupturing of a fallopian tube, there will be severe abdominal pain and pain in the shoulder tip. The condition is potentially life-threatening and requires urgent surgery.

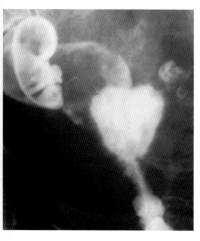

X-RAY OF AN ECTOPIC PREGNANCY
This X-ray shows an ectopic pregnancy at about 10–12 weeks in which the fetus is developing in the mother's right fallopian tube. Left untreated, the tube will rupture, causing bleeding into the abdomen.

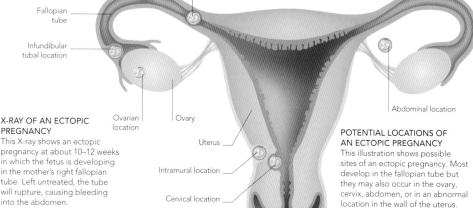

Ampullar tubal location

Isthmic tubal location

Fallopian tube

Infundibular tubal location

Ovarian location

Ovary

Uterus

Intramural location

Cervical location

Abdominal location

POTENTIAL LOCATIONS OF AN ECTOPIC PREGNANCY
This illustration shows possible sites of an ectopic pregnancy. Most develop in the fallopian tube but they may also occur in the ovary, cervix, abdomen, or in an abnormal location in the wall of the uterus.

MOLAR PREGNANCY

THIS OCCURS WHEN A SPERM FERTILIZES AN EGG BUT THE RESULTING SET OF CHROMOSOMES IS ABNORMAL SO THAT A NORMAL PREGNANCY CANNOT DEVELOP.

In a complete molar pregnancy, a mass of cysts forms in the uterus. In a partial molar pregnancy, an embryo and placenta start to grow, but the embryo does not survive. Symptoms include vaginal bleeding, which may begin from about six weeks, and nausea and vomiting, which may be severe. A molar pregnancy is treated by opening the cervix (under general anesthesia) so that the tissue can be removed. Rarely, the molar tissue becomes cancerous and further treatment, such as chemotherapy, is necessary.

Multiple cysts develop within uterus

COMPLETE MOLAR PREGNANCY
The mass of cysts formed in the uterus is sometimes known as hydatidiform mole (from Greek for "grapelike").

NORMAL EMBRYO DEVELOPMENT
Usually, a single egg and a single sperm, each with 23 chromosomes, combine at fertilization to give a normal embryo with 46 chromosomes.

Sperm

23 chromosomes from father

23 chromosomes from mother

Egg

Normal embryo with 46 chromosomes

COMPLETE MOLAR PREGNANCY
A sperm with 23 chromosomes fertilizes an empty egg with no chromosomes. The 23 chromosomes from the sperm duplicate, giving 46.

One sperm

Empty egg

23 chromosomes from father

No chromosomes from mother

Abnormal embryo with 23 pairs of duplicated paternal chromosomes

PARTIAL MOLAR PREGNANCY
Two sperm, each with 23 chromosomes, fertilize a single egg, also with 23 chromosomes, giving an abnormal embryo with 69 chromosomes.

Two sperm

Egg

46 chromosomes from father

23 chromosomes from mother

Resulting embryo is abnormal with 69 chromosomes

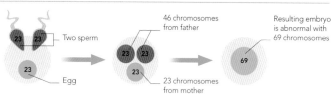

VAGINAL BLEEDING IN PREGNANCY

BLEEDING CAN OCCUR AT ANY TIME DURING PREGNANCY AND CAN BE DUE TO A WIDE VARIETY OF CAUSES. BLEEDING AT ANY STAGE IS POTENTIALLY SERIOUS AND REQUIRES IMMEDIATE SPECIALIST ATTENTION.

Vaginal bleeding during the first 14 weeks may indicate a miscarriage (see opposite page) or, less commonly, an ectopic

pregnancy (see above). In some cases, there may also be pain, which tends to be severe if the cause is an ectopic

pregnancy. Occasionally, light bleeding occurs for no apparent reason and the pregnancy continues. Between 14 and 24 weeks, bleeding may signify a late miscarriage, commonly due to cervical incompetence (see opposite page). Important causes of bleeding after 24 weeks include placental abruption (see

p.228), which is painful, and placenta previa (see p.228), which is painless. Certain conditions, such as cervical polyps (noncancerous growths on the cervix), can cause bleeding at any time. Investigations of the cause may include a cervical examination and an ultrasound scan. Treatment depends on the cause.

PLACENTA PREVIA

IF THE PLACENTA LIES LOW IN THE UTERUS AND PARTLY OR FULLY COVERS THE OPENING TO THE CERVIX, IT CAN INTERFERE WITH BIRTH. THE CONDITION AFFECTS ABOUT 1 IN 200 PREGNANCIES.

Placenta previa is a common cause of painless vaginal bleeding after the 24th week of pregnancy. Heavy bleeding can be potentially life-threatening for both fetus and mother. Risk factors include a previous cesarean section, multiple pregnancy, and several previous pregnancies. It is diagnosed by an ultrasound scan. Often, the placenta will move up as the uterus grows, but if it stays low and bleeding occurs, admission to the hospital is necessary. Hospital admission may be recommended from about 30 weeks for all women with complete placenta previa, with a cesarean section being planned for about 38 weeks. If severe bleeding occurs, an emergency cesarean section is needed. A cesarean section is also recommended for women with a partial placenta previa.

PLACENTAL POSITIONS
In placenta previa, the position of the placenta varies from lying low in the uterus without encroaching on the cervix to lying centrally across the cervix.

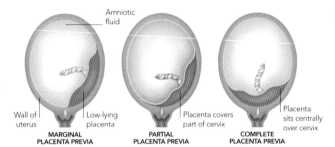

Amniotic
fluid

Wall of
uterus

Low-lying
placenta

**MARGINAL
PLACENTA PREVIA**

Placenta covers
part of cervix

**PARTIAL
PLACENTA PREVIA**

Placenta
sits centrally
over cervix

**COMPLETE
PLACENTA PREVIA**

PLACENTAL ABRUPTION

THIS IS A POTENTIALLY LIFE-THREATENING CONDITION IN WHICH PART OR ALL OF THE PLACENTA DETACHES ITSELF FROM THE WALL OF THE UTERUS BEFORE THE BABY IS BORN.

There are two forms of placental abruption: revealed abruption, a common cause of vaginal bleeding after 28 weeks; and concealed abruption, which does not cause bleeding as the blood remains in the uterus. Risk factors include long-standing high blood pressure, a previous abruption, and several previous pregnancies. Smoking, excessive drinking, and drug abuse also increase the risk. In contrast to bleeding in placenta previa, a placental abruption is always painful and causes the uterus to contract. An ultrasound scan will be done and the fetal heart checked. Induction may be recommended; in severe cases, emergency cesarean section may be necessary.

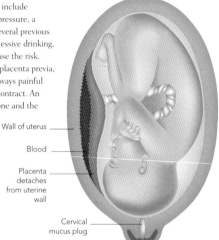

Wall of uterus

Blood

Placenta
detaches
from uterine
wall

Cervical
mucus plug

PLACENTAL DETACHMENT
In most cases, the placenta becomes partly detached, and blood either passes out through the vagina or collects between the placenta and uterine wall. Rarely, the entire placenta may become detached.

AMNIOTIC FLUID PROBLEMS

THE AMOUNT OF FLUID CONTAINED IN THE AMNIOTIC SAC CAN BE AFFECTED BY A NUMBER OF CONDITIONS, RESULTING IN EITHER AN ABNORMALLY LARGE VOLUME (POLYHYDRAMNIOS) OR AN ABNORMALLY SMALL VOLUME (OLIGOHYDRAMNIOS).

Polyhydramnios can cause maternal discomfort and is associated with premature rupture of the membranes and premature labor. Polyhydramnios also increases the risk of placental abruption (see above), postpartum hemorrhage (see p.240), cesarean section, and unstable lie (where the fetal position is constantly changing). The condition is managed to prolong the pregnancy and prevent complications for mother and fetus. Where possible, the underlying causes are treated. Oligohydramnios is often only noticed during prenatal checks. This condition, caused by the premature rupture of the membranes, is associated with premature labor and fetal growth restriction (see opposite). Regular assessment of fetal well-being should be performed.

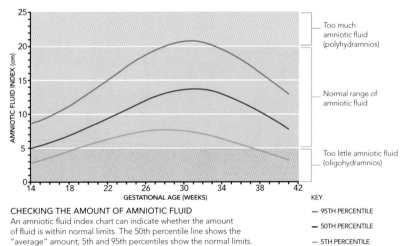

Too much amniotic fluid (polyhydramnios)

Normal range of amniotic fluid

Too little amniotic fluid (oligohydramnios)

CHECKING THE AMOUNT OF AMNIOTIC FLUID
An amniotic fluid index chart can indicate whether the amount of fluid is within normal limits. The 50th percentile line shows the "average" amount; 5th and 95th percentiles show the normal limits.

KEY
— 95TH PERCENTILE
— 50TH PERCENTILE
— 5TH PERCENTILE

CAUSES OF AMNIOTIC FLUID PROBLEMS

Excessive amniotic fluid (polyhydramnios) or too little (oligohydramnios) may be associated with factors in the mother or the fetus. Some common factors are below.

CAUSES OF OLIGOHYDRAMNIOS	CAUSES OF POLYHYDRAMNIOS
Premature rupture of membranes	Diabetes mellitus
Fetal growth restriction, for example due to preeclampsia	Gastrointestinal (bowel) obstruction
A fetal abnormality causing reduced urine production or obstruction of passage of urine	Impaired fetal swallowing due to fetal abnormalities, such as anencephaly
The use of drugs, such as nonsteroidal anti-inflammatory drugs	Heart failure due to congenital reasons or anemia
Twin–twin transfusion syndrome (an imbalance when one twin receives more blood than the other)	Increased fetal urine production (such as twin–twin transfusion syndrome)
Infection	Infection, such as syphilis or parvovirus
Chromosomal abnormalities, such as Down syndrome	Chromosomal abnormalities, such as Down syndrome
Postmaturity—a baby is overdue	Achondroplasia (a bone disorder causing short stature)

FETAL GROWTH RESTRICTION

ALSO KNOWN AS INTRA-UTERINE GROWTH RETARDATION, THIS CONDITION OCCURS WHEN A FETUS FAILS TO GROW SUFFICIENTLY IN THE UTERUS SO THAT IT IS THIN AND HAS A LOW BIRTH WEIGHT (LESS THAN 5½ LB/2.5 KG).

Fetal growth restriction has many possible causes, including long-standing high blood pressure, preeclampsia (see below), or a maternal infection, such as rubella. In some cases, it may occur because the placenta fails to supply sufficient nutrients to the fetus. Inherited fetal disorders, such as Down syndrome, are also possible causes. The risk of growth restriction is increased if the mother has a poor diet, smokes,

drinks excessively, or abuses drugs. Repeat ultrasound scans, and sometimes Doppler scans of blood flow in the umbilical artery, are used to monitor fetal growth. Hospital admission for bed rest and monitoring may be needed, and, when possible, any underlying causes are treated. An early delivery may be recommended if there are concerns about the baby's health.

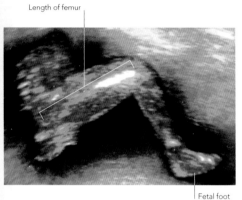

Length of femur

Fetal foot

MEASURING THE LENGTH OF THE FEMUR

The femur (thighbone) can be measured from ultrasound images. This measurement can be taken at intervals and, together with the abdominal circumference, used to monitor fetal growth.

MONITORING FETAL GROWTH

This graph shows fetal weight-increase curves in pregnancy; the 97th percentile and 3rd percentile lines show the upper and lower limits of the normal range. If weight begins to fall below the lower limit, fetal growth restriction is indicated.

(Graph: WEIGHT (KG) on vertical axis 0–5; GESTATIONAL AGE (WEEKS) on horizontal axis 22–42; curves labeled 97th percentile, 50th percentile, 3rd percentile)

KEY — ● NORMAL BABY — ● NORMAL SMALL BABY — ● GROWTH-RESTRICTED BABY

OBSTETRIC CHOLESTASIS

ALSO CALLED INTRAHEPATIC CHOLESTASIS OF PREGNANCY, OBSTETRIC CHOLESTASIS IS A CONDITION IN WHICH THERE IS ABNORMALLY SLOW FLOW OF BILE, A SUBSTANCE THAT THE LIVER PRODUCES TO AID DIGESTION.

Obstetric cholestasis causes a buildup of bile in the mother's blood, producing symptoms such as very itchy skin and jaundice. The itching is often most pronounced on the hands and feet but may affect the entire body; it is also often worse at night. Obstetric cholestasis can increase the risk of premature birth or stillbirth, so it is important that it be recognized and treated early. The cause of the condition is unknown, but there may be a genetic link because it is more common in women of South American, South Asian, or Scandinavian origin, and it can run in families. It is also more common in women who are pregnant with twins or other multiple fetuses. To diagnose obstetric cholestasis, the doctor may ask about the mother's family history and may order blood tests to assess liver function and levels of bile acids. The mother may be given a drug called rsodeoxycholic acid to reduce bile acid levels and help to relieve itching. The mother's liver function and levels of bile acids will be monitored throughout the pregnancy, and the fetus will also be monitored to detect any problems, such as an abnormal fetal heart rate. Severe obstetric cholestasis may interfere with normal blood clotting in the mother; vitamin K may be given to correct this problem. In some severe cases, labor may be induced if the fetus is older than 36 weeks' gestation, to prevent stillbirth.

PREECLAMPSIA AND ECLAMPSIA

THESE CONDITIONS ARE UNIQUE TO PREGNANCY AND ALWAYS IMPROVE FOLLOWING DELIVERY OF THE BABY.

In preeclampsia, the blood pressure increases, fluid is retained, and protein is lost in the urine. Symptoms occur quite late in the condition, including swelling of the hands, face, and feet, headache, visual disturbances, and abdominal pain. If untreated, high blood pressure leads to eclampsia (seizures) in 1 percent of women with preeclampsia. For this reason, every pregnant woman has her urine checked for the presence of protein and her blood pressure measured at each prenatal visit. Treatment aims to return the blood pressure to within the normal range. There may be fetal growth restriction (see above) and hospital monitoring, and early delivery of the baby may be necessary. Eclampsia is treated urgently and delivery by cesarean section usually follows once the mother has been stabilized.

Visual disturbance
Flashing lights, blurry vision, and sensitivity to light are common

Severe headache
Commonly experienced at the front of the head

Nausea and vomiting
May be experienced alongside some dizziness

Sudden weight gain
Unusually fast weight gain (more than 2lb (0.9kg) a week)

Abdominal pain
This pain tends to occur in the center of the upper part of the abdomen

PREECLAMPSIA SYMPTOMS
Preeclampsia may be mild and produce no symptoms. In many cases, symptoms do develop, affecting various parts of the body. Severe symptoms may herald life-threatening eclampsia.

Sudden edema
Sudden swelling of the feet (and/or face or hands) is a sign of preeclampsia

RISK FACTORS FOR PREECLAMPSIA

The underlying cause of preeclampsia is not fully understood, although it may be due to a problem with the placenta. However, various factors have been identified that increase the risk of developing the condition, and these are listed below.

Being overweight or obese

A family or personal history of preeclampsia

A multiple pregnancy

First pregnancy or first pregnancy with a new partner

Ten years or more have passed since the last pregnancy

Being over the age of 35

Preexisting kidney disease

Preexisting high blood pressure

Preexisting diabetes mellitus

Certain autoimmune disorders

GESTATIONAL DIABETES

DIABETES MELLITUS CAN DEVELOP IN PREGNANCY IF THE PANCREAS CANNOT MEET THE INCREASED NEED FOR THE BLOOD-GLUCOSE-REGULATING HORMONE INSULIN.

Gestational diabetes often causes no symptoms, but if they do occur they may include excessive thirst, tiredness, and passing large amounts of urine. It is diagnosed by blood tests. Treatment is by dietary control and, in a few cases, insulin injections.

THE CONSEQUENCES OF GESTATIONAL DIABETES

The typical result is a large baby. The mother's insulin and glucose usually normalize after the birth.

insulin injections. The baby may grow very large, which may necessitate a cesarean section. Gestational diabetes usually disappears after the birth but may recur.

A woman with gestational diabetes has poorly controlled blood glucose levels because insulin is not being made in sufficient amounts. Consequently, her blood glucose levels are high.

This blood high in glucose passes to the baby via the placenta. Blood glucose is the baby's main food source.

The baby increases insulin production to utilize this glucose; unused glucose is laid down as fat. Consequently, the baby grows larger than normal, which may pose problems during delivery.

HYPEREMESIS GRAVIDARUM

VOMITING IN EARLY PREGNANCY CAN BE SO SEVERE THAT NO FLUIDS OR FOOD CAN BE KEPT DOWN.

In contrast to women with normal morning sickness, who gain weight, those with hyperemesis gravidarum lose weight and may also become dehydrated. The cause is not fully understood, but very high levels of the hormone human chorionic gonadotropin (hCG), produced in pregnancy, may play a role. Having twins is associated with high levels of hCG and an increased risk of hyperemesis gravidarum. Stress may also worsen the condition. If the vomiting is extremely severe, hospital admission may be arranged, where blood tests will be done to assess the level of dehydration and ultrasound used to check the fetus. Intravenous fluids and antinausea drugs may be given. The condition usually clears up by about the 14th week of pregnancy but may recur in future pregnancies.

WHEN DOES MORNING SICKNESS BECOME HYPEREMESIS GRAVIDARUM?	
MORNING SICKNESS	**HYPEREMESIS GRAVIDARUM**
There is little, if any, loss of weight. In fact, there is usually weight gain.	A significant amount of weight—5–20 lb (2.2–9 kg) or sometimes even more—is lost.
Nausea and vomiting do not interfere with the ability to eat and drink.	Nausea and vomiting result in a poor appetite and dehydration.
Vomiting is infrequent, and nausea tends to be episodic and mild.	Vomiting is frequent and may contain bile or blood. Nausea is constant and moderate to severe.
Dietary and lifestyle changes are usually all that is required to improve well-being.	Intravenous fluid rehydration and antinausea medicine are necessary.
Typically, improvement is seen after the first trimester, but queasiness may occur at times.	Symptoms may wane during mid-pregnancy but nausea and vomiting may continue.
Usual tasks, such as work and looking after children, are possible on most days.	The mother may be unable to work for weeks or months and may need to be cared for.

RHESUS INCOMPATIBILITY

A MISMATCH BETWEEN THE RHESUS BLOOD GROUP OF A FETUS AND THE MOTHER CAN RESULT IN PROBLEMS IN A FUTURE PREGNANCY IF THE SAME MISMATCH OCCURS.

Blood is classed as Rh positive or Rh negative depending on whether the red blood cells have Rhesus proteins on their surface or not (Rhesus status). If an Rh-negative woman has a Rh-positive partner, she may have a Rh-positive baby. The baby's Rh-positive cells may trigger the formation of antibodies in the mother against the Rh-positive blood cells. This does not cause problems in the first pregnancy, but if the woman has a Rh-positive baby again, her antibodies if formed will cross the placenta and destroy fetal red blood cells. This will cause anemia in the fetus and jaundice (see p.235) after birth. In mild cases of Rhesus incompatibility, labor may be induced prior to 37 weeks; in more severe cases, it may be as early as 26 weeks. If a fetus is too ill or immature to be delivered, it may be given a transfusion of Rh-negative blood. Antibody injections are given to the mother during each pregnancy to destroy any fetal blood cells that enter her circulation, thereby preventing antibody formation, and prevent this complication.

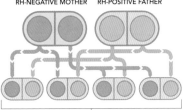

HOW RHESUS STATUS IS PASSED DOWN
Every person has two versions of the gene for Rhesus (Rh) status. If one is the Rh-negative version and the other Rh-positive, the Rh-positive version will prevail.

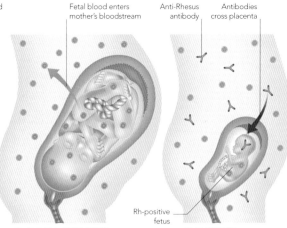

1 THE FIRST PREGNANCY
An Rh-negative mother will not have incompatibility problems with an Rh-positive baby. Problems arise in later pregnancies if the mother is Rh negative and the fetus Rh positive.

2 DURING CHILDBIRTH
Leakage of fetal red blood cells into the mother's circulation during childbirth will cause her to produce antibodies. These antibodies will react against Rh-positive red blood cells in any subsequent pregnancies.

3 THE NEXT PREGNANCY
The mother's antibodies move freely across the placenta and destroy Rh-positive fetal red blood cells, which causes anemia in the fetus.

URINARY TRACT INFECTIONS

BACTERIAL INFECTIONS OF THE URINARY TRACT ARE COMMON DURING PREGNANCY DUE TO THE DELAYED CLEARING OF URINE.

Hormonal changes in pregnancy and the enlarged uterus delay the urine flow, which makes pregnant women susceptible to urinary infections. Symptoms include a burning sensation when urinating, frequent urination, and pain in the lower abdomen, lower back, or on one side. Fever and pain in the kidney area may indicate that the infection has spread up the urinary tract. A urine test may be done to confirm the diagnosis; treatment is with antibiotics. Untreated, a urinary tract infection may lead to premature labor or a low-birthweight baby.

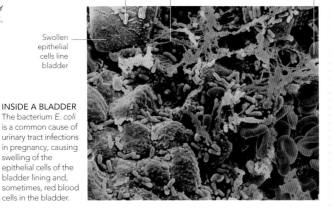

E. coli bacteria cover bladder's interior surface

Mucus filaments secreted by epithelial cells

Red blood cells result from bleeding caused by infection

Swollen epithelial cells line bladder

INSIDE A BLADDER
The bacterium *E. coli* is a common cause of urinary tract infections in pregnancy, causing swelling of the epithelial cells of the bladder lining and, sometimes, red blood cells in the bladder.

CARPAL TUNNEL SYNDROME

THIS TINGLING, NUMBNESS, AND PAIN IN THE HAND IS DUE TO COMPRESSION OF A NERVE IN THE WRIST.

One of the nerves to the hand passes through a small gap (the carpal tunnel) between the wrist bones and the ligament over them. In pregnancy, swelling of the tissues can reduce this gap, compressing the nerve and causing tingling, numbness, and sometimes pain in the hand. Bending and straightening the wrist and fingers may help relieve the symptoms. The condition usually clears up after delivery, but sometimes surgery may be needed to release pressure on the nerve and relieve the symptoms.

SCIATICA

THIS PAIN SPREADS FROM THE BUTTOCK DOWN THE BACK OF THE LEG DUE TO PRESSURE ON THE SCIATIC NERVE.

The changes in posture during pregnancy can put pressure on the sciatic nerve, which runs down the back of the leg and divides at the knee to go to the outer border and sole of the foot. In addition to pain, sciatica may also make it difficult to stand upright, and even to walk if the condition is severe. The symptoms tend to be intermittent and usually clear up after birth. In the meantime, they may be alleviated by adopting good posture, with the shoulders pulled back, the spine kept straight, the bottom tucked under, the abdomen tucked in, and the knees kept relaxed.

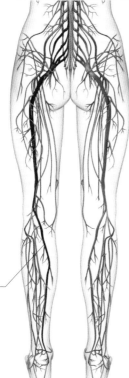

Sciatic nerve runs from buttock down back of thigh and divides at knee to supply foot

THE COURSE OF THE SCIATIC NERVE
The sciatic nerve, the largest nerve in the body, is formed when nerves from the lower spine combine to form one thick nerve. The nerve and its branches extend along the length of the leg.

EDEMA

SWELLING DUE TO FLUID BUILDUP IS COMMON IN PREGNANCY. THE FEET, LEGS, AND HANDS MAY BE AFFECTED.

Fluid retention is particularly common in the last few months of pregnancy, affecting up to 80 percent of healthy pregnant women. The accumulated fluid causes swelling, which tends to disappear during the night after lying in bed and then gradually worsens through the day. It may be improved by raising the legs when sitting and by staying mobile—for example, by walking or swimming—which aids the circulation; support hose can also be helpful. The fluid retention is usually no cause for concern, although it may occur as a symptom of preeclampsia (see p.229).

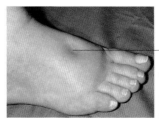

Pressure causes depression in skin that disappears only gradually when presssure is removed

SWOLLEN FOOT
Fluid usually accumulates first in the feet and, if severe, may extend up the leg; the hands may also be affected. Pressing the swelling causes a long-lasting depression.

VARICOSE VEINS

SWOLLEN VEINS IN THE LEGS MAY DEVELOP OR WORSEN DURING PREGNANCY DUE TO PRESSURE FROM THE ENLARGED UTERUS.

In the later stages of pregnancy, the enlarged uterus places pressure on the deep veins that carry blood away from the legs, causing the valves that normally prevent backflow in them to become leaky. As a result, blood builds up in superficial veins, which drain into the deep veins, causing the superficial veins to become swollen and distorted (varicosed). The problem may be helped by keeping mobile, raising the legs when sitting, and wearing support hose. Injection therapy and surgery may be options after pregnancy, if treatment is required.

PROBLEMS SPECIFIC TO MULTIPLE PREGNANCIES

A multiple pregnancy increases the risk of problems for the mother and her babies. The normal complaints, such as morning sickness, are often worse due to the higher hormone levels and the larger uterus. There is also an increased risk of developing medical problems, such as iron-deficiency anemia, high blood pressure, preeclampsia (see p.229), hyperemesis gravidarum (see opposite), placenta previa (see p.228), polyhydramnios (see p.228), and miscarriage (see p.226). Babies tend to be small, and premature labor is more likely. A multiple pregnancy requires special monitoring but most have a good outcome.

THREE IN A WOMB
Triplets occur in only 1 in 8,000 births. Multiple pregnancies from assisted conception are now less common due to restrictions on the number of embryos used.

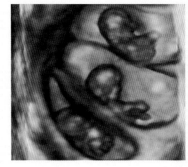

LABOR AND DELIVERY PROBLEMS

FOR MANY WOMEN, LABOR AND DELIVERY ARE UNPROBLEMATIC, INTENSE, AND JOYFUL. BUT FOR SOME, PROBLEMS ARISE EITHER FOR THE FETUS OR FOR THE MOTHER. FOR INSTANCE, LABOR SOMETIMES STARTS BEFORE PREGNANCY HAS REACHED ITS FULL TERM, OR THE FETUS MAY SHOW SIGNS OF DISTRESS OR BECOME STUCK AND NEED URGENT DELIVERY. FROM THE MOTHER'S POINT OF VIEW, TISSUES AROUND THE VAGINAL OPENING CAN TEAR DURING CHILDBIRTH AND SOMETIMES AN ASSISTED DELIVERY MAY RESULT IN INJURY.

PREMATURE LABOR

THIS TERM DESCRIBES LABOR THAT BEGINS BEFORE 37 WEEKS. BABIES BORN PREMATURELY MAY HAVE ASSOCIATED COMPLICATIONS (SEE P.234).

Causes of early labor include multiple pregnancy and maternal infections. Often no reason is found. Several factors can increase the risk, including smoking or drinking alcohol when pregnant, stress, and a previous premature labor. Tightenings in the abdomen, normally painless become painful and regular with bloody mucus discharge and pains in the lower back. If the onset of labor is very premature, the doctor may try to halt its progress with drugs given to the mother intravenously. If this is not possible, she may be given corticosteroid injections to help the fetal lungs mature. Premature babies, depending on how early they arrive, may need care in a special unit while their organs mature.

PREMATURE TRIPLETS
Women who have multiple pregnancies are more likely to go into labor early. It may be because the uterus is overly stretched.

FETAL DISTRESS

PARTICULAR SIGNS INDICATE WHEN A FETUS IS NOT WELL OR IS NOT RESPONDING AS NORMAL OR EXPECTED DURING PREGNANCY OR LABOR.

Fetal distress may be suggested by reduced fetal movements felt by the mother, meconium (fetal feces) in the amniotic fluid, and problems with the fetal heart rate, which may be faster than it should be (tachycardia), slower than it should be (bradycardia), or not showing as much variability as it should (the fetal heart rate usually fluctuates, with marked increases with maternal contractions). Possible causes include placental abruption (see p.228), but a cause may not be found. If necessary, the baby is delivered immediately either vaginally or by cesarean section.

USING CARDIOTOCOGRAPHY TO MONITOR FETAL WELL-BEING
Cardiotocography records the fetal heart rate and frequency of maternal contractions continuously. The heart rate should rise briefly with each contraction, which can be checked on the printed traces.

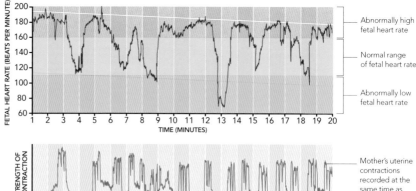

Abnormally high fetal heart rate

Normal range of fetal heart rate

Abnormally low fetal heart rate

Mother's uterine contractions recorded at the same time as fetal heart rate

CORD PROLAPSE

THIS EMERGENCY SITUATION ARISES WHEN THE UMBILICAL CORD APPEARS THROUGH THE CERVIX BEFORE THE PRESENTING PART OF THE FETUS (THE PART NEAREST THE CERVIX), WHICH CAN COMPROMISE ITS BLOOD SUPPLY.

Cord prolapse usually occurs during labor, but occasionally it happens when the waters break during pregnancy. The fetus may compress the cord, reducing the blood supply it receives. A cord prolapse may occur when a fetus is not engaged (see p.189), when it is not head-down (particularly if it is lying across the uterus), when there is a multiple pregnancy, or when there is excess amniotic fluid (polyhydramnios, see p.228). It is essential for the mother to move into the correct position (see below) if a cord prolapse is present. An urgent vaginal delivery may be possible (if necessary using forceps or vacuum suction) if the cervix is fully dilated; otherwise an emergency cesarean section will be performed immediately.

RELIEVING PRESSURE
The mother kneels on all fours. The physician or midwife may insert a hand into the vagina to hold the baby away from the umbilical cord.

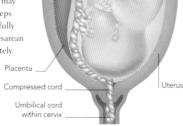

Placenta

Compressed cord

Uterus

Umbilical cord within cervix

BABY PRESSING ON CORD
If the baby presses on the umbilical cord, compressing the blood vessels it contains, it can reduce the flow of blood, and oxygen, from the placenta to the baby.

RETAINED PLACENTA

SOMETIMES THE PLACENTA OR MEMBRANES DO NOT DETACH FROM THE WALL OF THE UTERUS AS THEY SHOULD AFTER THE BABY HAS BEEN DELIVERED.

A retained placenta may occur for a number of reasons, including uterine atony (the uterus stops contracting as it should to expel the placenta) or the uncommon condition placenta accreta, present when part of the placenta cannot detach itself because it is deeply embedded in the uterine wall. If part or all of the placenta or membranes are retained, the uterus cannot contract effectively so that bleeding from the uterine blood vessels persists. If the placenta remains stuck in the uterus, it will need to be delivered by hand, using a regional anesthetic (epidural or spinal, see pp.196–97) or a general anesthetic.

SHOULDER DYSTOCIA

THIS MEDICALLY URGENT SITUATION ARISES WHEN A BABY'S HEAD IS DELIVERED BUT ONE OF ITS SHOULDERS GETS STUCK BEHIND THE PUBIC SYMPHYSIS, THE JOINT AT THE FRONT OF THE MOTHER'S PELVIS.

Shoulder dystocia occurs unexpectedly during a normal vaginal delivery or an assisted delivery (helped by forceps or vacuum suction, see p.202). It causes problems because the baby cannot start to breathe and it may compress the umbilical cord. This is an emergency situation. The physician or midwife asks the mother to stop pushing and may also ask her to change her position to make more room for the baby to come out. He or she may press on the lower abdomen to dislodge the shoulder and may also try to reposition the baby vaginally. An episiotomy may be made to provide more room for the delivery. Shoulder dystocia may result in damage to the network of nerves that supply the arm (the brachial plexus).

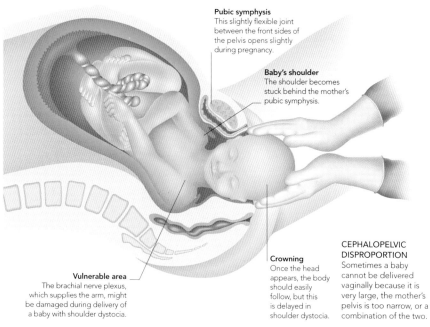

Pubic symphysis
This slightly flexible joint between the front sides of the pelvis opens slightly during pregnancy.

Baby's shoulder
The shoulder becomes stuck behind the mother's pubic symphysis.

Vulnerable area
The brachial nerve plexus, which supplies the arm, might be damaged during delivery of a baby with shoulder dystocia.

Crowning
Once the head appears, the body should easily follow, but this is delayed in shoulder dystocia.

CEPHALOPELVIC DISPROPORTION
Sometimes a baby cannot be delivered vaginally because it is very large, the mother's pelvis is too narrow, or a combination of the two.

GROUP B STREP TRANSFER

THIS BACTERIAL INFECTION CAN CAUSE PROBLEMS FOR A NEWBORN IF IT IS TRANSFERRED FROM MOTHER TO BABY DURING PREGNANCY OR DELIVERY.

Group B streptococcus is present normally in the intestine and vagina of many women (up to about one-third). In some of these women, the infection will be transferred to the fetus either in utero or during the delivery. Certain factors increase the risk of passing the bacteria on to the baby, such as premature labor (that is, before 37 weeks) or a urinary tract infection caused by Group B streptococcus. Symptoms in an affected baby may include fever, breathing problems, problems with feeding, and fits. Babies may have blood tests to detect the infection, which is treatable with antibiotics.

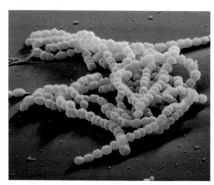

GROUP B STREPTOCOCCAL CHAINS
These bacteria, which can be present in the bowel and others areas in healthy adults without causing any problems, can have very serious effects if transferred to a newborn.

PERINEAL TEARS

WHEN TISSUES UNDERGO EXTREME STRETCHING AS A BABY PASSES DOWN THE BIRTH CANAL, TEARS CAN OCCUR BETWEEN THE VAGINAL OPENING AND THE ANUS.

Tears may range from a tiny one involving the edge of the vaginal opening to those that involve deeper layers of muscle or extend to the anus. Small tears may also occur in the upper vagina. Rarely, the cervix or the labia are involved. A number of factors put women at increased risk of a tear, including a first vaginal delivery, a previous severe tear, an assisted delivery, a big baby, or a baby facing forward rather than downward. Sometimes, episiotomy cuts (see p.202) tear further. Stitches may be needed to bring the torn layers of a tear back together to heal.

TISSUES INVOLVED IN PERINEAL TEARS
Tears can extend from the edge of the vagina toward the anus. If tears extend to the deeper tissues, they may take some weeks to heal.

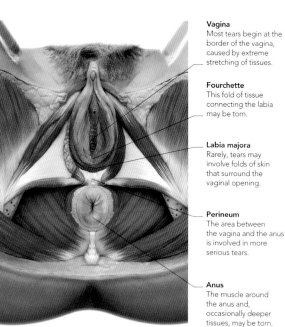

Vagina
Most tears begin at the border of the vagina, caused by extreme stretching of tissues.

Fourchette
This fold of tissue connecting the labia may be torn.

Labia majora
Rarely, tears may involve folds of skin that surround the vaginal opening.

Perineum
The area between the vagina and the anus is involved in more serious tears.

Anus
The muscle around the anus and, occasionally deeper tissues, may be torn.

CLASSIFICATION OF PERINEAL TEARS	
DEGREE	**DETAILS OF INVOLVEMENT**
First	The most common tears, these affect skin and tissues around the vaginal opening but no muscles are involved. They may need stitches or may heal themselves.
Second	When muscle around the vagina is involved, tears tend to be quite sore. Dissolvable stitches are used to repair the layers, and healing takes a few weeks.
Third	Vaginal tissue, perineal skin, and muscles beneath, as well as muscle around the anus (anal sphincter), are involved in a third-degree tear. All layers need stitching.
Fourth	When a third-degree tear extends into the tissues beneath the anal sphincter, it becomes a fourth-degree tear. Many stitches are needed to reposition all the tissues.

PROBLEMS IN NEWBORNS

NEWBORN BABIES ARE AT RISK OF VARIOUS ILLNESSES, SUCH AS INFECTIONS PASSED ACROSS THE PLACENTA OR DURING BIRTH. THESE ILLNESSES MAY ARISE AS A RESULT OF PREMATURE BIRTH OR PROBLEMS DURING PREGNANCY OR DELIVERY—OR OCCUR FOR NO OBVIOUS REASON. PEDIATRICIANS ARE SKILLED IN MANAGING THESE CONDITIONS, WHICH SOMETIMES REQUIRE CARE IN A NEONATAL INTENSIVE CARE UNIT.

PREMATURE BIRTH COMPLICATIONS

Premature babies are particularly susceptible to certain medical problems, especially if they are born very early or have a very low birth weight. This is because they have had less time to develop; this lack of maturity can be seen particularly clearly in respiratory distress syndrome (see lung problems, below). Continuing advances in the management of premature babies have improved the outlook for them, but nevertheless they are still at increased risk of certain chronic problems that require long-term treatment.

LUNG PROBLEMS

PREMATURITY IS ASSOCIATED WITH A NUMBER OF RESPIRATORY PROBLEMS IN THE NEWBORN BABY, SUCH AS RESPIRATORY DISTRESS SYNDROME AND EPISODES IN WHICH THE BABY'S BREATHING IS ABNORMALLY SLOW OR EVEN STOPS COMPLETELY.

Respiratory distress syndrome is most likely to affect babies born before 28 weeks' gestation. It is due to lack of a substance called surfactant, which helps to keep the lungs' tiny air sacs (alveoli) open. As a result, the surface area of the lungs is reduced, causing the baby's breathing to be labored and the breathing rate to be higher than normal. If a premature birth is anticipated, steroids may be given to the woman during pregnancy to help the baby's lungs mature. After delivery, surfactant may be given directly into the baby's lungs through a tube. A chest X-ray may be taken to confirm the diagnosis. Oxygen is given, and assisted ventilation may be needed, either in the form of CPAP (continuous positive airways pressure), which maintains the pressure in the airways between breaths, or mechanical ventilation, in which a machine takes over breathing for the baby. Episodes of slow

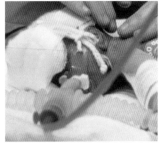

BREATHING ASSISTANCE
A premature baby may need help with breathing, either to keep the airways open or to take over breathing while the lungs mature.

or absent breathing are also common in premature infants. Possible causes include low oxygen levels or low blood-sugar levels, but in many cases no cause can be found. A respiratory stimulant drug may be needed, and, in some cases, CPAP.

BRAIN HEMORRHAGE

BLEEDING INTO THE BRAIN IS COMMON IN VERY PREMATURE BABIES, USUALLY OCCURRING WITHIN THE FIRST 72 HOURS AFTER THE BIRTH. THE PROBLEMS CAUSED VARY GREATLY, DEPENDING ON THE SEVERITY AND SITE OF THE BLEEDING.

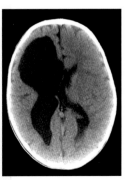

Brain hemorrhages are more common in babies with severe respiratory distress syndrome (see above) and those who have been deprived of oxygen around the time of birth. Some hemorrhages may result in cerebral palsy (see opposite page), in which problems arise from damage to the brain's nerve tissue and a buildup of fluid in the brain (hydrocephalus). CT or ultrasound scanning is used to assess the location and size of the hemorrhage. With hydrocephalus, excess fluid may be removed or a permanent shunt may be inserted that diverts the excess fluid from the brain into the abdomen.

BLEEDING IN THE BRAIN
In this CT brain scan of a preschool child, there is partial disappearance of the brain cavities due to a hemorrhage.

RETINOPATHY OF PREMATURITY

THIS CONDITION (OFTEN SIMPLY CALLED ROP) AFFECTS THE DEVELOPMENT OF BLOOD VESSELS IN THE RETINA, THE INNERMOST LAYER OF THE EYE THAT CONTAINS LIGHT-SENSITIVE CELLS AND NERVE CELLS THAT SEND MESSAGES TO THE BRAIN TO FORM IMAGES.

This condition affects about 20 percent of very low birth weight babies who are born before 31 weeks' gestation. There is abnormal development of the retinal blood vessels, which grow excessively in some areas of the retina but do not reach others. The abnormal vessels are fragile and can leak, damaging the retina and impairing vision. In severe cases, the condition may progress to detachment of the retina from the underlying tissue layers and loss of vision. ROP is diagnosed and assessed by imaging the retina. Mild cases may clear up by themselves, but in more severe cases, laser treatment may be needed to reduce visual impairment.

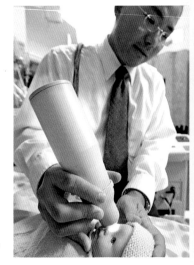

IMAGING THE RETINA
Premature babies can be checked for retinopathy using a retinal camera (shown here being demonstrated on a doll).

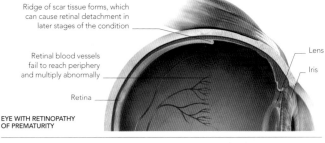

Ridge of scar tissue forms, which can cause retinal detachment in later stages of the condition

Retinal blood vessels fail to reach periphery and multiply abnormally

Retina

Lens

Iris

EYE WITH RETINOPATHY OF PREMATURITY

HEALTHY EYE

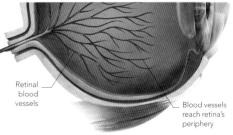

BLOOD VESSEL DEVELOPMENT
In retinopathy of prematurity some areas of the retina do not have blood vessels and so do not receive the oxygen and nutrients they need.

Retinal blood vessels

Blood vessels reach retina's periphery

MEDICAL CONDITIONS

What happens to babies during pregnancy, delivery, and the time soon after birth affects their health. Certain infections passed to a baby during pregnancy or on the journey down the birth canal can cause problems, as can excessive drinking by the mother during pregnancy. Damage to the brain at or around delivery can cause cerebral palsy. Jaundice is another common problem in the newborn.

NEONATAL JAUNDICE

A COMMON SYMPTOM IN NEWBORN BABIES, JAUNDICE IS A YELLOW COLORATION OF THE SKIN AND WHITES OF THE EYES. THIS IS USUALLY CONSIDERED NORMAL AND CLEARS UP BY ITSELF IN A FEW DAYS.

Jaundice is due to high levels of the pigment bilirubin, which is formed naturally in the body. The liver, which normally disposes of bilirubin, may not function properly at first, causing bilirubin levels to rise, but this usually corrects itself in a few days, although phototherapy may be needed. Occasionally jaundice may be due to an underlying problem, such as Rhesus incompatibility (see p.230), an infection, or a liver abnormality. In such cases, jaundice may be severe and, if left untreated, may affect hearing and brain function.

PHOTOTHERAPY
A newborn baby is given light therapy, during which the light waveform breaks down bilirubin, reducing the jaundice.

CONGENITAL INFECTIONS

THESE ARE INFECTIONS THAT A BABY HAS WHEN IT IS BORN, WHICH HAVE BEEN TRANSFERRED FROM THE MOTHER EITHER IN PREGNANCY OR DURING DELIVERY.

In early pregnancy, fetal development may be disrupted by infections such as rubella, which can cause heart defects. Some infections in early pregnancy can also lead to miscarriage. Later in pregnancy, certain infections may lead to premature labor and illnesses in the newborn. Infections that can be passed on during delivery include streptococcus and herpes. Preventive measures include immunization against rubella and taking care with food hygiene. Cesarean section may be recommended for some women, including those with HIV or genital herpes.

FETAL ALCOHOL SYNDROME

DRINKING TOO MUCH ALCOHOL IN PREGNANCY MAY LEAD TO FETAL ALCOHOL SYNDROME, WITH FEATURES INCLUDING HEART PROBLEMS, LEARNING DIFFICULTIES, AND CERTAIN DISTINCTIVE FACIAL FEATURES.

The features of fetal acohol syndrome (FAS) vary from one individual to another but typically include reduced growth, developmental delay, heart abnormalities, and the presence of certain facial characteristics. Diagnosis of the condition is based on the features present. Affected children may require surgery to treat heart defects and special help at school for learning difficulties. There are also often behavioral problems. The syndrome is a lifelong condition, and affected individuals may be unable to live independently later in life.

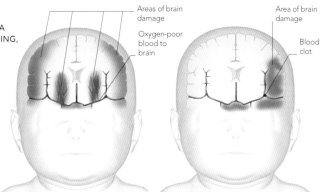

Small eyes
Skin folds under eyes
Low-set ears
Flat midface
High-arched eyebrows
Eyelid droop (ptosis)
Saddle-shaped nose
Smooth, indistinct philtrum (the groove between nose and upper lip)
Thin upper lip
Short jaw (micrognathia)

CHARACTERISTIC FACIAL FEATURES
Children affected by fetal alcohol syndrome tend to have characteristic facial features.

Corpus callosum

MRI SCAN OF CHILD'S BRAIN
The corpus callosum (colored purple), which connects the two hemispheres of the brain, is commonly affected in fetal alcohol syndrome.

CEREBRAL PALSY

THIS MOVEMENT DISORDER OCCURS AS A RESULT OF BRAIN DAMAGE BEFORE, DURING, OR IN THE EARLY YEARS AFTER BIRTH.

Cerebral palsy may occur for no apparent reason or it may result from a congenital infection (see above) or from oxygen deprivation during birth. Very premature babies are particularly at risk because they are prone to bleeding into the brain. Meningitis or a head injury in early life can also be a cause. Symptoms only tend to become apparent after several months, and may include limb weakness, lack of control of movement, swallowing problems, developmental delay, and vision and hearing problems. About a quarter of affected children have learning difficulties. Cerebral palsy is lifelong but does not progress. Treatment and support are tailored to meet individual needs.

Areas of brain damage
Oxygen-poor blood to brain
Area of brain damage
Blood clot

OXYGEN STARVATION AT BIRTH
If the brain is deprived of oxygen during birth, there will be generalized brain damage, which may cause a wide range of symptoms.

STROKE IN A NEWBORN
If a clot deprives one brain area of blood, damage is localized and only the actions controlled by that area will be affected.

CONGENITAL HYPOTHYROIDISM

IN A BABY BORN WITH AN UNDERACTIVE THYROID GLAND, INSUFFICIENT THYROID HORMONES ARE PRODUCED.

Thyroid hormones regulate the body's metabolism. Symptoms of deficiency tend to be noticed only as the child gets older. They include a failure to grow and put on weight, feeding problems, prolonged jaundice, dry, mottled skin, a large tongue, and a hoarse cry. There are also learning difficulties. Screening is carried out on all newborn babies so that treatment can be started as early as possible to prevent problems from developing. Treatment is with thyroid hormone supplements and is lifelong. In most cases, children treated early develop normally and do not have learning difficulties.

235

CHROMOSOMAL AND GENETIC DISORDERS

The way in which the body develops, grows, and functions is determined by the 20,000–25,000 pairs of genes arranged in 23 pairs of chromosomes in the body cells. Gene and chromosomal abnormalities may sometimes cause no noticeable problems, but they can also produce a wide range of disorders, each one being rare and affecting one or more body systems. These disorders may develop as a result of an incorrect number of one of the chromosomes—as occurs, for example, in Down syndrome and Turner syndrome—or as a result of a defect in one of the genes, as is the case with cystic fibrosis.

NEUROFIBROMATOSIS

THIS IS A GENETIC DISORDER IN WHICH NON-CANCEROUS GROWTHS (NEUROFIBROMAS) DEVELOP ON NERVE FIBERS THROUGHOUT THE BODY.

Symptoms usually develop in childhood and include flat, brown patches and freckles on the skin, and soft swellings under the skin, which may be small or large and disfiguring. Further problems may develop if the swellings press on nearby tissues. There may also be learning difficulties, and some children develop epilepsy. Rarely, the neurofibromas become cancerous. A rarer form of the condition affects adults. In this form, tumors do not develop under the skin but the inner ear is often affected, which can cause hearing problems. In both cases, CT or MRI scanning may be arranged to look for tumors. There is no cure, but large tumors may be removed if they cause problems. Educational support may be needed for children with learning difficulties.

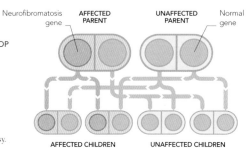

Neurofibromatosis gene — AFFECTED PARENT | UNAFFECTED PARENT — Normal gene

AFFECTED CHILDREN | UNAFFECTED CHILDREN

AUTOSOMAL DOMINANT INHERITANCE PATTERN
Neurofibromatosis is inherited in an autosomal dominant way. If both a neurofibromatosis and a normal gene are present, the neurofibromatosis gene will override the normal one.

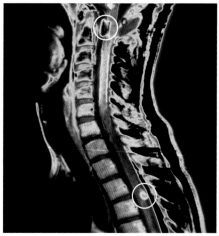

SPINAL NEUROFIBROMAS
This color-enhanced MRI scan shows the presence of two large neurofibromas (colored green) in the spinal cord (colored purple) of the chest and lower back.

PHENYLKETONURIA

IN THIS GENETIC DISORDER THERE IS LACK OF AN ENZYME THAT BREAKS DOWN PHENYLALANINE IN PROTEIN, WHICH MAY LEAD TO BRAIN DAMAGE.

Phenylketonuria (PKU) is a rare autosomal recessive disorder in which the body does not produce an enzyme that breaks down phenylalanine, a substance in protein-containing foods. Instead, it is broken down into harmful chemicals. Symptoms usually develop between six and 12 months and may include developmental delay, vomiting, and seizures. If PKU is not treated, brain damage can result, causing learning difficulties. Treatment is with special milk containing enough protein but little phenylalanine, and later a diet low in phenylalanine. With early treatment, children develop normally. All babies are screened for PKU shortly after birth.

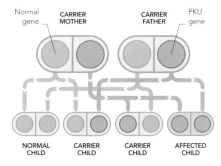

Normal gene — CARRIER MOTHER | CARRIER FATHER — PKU gene

NORMAL CHILD | CARRIER CHILD | CARRIER CHILD | AFFECTED CHILD

AUTOSOMAL RECESSIVE INHERITANCE PATTERN
A child must inherit the PKU gene from both parents to develop the condition. If one PKU and one normal gene are inherited, the child will not have PKU but will be a carrier.

CYSTIC FIBROSIS

THIS GENETIC DISORDER AFFECTS MUCUS-PRODUCING GLANDS AROUND THE BODY, CAUSING THEM TO PRODUCE ABNORMALLY THICK MUCUS.

Cystic fibrosis (CF) is one of the more common genetic disorders: about 1 in 2,500 babies is born with the condition and 1 in 25 people carry the CF gene. It is inherited in an autosomal recessive manner, so a child must inherit two copies of the CF gene to be affected. The disorder affects all the mucus-secreting glands but the lungs and pancreas are particularly affected; the latter often produces insufficient digestive enzymes due to obstruction by the thick mucus. A newborn baby with CF may have a distended abdomen and fail to pass feces for a few days. Later, the infant may grow slowly, fail to put on weight, suffer from recurrent chest infections, and produce pale, greasy feces. Permanent lung damage, liver damage, and diabetes mellitus may also develop. High levels of salt are found in the sweat, and this may be used to make the diagnosis. Regular physical therapy is needed to clear mucus in the airways, and antibiotics are given for chest infections. Other treatments include a high-calorie diet, vitamins, and enzymes to aid digestion. A heart–lung transplant may be possible in some cases. All babies are tested for CF shortly after birth.

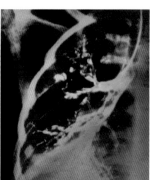

MUCUS-FILLED LUNGS
This colored chest X-ray of a person with CF shows that some of the airways are filled with mucus (green), which will cause breathing difficulty as well as a persistent cough.

EFFECTS OF CYSTIC FIBROSIS
CF can affect various parts of the body, but the main areas involved are the lungs and pancreas, which produces digestive enzymes.

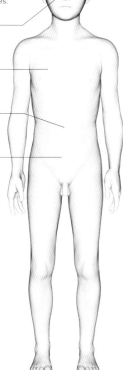

Sinuses
These cavities in the skull become inflamed (sinusitis).

Lungs
Mucus buildup in the lungs leads to coughing, breathing difficulty, and infections.

Pancreas
Digestion is inefficient because the pancreas fails to produce sufficient enzymes.

Intestines
Problems with the absorption of nutrients occur.

DOWN SYNDROME

THIS GROUP OF PHYSICAL AND MENTAL PROBLEMS IS CAUSED BY THE PRESENCE OF AN EXTRA COPY OF ONE OF THE CHROMOSOMES (CHROMOSOME 21).

Down syndrome is the most common chromosomal abnormality, and maternal age is a major risk factor for having an affected baby. The features and severity of the condition vary from one individual to another but typically include short stature, characteristic facial features, and learning difficulties. Children with Down syndrome are at increased risk of congenital heart defects, respiratory problems, leukemia, vision and hearing problems, and an underactive thyroid gland. They are also at increased risk of developing dementia from the age of 40. During pregnancy, tests are offered to give an indication of the risk of having an affected child and, if necessary, amniocentesis or chorionic villus sampling will be offered to make a definitive diagnosis. If the condition is not detected before birth, it can be confirmed later by chromosome analysis. A child with Down syndrome may need long-term specialized care and treatment; parents may also need support.

BABY WITH DOWN
This baby has the round face, almond-shaped eyes, flat-bridged nose, small chin, and protruding tongue that are typical of children with Down syndrome.

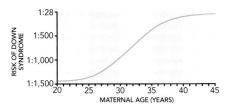

RISK OF HAVING A BABY WITH DOWN SYNDROME
Maternal age is the most important risk factor for having a baby with Down syndrome. The risk increases with a woman's age, reaching about 1:900 by age 30 and 1:28 by age 45.

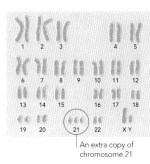

TRISOMY 21
This picture of an individual's chromosomes shows that there are three copies of chromosome 21 (trisomy 21), revealing that the person has Down syndrome.

An extra copy of chromosome 21

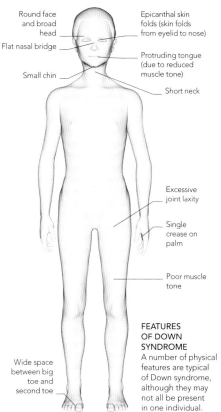

Round face and broad head

Flat nasal bridge

Small chin

Epicanthal skin folds (skin folds from eyelid to nose)

Protruding tongue (due to reduced muscle tone)

Short neck

Excessive joint laxity

Single crease on palm

Poor muscle tone

Wide space between big toe and second toe

FEATURES OF DOWN SYNDROME
A number of physical features are typical of Down syndrome, although they may not all be present in one individual.

TURNER SYNDROME

A RARE CHROMOSOMAL DISORDER, THIS OCCURS WHEN A GIRL HAS ONLY ONE COPY OF THE FEMALE X CHROMOSOME INSTEAD OF THE USUAL TWO.

At birth, features of Turner syndrome may include puffy feet, a broad chest, low-set ears, a short, wide neck, and feeding difficulties. However, there may be no signs until later in childhood, when short stature becomes apparent or when there is delay in the onset of puberty. Other problems may include abnormal narrowing of the aorta, kidney abnormalities, hearing problems, and, later, infertility. Chromosome analysis is used to confirm the diagnosis. Estrogen and growth-hormone supplements may be given to stimulate growth and help bring on normal puberty; estrogen is continued for life. Other disorders are treated as appropriate—for example, surgery to treat narrowing of the aorta.

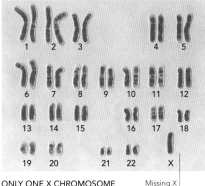

ONLY ONE X CHROMOSOME
This set of chromosomes from a woman with Turner syndrome shows an X chromosome is missing.

Missing X chromosome

NEONATAL SCREENING

PHYSICAL EXAMINATION
Babies are checked for a number of conditions soon after birth and again at six weeks. Their appearance is checked, and hearing tests are also offered. The checks include:

CONDITION	WHAT IS DONE
Physical abnormalities	The physical appearance is checked carefully for signs of conditions such as spina bifida and cleft palate. The reflexes may also be checked.
Congenital dysplasia of the hips	The hip joints are manipulated to check that the top of the femur (thighbone) is positioned securely in the socket of the pelvis.
Abnormal positioning of the testes	A boy's testes are examined to check that they sit within the scrotum.
Congenital cataracts	A light is shone into the eyes to check whether there are any opacities in the lens.
Congenital heart disease	The heart is checked for various structural abnormalities by listening for heart murmurs with a stethoscope.

BLOOD TESTS
These tests check for certain genetic disorders using a heel-prick blood sample taken from babies within a week of birth. The tests vary by state, but all test for phenylketonuria and congenital hypothyroidism.

CONDITION	WHAT IS DONE
Phenylketonuria (PKU)	The level of phenylalanine is checked. In PKU, harmful breakdown products of phenylalanine build up and can cause brain damage.
Congenital hypothyroidism	Levels of thyroid hormones are checked. A lack of thyroid hormones can lead to feeding problems, poor growth, and developmental delay.
Cystic fibrosis (CF)	The level of trypsinogen (an enzyme produced by the pancreas) is measured. CF causes recurrent chest infections, slow growth, and digestive problems.
Sickle cell disease	Levels of abnormal hemoglobin are checked. Sickle cell disease affects the red blood cells and can be associated with anemia and delayed growth.
MCADD (medium-chain acyl-CoA dehydrogenase deficiency)	Levels of an enzyme (MCAD) needed to metabolize fats properly are checked. Deficiency of the enzyme can lead to a harmful buildup of toxins.

ANATOMICAL PROBLEMS

Problems can occur at any stage in the development of a fetus and may affect the structure of one or several areas of the body. Some anatomical problems are immediately obvious at birth because they are clearly visible, such as a cleft lip. Other problems are internal—heart defects, for example—and may take time to show themselves through symptoms developing or signs being detected during routine newborn examinations. It is usually possible to treat most anatomical problems.

HEART DEFECTS

A NUMBER OF STRUCTURAL HEART ABNORMALITIES MAY BE PRESENT AT BIRTH; SOME MAY RESOLVE ON THEIR OWN, BUT SOME REQUIRE SURGICAL CORRECTION.

Heart defects may be due to persistence of the special features of the fetal heart that normally disappear at birth, such as an open foramen ovale and patent (open) ductus arteriosus. Alternatively, they may result from failure of the fetal heart to develop normally during pregnancy, as with coarctation of the aorta (narrowing of the body's main artery close to the heart) and valve defects. Sometimes, several problems are present. Heart defects may cause shortness of breath, which may affect feeding and thereby impair growth. They may be detected as murmurs during a routine examination, or may be found when investigating symptoms. If a defect is suspected, the heart can be examined using echocardiography (ultrasound scanning of the heart). Many defects clear up without treatment, but about a third need corrective surgery.

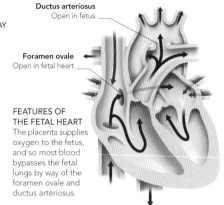

Ductus arteriosus
Open in fetus

Foramen ovale
Open in fetal heart

FEATURES OF THE FETAL HEART
The placenta supplies oxygen to the fetus, and so most blood bypasses the fetal lungs by way of the foramen ovale and ductus arteriosus.

KEY
← OXYGEN-RICH BLOOD
← OXYGEN-POOR BLOOD
← MIXED BLOOD

Ductus arteriosus
Has closed

Aorta

Mixing of blood
Oxygen-rich blood mixes with oxygen-poor blood

Ductus arteriosus
Is still open; it should have closed

Foramen ovale
Has closed

Foramen ovale
Is still open; it should have closed

Left ventricle

HEALTHY NEWBORN HEART
With the first breath, the newborn lungs are inflated, triggering changes in the heart that allow it to work independently of the placenta. The foramen ovale and the ductus arteriosus both close.

HEART WITH OPEN FORAMEN OVALE
If the foramen ovale fails to close, oxygen-rich blood is able to travel to the right side of the heart, where it recirculates to the lungs. This results in a less efficient circulation.

HEART WITH OPEN DUCTUS ARTERIOSUS
If this small duct persists in the newborn, oxygen-poor blood passes into the aorta, where it joins oxygen-rich blood from the left ventricle.

PYLORIC STENOSIS

IN THIS CONDITION, THE OUTLET OF THE STOMACH IS NARROWED, WHICH PREVENTS FOOD FROM PASSING FROM THE STOMACH INTO THE SMALL INTESTINE.

Pyloric stenosis is about five times more common in boys than in girls, but its cause is not known. Symptoms tend to develop between three and eight weeks after birth. The main symptom is persistent vomiting, which may be very forceful (projectile vomiting) and cause hunger immediately afterward. Admission to the hospital for intravenous fluids may be necessary because affected babies tend to become dehydrated. The doctor will examine a baby's abdomen, sometimes during a feeding, and an ultrasound scan or special X-rays may be taken to confirm the diagnosis. The condition is treated by a surgical procedure to widen the stomach outlet, which usually cures the problem completely.

NEURAL TUBE DEFECTS

DEFECTS OF THE SPINAL CORD (SPINA BIFIDA) AND BRAIN ARISE DUE TO ABNORMAL DEVELOPMENT OF THE NEURAL TUBE DURING EARLY PREGNANCY.

If the neural tube (see p.99) fails to close properly, brain and spinal cord defects may be present at birth, ranging from a minor abnormality signified only by a dimple or tuft of hair on the lower back, to part of the spinal cord being exposed; rarely, the brain is affected. In severe cases, leg movement and sensation may be affected as well as bowel and bladder control. A fetal anomaly scan (see p.139) and blood tests may detect the condition in pregnancy. Taking folic acid supplements before and during pregnancy reduces the risk of neural tube defects.

Ribcage

Spinal cord bulges out of fetus's back

SPINA BIFIDA IN A FETUS
This 3-D ultrasound scan shows a bulge in the lower back where the spinal cord bulges through a gap in the spinal column.

HERNIA

HERE PART OF AN ORGAN, MOST COMMONLY THE GUT, PROTRUDES THROUGH A WEAKENED AREA OF MUSCLE, SOMETIMES CAUSING A VISIBLE BULGE.

A hernia may occur at various sites, but an inguinal hernia is particularly common in babies, especially boys. It typically causes an intermittent swelling in the groin or scrotum that appears when a baby cries. A hernia may become trapped (strangulated), in which case it appears as a persistent lump and is accompanied by vomiting and being unwell. A strangulated hernia is a serious condition that requires emergency treatment. To avoid this, early surgery is generally recommended for an inguinal hernia.

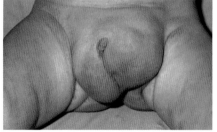

INGUINAL HERNIA ON EACH SIDE
This six-month-old boy has a bilateral inguinal hernia (a hernia on each side of the groin) that is so large it has extended down into the scrotum and obscured the genitals.

CONGENITAL DYSPLASIA OF THE HIP

PRESENT AT BIRTH, THIS PROBLEM OCCURS WHEN THE UPPER END OF THE FEMUR (THE "BALL" OF THE THIGHBONE) DOES NOT FIT PROPERLY INTO THE SOCKET OF THE PELVIS. IF UNTREATED, CONGENITAL HIP DYSPLASIA MAY CAUSE PROBLEMS WHEN WALKING STARTS.

Congenital hip dyplasia ranges from mild instability of the hip joint, through subluxation (in which the ball slips out of the socket but can be maneuvered back into position), to full dislocation (in which the ball of the femur lies completely outside the socket in the pelvis). The mildest forms may result from loose ligaments that allow the ball to move excessively. The more severe forms are due to failure of the hip socket to develop normally. Early detection of the condition can prevent other problems from developing and reduce the likelihood of surgical treatment being needed. Consequently, the condition is checked for during neonatal screening (see p.237), and in some cases ultrasound scanning may also be carried out. Left untreated, congenital hip dysplasia may lead to restricted leg movement, shortening of the affected leg, or a limp. If the condition is suspected, an orthopedic specialist will make an assessment. The condition may be

treated by placing the baby in a splinting device for several months to hold the ball of the femur in the socket of the pelvis. Progress is monitored with X-rays or ultrasound scans. If this treatment is unsuccessful, surgery to correct the hip dysplasia may be recommended.

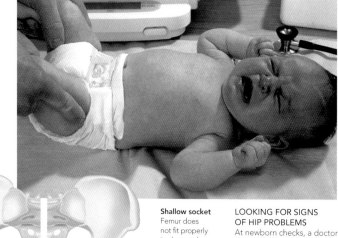

Pelvis
Provides the "socket" part of the hip's ball-and-socket joint

Head of femur
Provides the "ball" that fits into the socket

NORMAL HIP
The ball-shaped end of the upper femur fits snugly into the cup-shaped socket of the pelvis. This arrangement allows a wider range of movement than any other joint in the body.

Shallow socket
Femur does not fit properly in abnormal hip socket

LOOKING FOR SIGNS OF HIP PROBLEMS
At newborn checks, a doctor bends a baby's knees and manipulates the legs to see if the hip joints are stable or if the ball can be moved in and out of the socket.

HIP WITH POTENTIAL PROBLEMS
If the socket fails to develop properly in pregnancy, it does not form the cup needed to hold the ball securely. The surrounding tissues cannot hold the ball in the socket, and problems can develop.

CLEFT LIP AND PALATE

FAILURE OF THE UPPER LIP AND ROOF OF THE MOUTH TO CLOSE PROPERLY DURING DEVELOPMENT, THIS CONDITION SOMETIMES RUNS IN FAMILIES.

A cleft lip and palate are among the most common congenital defects. They may occur singly or together, and one or both sides may be affected. Risk factors for the condition include taking certain drugs (in particular, some anticonvulsants) during pregancy and drinking excessive amounts of alcohol when pregnant. Cleft lip and palate may cause problems with feeding, and speech may be affected if treatment is delayed. A buildup of fluid in the middle ear may also occur. The usual treatment is surgery. Surgical repair of the cleft lip is usually performed first, with the cleft palate repair being performed later. A plate may be fitted to cover the gap in the palate and help with feeding until surgery. Corrective surgery often achieves good results and makes it possible for speech to develop normally.

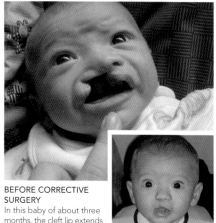

BEFORE CORRECTIVE SURGERY
In this baby of about three months, the cleft lip extends up to affect the nostril and septum (division) of the nose.

TWO WEEKS AFTER SURGERY

DIGIT ABNORMALITIES

IN POLYDACTYLY, THERE ARE MORE THAN THE NORMAL NUMBER OF DIGITS. IN SYNDACTYLY, TWO OR MORE DIGITS ARE FUSED TOGETHER, GIVING A WEBBED APPEARANCE.

Polydactyly may occur by itself, or occasionally it may be a feature of a genetic disorder. It may affect the fingers or toes or both. The additional digits are often poorly developed, but sometimes they are fully formed and functional. Poorly developed digits are usually removed surgically.

Syndactyly, in which the normal webbing at the base of the digits extends farther along the digits, may likewise affect the hands and the feet. When it affects the feet, it tends to occur between the second and third toes. Often, no treatment is necessary, but sometimes surgery is recommended to release the fingers if the webbing restricts movement.

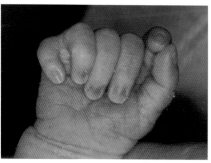

AN EXTRA FINGER
A sixth finger can clearly be seen on this baby's hand. This condition, known as polydactyly, can be passed down in families or may occur with no family history.

1 TRIMMING
The edges of the cleft, which here extend up from the lip and into the nose, are carefully trimmed.

2 NOSTRIL REPAIR
The bottom of the nose is stitched to form a complete nostril as similar as possible to the one on the other side.

3 CLOSING THE LIP
The edges in the lip area are brought together carefully with multiple stitches to form the upper lip.

4 COMPLETION
The stitches now close the openings entirely, and the procedure is complete. Healing will take several weeks.

PROBLEMS AFFECTING THE MOTHER AFTER DELIVERY

FOR MOST WOMEN, CHILDBIRTH PROGRESSES WITHOUT ANY MAJOR ISSUES. BUT EVEN IN THESE WOMEN, PROBLEMS MAY DEVELOP AFTER THE BIRTH. EVENTS THAT OCCUR DURING DELIVERY OR OTHER FACTORS, SUCH AS A PREEXISTING CONDITION, MAY INCREASE THE LIKELIHOOD OF PROBLEMS.

MOST POST-DELIVERY PROBLEMS CAN BE RESOLVED AND ARE NOT SERIOUS. HOWEVER, SOME PROBLEMS, SUCH AS DEEP VEIN THROMBOSIS, REQUIRE URGENT TREATMENT BECAUSE THEY CAN BE LIFE THREATENING. OTHERS, SUCH AS INCONTINENCE, THOUGH NOT SERIOUS, CAN BE DIFFICULT TO TREAT.

POSTPARTUM HEMORRHAGE

THIS IS DEFINED AS THE LOSS OF MORE THAN 18 FL OZ (500 ML) OF BLOOD WITHIN ONE DAY OR SIX WEEKS OF DELIVERY. SUCH BLEEDING CAN BE LIFE-THREATENING AND NEEDS URGENT TREATMENT.

Postpartum hemorrhage (PPH) can be primary (within 24 hours of delivery) or secondary (between 24 hours and six weeks after delivery). The most common causes of primary PPH are uterine atony (the uterus can no longer contract) and retained placental tissue. Very heavy bleeding may lead to life-threatening shock. If a primary PPH occurs, careful examination is needed; blood loss and blood pressure are closely monitored. A blood transfusion and drugs to help the uterus contract may be given. Surgery may be needed. Two common causes of secondary PPH are infections of the uterine lining and retained tissue. Any causes are investigated and treated.

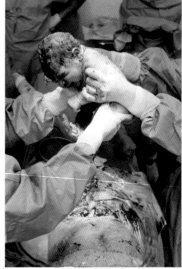

EMERGENCY CESAREAN SECTION
This surgical procedure to deliver a baby is associated with an increased risk of both primary and secondary PPH.

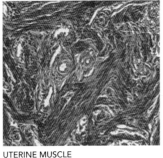

UTERINE MUSCLE
This micrograph shows muscle from the uterine wall. Atony of this muscle (where it cannot contract properly) is a cause of bleeding after delivery.

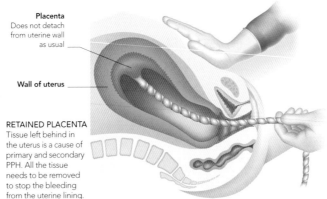

Placenta
Does not detach from uterine wall as usual

Wall of uterus

RETAINED PLACENTA
Tissue left behind in the uterus is a cause of primary and secondary PPH. All the tissue needs to be removed to stop the bleeding from the uterine lining.

PROLAPSE OF THE UTERUS AND VAGINA

IF THE MUSCLES AND LIGAMENTS THAT SUPPORT THE UTERUS AND VAGINA ARE WEAKENED, A PROLAPSE MAY OCCUR WHERE THEY BECOME DISPLACED.

The tissues supporting the uterus and vagina can be weakened by childbirth combined with other risk factors (see table, right). The degree of uterine prolapse varies from slight displacement to the uterus protruding from the vagina. Symptoms of uterine and vaginal prolapse can include problems with defecation or urination, including frequent urination. There is often a feeling of something in the vagina, and in severe cases a lump may be felt beneath the vagina. Stress incontinence, in which there is urine leakage when abdominal pressure is increased, for example when laughing, is often associated with a cystocele (a prolapse affecting the bladder) and is a common symptom after childbirth. Kegel exercises may be helpful in mild cases. After menopause, estrogen supplements may be used to help strengthen the supporting tissues. A vaginal ring pessary may be inserted to keep the uterus in position. In older women, corrective surgery may be considered.

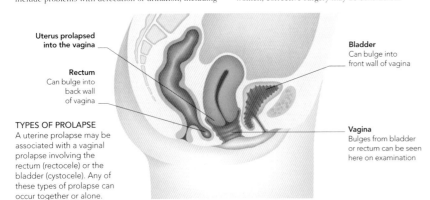

Uterus prolapsed into the vagina

Rectum
Can bulge into back wall of vagina

TYPES OF PROLAPSE
A uterine prolapse may be associated with a vaginal prolapse involving the rectum (rectocele) or the bladder (cystocele). Any of these types of prolapse can occur together or alone.

Bladder
Can bulge into front wall of vagina

Vagina
Bulges from bladder or rectum can be seen here on examination

RISK FACTORS FOR UTERINE AND VAGINA PROLAPSE
RISK FACTORS
Increasing age (the risk doubles with each decade of life)
Having a vaginal delivery
Having several vaginal deliveries (the number increases the risk)
Being overweight or obese
Having a family history of prolapse
Carrying a large fetus during pregnancy
Pushing for a long time (prolonged second stage of labor)
Having an episiotomy
Having an assisted delivery, for example by forceps
Receiving the drug oxytocin during labor
Having reduced levels of estrogen after menopause
Suffering from a chronic cough or chronic constipation

URINARY INCONTINENCE

LEAKS OF URINE DUE TO RAISED PRESSURE IN THE ABDOMEN WHEN COUGHING OR LAUGHING ARE COMMON AFTER CHILDBIRTH.

Problems with urine leakage during pregnancy make stress incontinence after childbirth more likely. The muscles of the pelvic floor are put under pressure during pregnancy and childbirth (and hormonal changes during pregnancy make the muscles looser). A prolapse affecting the bladder (called a cystocele, see opposite) is a particular cause of urine leakage. Stress incontinence may be temporary, last for several weeks, or persist for longer periods. Kegel exercises can help, but for a few women surgery may be offered to tighten the bladder supports and to correct a prolapse, if necessary.

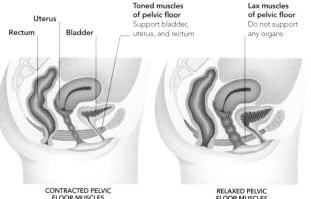

Uterus
Rectum
Bladder
Toned muscles of pelvic floor
Support bladder, uterus, and rectum
Lax muscles of pelvic floor
Do not support any organs

CONTRACTED PELVIC FLOOR MUSCLES

RELAXED PELVIC FLOOR MUSCLES

PELVIC FLOOR MUSCLES AND INCONTINENCE
Looseness of the pelvic floor muscles, which support the uterus and bladder, predisposes a woman to incontinence. Kegel exercises performed regularly during pregnancy and after childbirth can help prevent or lessen this.

FECAL INCONTINENCE

CONTROLLING THE PASSAGE OF FECES OR GAS MAY BE MORE DIFFICULT THAN NORMAL AFTER CHILDBIRTH.

Fecal incontinence may result from weakness in the pelvic floor muscles, which may also cause a prolapse of the rectum, or an injury to the ring of muscle around the anus, perhaps as a result of a tear (see p.233). Tears are more likely if the baby delivered is big, the pushing (second) stage of labor is prolonged, or the baby is born facing upward. Fecal incontinence may persist for a few months or may clear up very quickly. For a few women, it persists in the longterm. Kegel exercises can help, but surgery may be offered if there is a persistent problem.

WOUND INFECTION

THE WOUND FOLLOWING A CESAREAN SECTION, AN EPISIOTOMY, OR A TEAR MAY BECOME INFECTED AND REQUIRE TREATMENT WITH ANTIBIOTICS.

The area around any such wound from childbirth will be reddened and may feel warm if infected; there may also be tenderness or pain. If discharge is present, a swab will be taken and sent off for laboratory analysis for the presence of bacteria. Once the swab has been taken, antibiotics may be given based on those bacteria that are likely to be present. The prescription may then be amended once the swab results are back. The antibiotics should clear the infection so that healing can take place.

INFECTION OF THE UTERUS

ALSO KNOWN AS ENDOMETRITIS, AN INFECTION OF THE UTERINE LINING AFTER DELIVERY IS NOT UNCOMMON AND CAN BE PAINFUL.

If labor is prolonged or there is a long time between membrane rupture and delivery, the likelihood of endometritis is increased. Women who have a cesarean section are at increased risk, particularly if the procedure is performed after the membranes have ruptured or if labor has already started. Endometritis causes pain in the lower abdomen. Body temperature may be raised, causing a fever and chills. In addition, the fluid that is normally

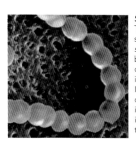

STREPTOCOCCUS A
This SEM image shows a chain of *streptococcus A* bacteria. This bacteria can cause inflammation in the endometrium. It is also a possible cause of wound infections. Such infections are usually treated with antibiotics.

passed vaginally after delivery (the lochia) will have an unpleasant odor. Swabs of the lochia will be taken and tested to look for infections. Antibiotics should clear the condition.

DEEP VEIN THROMBOSIS

WHEN A CLOT FORMS IN ONE OF THE DEEP VEINS OF THE LEG, FRAGMENTS OF THE CLOT CAN BREAK OFF AND TRAVEL TO THE LUNGS.

Women are at increased risk of deep vein thrombosis (DVT) after childbirth because there is an increased tendency for the blood to clot. Women who have a cesarean section are also at higher risk and may be given special hose to wear for a day or two after surgery. An affected leg may feel painful and warm; it may also be swollen and reddened. Body temperature may rise slightly. Clots that travel and cause a blockage in the lungs lead to a condition called pulmonary embolism (PE), which is life-threatening and may cause persistent shortness of breath and chest pain. If DVT is suspected, urgent tests will be arranged, such as Doppler scanning to check blood flow in the legs' deep veins. Drugs will be given to reduce the tendency of the blood to clot and so reduce the risk of a PE.

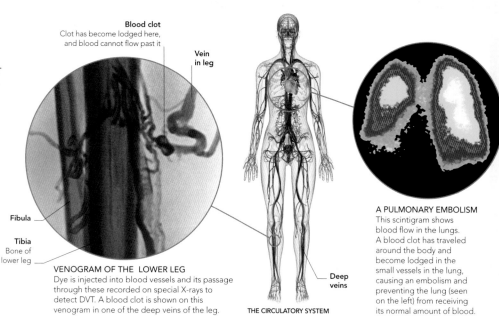

Blood clot
Clot has become lodged here, and blood cannot flow past it

Vein in leg

Fibula

Tibia
Bone of lower leg

VENOGRAM OF THE LOWER LEG
Dye is injected into blood vessels and its passage through these recorded on special X-rays to detect DVT. A blood clot is shown on this venogram in one of the deep veins of the leg.

Deep veins

THE CIRCULATORY SYSTEM

A PULMONARY EMBOLISM
This scintigram shows blood flow in the lungs. A blood clot has traveled around the body and become lodged in the small vessels in the lung, causing an embolism and preventing the lung (seen on the left) from receiving its normal amount of blood.

DEPRESSION AFTER PREGNANCY

THE HORMONAL AND LIFE CHANGES THAT FOLLOW THE DELIVERY OF A BABY CAN BE ASSOCIATED WITH LOW MOOD AND TEARFULNESS. POSTPARTUM SUPPORT, FROM FAMILY AND MEDICAL PROFESSIONALS, IS VITAL TO HELP WOMEN THROUGH THIS EMOTIONAL TIME.

The mood changes that often follow childbirth vary: from mild and transient in most cases to severe and debilitating in a few. Any symptoms of low mood, whether they are mild or severe, should be noted so that the appropriate help can be given.

Baby blues

Known as the baby blues, feelings of sadness often accompanied by weeping are very common and begin within a few days of the birth. There can be marked mood swings with feelings of sadness one minute followed by elation the next. New mothers may also be irritable and tired, partly due to hormonal fluctuations but also due to an inevitable lack of sleep. The baby blues usually settle within a few weeks.

Postpartum depression

This condition is thought to relate to the hormonal changes that also cause the baby blues—the fall in progesterone and estrogen after delivery. Postpartum depression tends to recur, and a family history of the condition puts women at increased risk. Other factors, including lack of sleep, relationship problems, and a difficult labor, can play a role. Depression can develop within the first six months after delivery with various symptoms including feelings of exhaustion, having little interest in the baby, guilt, loss of appetite, signs of anxiety, and problems sleeping. Antidepressant drugs may be recommended, and an improvement in symptoms should be seen within a few weeks.

Puerperal psychosis

A personal or family history of mental illness increases the risk of this condition developing. Symptoms appear within about three weeks of childbirth and include hallucinations, problems sleeping, and periods of mania alternating with depression. This serious illness requires prompt and specialist help and hospital treatment.

NOT BONDING WITH A BABY
New mothers with depression may have little interest in their babies and feel they are not forming a close bond. This can worsen the feelings of sadness and guilt they already have.

Postpartum depression
Serious mental health condition that affects 1 in 10 women after having a baby

Puerperal psychosis
Rare but serious condition; 1 in 1,000 women affected after having a baby

Baby blues
Experienced by most new mothers to some degree

HOW COMMON IS DEPRESSION?
Baby blues are very common, affecting most new mothers. Postpartum depression is less common; and puerperal psychosis affects a small number of women.

COPING STRATEGIES	
A few simple measures can help new mothers who may be feeling low after the birth of a baby. Feelings of isolation are common, so it is vital to spend	time with those who offer emotional support as well as practical help. Reassurance from midwives and physicians is helpful too.
New mothers should be encouraged to accept the help and support of others, whether it is by spending time talking or allowing them to help with the baby.	Seeing the outside world, when they feel able, and talking to other people helps achieve a positive outlook, making having a new baby more enjoyable.
Finding "me" time can make a big difference, as can seizing opportunities to sleep, following the age-old rule of sleeping whenever the baby sleeps.	It is important to resist being self-critical and to be proud of achievements, no matter how small. This is a steep learning curve, especially with the first baby.
Friends and family can be a great source of support and encouragement. New mothers should try to maintain regular contact and avoid isolation.	New mothers need to avoid having unreasonable expectations, for example, by learning to accept that it does not matter if household chores are left undone.

BREAST ENGORGEMENT

BEFORE FEEDING IS ESTABLISHED, MILK CAN ACCUMULATE IN THE BREASTS, CAUSING PAIN AND SWELLING.

Engorgement can also occur if a woman has to stop breastfeeding. The condition predisposes to infection in the breast (mastitis, see opposite). It is essential to wear a supportive bra. Acetaminophen may help ease any pain. The problem should settle within a few days once a baby is latching on and feeding well. When breastfeeding is being stopped, the number of feedings should be reduced over a week or two; in this way, the breasts become acclimatized to producing less milk.

CRACKED NIPPLES

THE SKIN OF THE NIPPLES MAY BECOME CRACKED, PARTICULARLY IN THE EARLY DAYS OF BREASTFEEDING.

Failing to latch on properly is the main cause of this painful condition, which is worse as the baby latches on or comes off the nipple. It is important to ensure that the baby is positioned correctly (see p.207). Emollient creams may provide relief, but need to be washed off before feedings. The problem should slowly resolve as feeding techniques improve. If the breasts continue to be painful, medical help should be sought because an infection requiring antibiotics may have developed.

BLOCKED MILK DUCT

A DUCT THAT DRAINS MILK FROM THE BREAST MAY BECOME BLOCKED, CAUSING AN AREA OF BREAST TISSUE TO BECOME PAINFUL AND SWOLLEN. THIS PROBLEM IS RELATIVELY COMMON.

Milk may accumulate in the affected area, causing it to be tender and painful. In some cases, the duct or the swollen area becomes infected, causing mastitis (see opposite). A blocked duct usually clears up within a day or two. Making sure that the baby latches on properly will help relieve the problem. It is essential to continue breastfeeding even when this is painful.

OBSTRUCTED FLOW
If the flow of milk out of the breast is blocked in a particular milk duct, the milk will accumulate in that area of the breast.

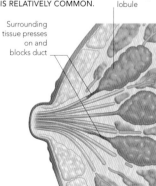

Milk-producing lobule

Surrounding tissue presses on and blocks duct

MASTITIS

THIS IS A COMMON PROBLEM IN THE FIRST SIX WEEKS OF BREASTFEEDING, IN WHICH AN AREA OF BREAST TISSUE BECOMES INFLAMED AND TENDER. ONE BREAST, OR LESS COMMONLY BOTH, MAY BE AFFECTED.

This common condition, which affects around 1 in 10 women who breastfeed, results from an infection in the breast tissue. A common cause is the bacterium *Staphylococcus aureus*. The affected area will be reddened, swollen, and painful. There may be associated flu-type symptoms, including a high body temperature and chills. A heated pad on the affected area may promote the flow of milk and provide some pain relief. Antibiotics are usually needed to treat the infection, and improvement should be seen within two or three days of taking them. Without treatment, an abscess (a collection of pus) may form, in which case a firm and painful lump will be felt in the affected area; fortunately, such abscesses are rarely seen today.

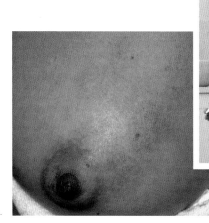

LOCALIZED AREA OF REDNESS
An area of the breast affected by mastitis becomes increasingly painful, red, and swollen, extending out from the nipple area.

EXPRESSING MILK
To prevent the buildup of milk, breastfeeding should be continued. Any extra milk can be expressed using a pump.

ALLEVIATING COMMON PROBLEMS

A number of problems may develop during the early days following childbirth that are part of the normal recovery process. At six weeks after delivery, the physician will check that all is well, including that the uterus is shrinking as it should. Mood changes, including the baby blues (see opposite), are some of the most noticeable features of early motherhood. There are many ways to alleviate the various symptoms after childbirth. Talking to the midwife or sharing problems at a postnatal group can be helpful. If a woman suspects she has problems requiring treatment, such as a urinary tract infection, medical advice should be sought promptly.

VAGINAL SORENESS
Tiny tears and grazes of the vagina or perineum (the area between vagina and anus) can cause soreness. These areas heal quickly, however, and discomfort should be short-lived. Stitched wounds may cause tenderness for a few weeks. Warm baths provide relief.

VAGINAL DISCHARGE
After the delivery, there will be bloody discharge from the vagina (called lochia). Initially, this is like having a period, and then continues in lesser amounts for up to six weeks. If it becomes foul-smelling or contains pus, medical advice should be sought to rule out infection.

CONTRACTIONS
Such "afterpains" may be felt as the uterus starts shrinking. They may be more noticeable when breastfeeding due to the release of the hormone oxytocin, which causes the muscles of the uterus to contract. These mild contractions gradually subside.

PROBLEMS WITH URINATION
Some degree of leakage of urine is common after pregnancy, particularly when coughing and laughing (stress incontinence), and may be improved by doing Kegel exercises frequently through the day. If incontinence persists, medical advice should be sought.

HEMORRHOIDS
These often develop during pregnancy. They can be eased by taking measures to avoid constipation (because straining can make them worse), having warm baths, and washing the area carefully after passing a motion. Creams and suppositories are also available.

SORE BREASTS AND LEAKING MILK
These problems are common before breastfeeding is established. They will be improved by wearing a supportive nursing bra, feeding the baby on demand, massaging the breast to help the flow of milk, drinking plenty of fluids, and ensuring the baby latches on well.

PROBLEMS WITH BOWEL MOVEMENTS
Constipation is a common problem. Keeping mobile, drinking plenty of fluids, and eating a healthy diet can improve symptoms. If a new mother had an episiotomy or tear repaired, she may be reluctant to try to have a bowel movement but these areas will not be affected.

SKIN CHANGES
Some women have acne in the first few weeks after their delivery, while others suffer from dry skin. The darkened patches that sometimes develop during pregnancy should fade gradually; exposure to the sun should be avoided to help this.

WEIGHT LOSS
In the first few days after childbirth, weight loss tends to be rapid due to losing both the weight of the baby and any retained fluid, passed out in urine. After this, weight loss will slow; sensible exercise and healthy eating can help shed weight gradually.

POSTNATAL CARE PLAN
The midwife visits frequently in the first few days, and there is a postnatal check after six weeks. At other times, it is important to seek help if any problems arise.

RESUMING EXERCISE
Gentle exercise after childbirth has physical and emotional benefits. Many types of strenuous exercise should be avoided until after the six-week postnatal check.

FADING STRETCHMARKS
The marks caused by the stretching of the skin and hormonal changes of pregnancy never disappear but do fade over time.

GLOSSARY

A

allele
A particular version of a gene. Different alleles of the same gene often have different effects.

amniocentesis
A technique used to obtain a sample of amniotic fluid. A hollow needle is passed through the abdominal and uterine walls, avoiding the placenta and fetus. It is performed from the 15th week of pregnancy onward.

amino acid
Any of around 20 different kinds of small molecules that are the basic building blocks of proteins. A single protein molecule may comprise hundreds or thousands of amino acids joined together.

amnion
A membrane that grows from the embryonic blastocyst and expands to surround the developing fetus in the uterus.
See also blastocyst.

amniotic fluid
The fluid enclosed by the amnion, which surrounds and protects the developing embryo and fetus.

Apgar score
A rating used to assess the health of a newborn baby in the first few minutes after delivery. The pulse, reflexes, breathing, movement, and skin color are rated as 0, 1, or 2, and the scores added together to give the overall Apgar score.

areola (plural: areolae)
The circular area of pigmented skin that surrounds the nipple.

B

blastocyst
The stage in embryonic development that follows the morula stage. The blastocyst comprises a hollow ball of cells (the trophoblast) that develops into protective membranes around the embryo, and a group of cells within the trophoblast called the embryoblast that will form the embryo itself.
See also morula.

blastomere
Any of the early cells that result from cleavage in the early embryo.
See also cleavage.

Braxton Hicks' contractions
The irregular contractions of the uterus that occur during pregnancy. They do not indicate that labor is about to begin.
See also contractions.

breech presentation
The term used to describe the position of the fetus when, rather than its head facing downward, its buttocks or feet are facing the cervix when birth is about to begin. Referred to as a "breech birth," this is more difficult to manage than the more common head-first birth.

C

cesarean section
A surgical operation that is performed in order to remove a baby from the uterus by cutting through the abdominal and uterine walls. It is often performed if there are actual or expected complications with a normal birth.

cervix
The lowest part of the uterus. It consists mainly of a ring of connective tissue surrounding a narrow, mucus-filled canal that connects the rest of the uterus to the vagina. During labor, the canal stretches and widens to allow the baby to pass through it.

chorion
The outermost of the membranes that surround the developing embryo and fetus. Part of it (the villous chorion) contributes to forming the placenta.
See also villi.

chorionic plate
The part of the chorion involved in attaching to the wall of the uterus. It is part of the placenta.

chorionic villus sampling (CVS)
A technique for obtaining small samples from the villi of the placenta, whose tissue is derived from the fetus and can be checked for fetal genetic abnormalities. CVS can be carried out at an earlier stage than amniocentesis.
See also amniocentesis, villi.

chromosomes
The structures within a cell's nucleus that contain the organism's genes. Humans have 23 pairs of chromosomes in a complete set of 46, with a set present in nearly every cell of the body. Each chromosome consists of a single long DNA molecule combined with various proteins. One pair of the 23 are the sex chromosomes, which come in two varieties—X and Y. Women have two X chromosomes, whereas men have one X and one Y.

cilia (singular: cilium)
The minute, beating hairs on the cell surfaces of some tissues, such as those lining the fallopian tubes.

cleavage
The early stage divisions of the fertilized egg cell, when it cleaves (divides) into a number of smaller cells without growing in total size.

clitoris
The structure of erectile tissue, part of the female genitals, that provides pleasurable sensations during sex. Its head is visible as a small projection but it also extends inward behind the wall of the vagina. It shares the same embryonic origin as parts of the penis.

colostrum
The milk produced by the breasts immediately after a baby is born. It is different in appearance and composition from milk produced subsequently.

contractions
The term used to describe the regular shortening of the strong muscle of the uterus that marks the beginning of labor. Contractions become stronger and more frequent with time. They serve first to stretch and dilate the cervix, and then to expel the baby from the uterus.
See also Braxton Hicks' contractions.

corpus hemorrhagicum
The mature follicle immediately after ovulation, before it develops into a corpus luteum.

corpus luteum (adjective: luteal)
A structure in the ovary formed from the remains of a mature follicle after ovulation. It produces progesterone, which keeps the uterus in a condition that allows it to support a potential pregnancy, but if no egg implants, the corpus luteum breaks down after a few days as part of the normal menstrual cycle.
See also follicle.

cotyledon
Any of the 15–20 lobes of the placenta that protrude into the lining of the uterus.

cytotrophoblast
A group of cells forming the inner cell layer of the trophoblast that plays a role in implantation.
See also blastocyst, implantation, syncytiotrophoblast.

D

decidua

The endometrial tissues of the pregnant uterus, some of which contribute to the placenta. They are shed after birth. *See also* endometrium.

diploid

Having two copies of each chromosome. Nearly all body cells, except gametes (sex cells), are diploid. *See also* haploid.

DNA

Short for deoxyribonucleic acid, a very long molecule made up of small individual units. DNA is found in the chromosomes of living cells, and the order of the small units "spells out" the instructions that determine the characteristics of the organism. *See also* gene.

E

ectoderm

The uppermost of the three layers of tissue into which the embryonic disk becomes divided. It later gives rise to the skin and nervous system. *See also* embryonic disk, endoderm, mesoderm.

ectopic pregnancy

A situation in which an early embryo implants outside the uterus, usually in a fallopian tube. Such pregnancies cannot succeed and therefore require medical intervention.

egg

In humans, the single yolk-containing cell that can potentially be fertilized by a sperm cell to result in a new individual. *See also* gamete, ovum.

embryo

The earliest stage of development of a human, covering approximately the first eight weeks after the egg is fertilized. (Sometimes the very earliest stages are termed the preembryo.) *See also* fetus.

embryonic disk

A disk-shaped region of tissue that appears within the blastocyst after implantation and that develops into the embryo. *See also* blastocyst.

endoderm

The lower of the three layers of tissue into which the embryonic disk becomes divided. It later gives rise to the gut and associated organs. *See also* ectoderm, embryonic disk, mesoderm.

endometrium

The inner lining of the uterus. It grows in thickness during each menstrual cycle, but if no pregnancy occurs it breaks down and some of its tissues and blood are expelled at menstruation. The early embryo implants in the endometrium and later the placenta also develops here. *See also* implantation, myometrium, perimetrium.

epididymis (plural: epididymides)

The long, highly coiled tube into which sperm pass after leaving the testis. Sperm become fully capable of fertilization only after maturing for several days in the epididymis.

epidural

Short for epidural anesthesia, a technique by which part of the body is made numb by applying anesthetic to the outer membrane (dura) of the spinal cord, usually in the lower-back region. It allows a mother to remain conscious during a potentially painful childbirth, or one that may involve surgery.

episiotomy

A surgical operation in which a cut is made into the perineum during childbirth to enlarge the opening of the vagina, in cases where the baby's head might otherwise tear the mother's tissues.

estrogen

Any of various natural or synthetic female sex hormones. Natural estrogens are produced by follicle cells of the ovary from puberty onward. They promote female characteristics such as breast development, and are essential to the menstrual cycle and female fertility.

F

fallopian tube

Either of the tubes leading to the uterus from the ovaries, down which an ovum travels after ovulation.

fetus (adjective: fetal)

The unborn baby in the uterus, from the time when it begins to show a recognizably human appearance, at approximately eight weeks after fertilization or ten weeks after the start of the mother's last menstrual period. *See also* embryo.

fimbria (plural: fimbriae)

Any of the several fingerlike projections at the end of each fallopian tube that help collect an ovum released by the ovary, so that it will be conveyed toward the uterus.

follicle (adjective: follicular)

A small cavity lined with cells. In reproductive contexts, the term refers to an ovarian follicle, a structure within the ovary comprising an oocyte (immature egg cell) surrounded by a covering of other specialized cells. Small primordial follicles form in the fetal ovary before birth, but remain inactive until puberty. After puberty, a few follicles every month start developing into primary and then secondary follicles, although usually only one develops fully into a tertiary or Graafian follicle, a fluid-containing structure that releases a mature ovum at ovulation. *See also* ovum.

follicle stimulating hormone (FSH)

A hormone secreted by the pituitary gland that affects the ovaries and testes. Increasing levels of FSH are necessary for puberty to occur in both sexes and, in women, the hormone also stimulates follicular development during the menstrual cycle. *See also* follicle.

folliculogenesis

The development of a primordial follicle into one that is fully mature. *See also* follicle.

fontanelles

The "soft spots" on the head of a baby where the underlying skull bones have not yet fused together.

forceps (obstetrical)

An instrument whose ends can be eased around a baby's head during labor if necessary: gentle pulling on the forceps then aids the baby's exit from the birth canal.

fundus

The top part of the uterus, whose position can be felt from the outside of a woman's body in the later stages of pregnancy. It is the usual site of the placenta.

G

gamete

A haploid sex cell, i.e. a sperm cell or an unfertilized egg cell. *See also* haploid, zygote.

gene
A length of a DNA molecule containing a particular genetic instruction. Many genes are blueprints for making particular protein molecules, while others have a role in controlling other genes. Nearly every cell in the human body contains a complete set of genes (the genome), although different genes are "switched on" in different cells.

genome
The complete set of genes found in the cell of a human or other living species.

germ cell
A stem cell from which gametes are derived; also, an immature or mature gamete. *See also* stem cell.

germ layer
Any of the basic cell layers into which the embryonic disk becomes divided. *See also* ectoderm, endoderm, mesoderm.

goblet cells
The mucus-secreting cells found in the surface linings of some tissues, such as those of the fallopian tubes.

H
haploid
Having only one copy of each chromosome, rather than a pair. Gametes (sex cells) are haploid, allowing them to re-create a normal diploid individual when they unite at fertilization. *See also* diploid.

human chorionic gonadotropin (hCG)
A hormone produced by the placenta that causes the corpus luteum of the ovary to continue producing progesterone, so that pregnancy continues.

hypothalamus
A control center at the base of the brain, close to the pituitary gland. Its many functions include stimulating the pituitary gland to produce luteinizing hormone and follicle stimulating hormone. *See also* follicle stimulating hormone, luteinizing hormone.

I
implantation
The process in which the early embryo (at the blastocyst stage) attaches to and becomes incorporated within the endometrium of the uterus. *See also* blastocyst, endometrium.

in vitro fertilization (IVF)
An assisted-conception technique that involves removing some of a woman's unfertilized eggs from her ovaries, fertilizing them with sperm in a laboratory, culturing them until the blastocyst stage, and then introducing them into the uterus so that they can implant. The technique may be used, for example, when a woman is infertile because her fallopian tubes have become blocked. *See also* blastocyst, implantation.

induction
The process of artificially starting labor by various means, in cases where the onset of natural labor is overdue.

intervillous space
Any of the spaces between the villi of the placenta through which the mother's blood circulates. It is here that gases are exchanged between the blood of mother and fetus. *See also* villi.

L
labia
Either of the two pairs of folds that form part of a woman's vulva (external genitals), comprising the labia majora (the outer labia) and the more delicate labia minora (the inner labia).

labor
The process of giving birth. In the first stage of labor, regular contractions of the uterus stretch the cervix and widen (dilate) its opening, until it is wide enough for the baby's head to pass through. In the second stage, the baby is born. In the third stage, the placenta and other materials are expelled.

lactation
A term used to describe the process of milk production by the breasts.

lanugo
The fine hair covering the skin of a fetus.

laparoscopy
The technique of viewing internal abdominal organs by inserting an instrument (laparoscope) through the body wall. The laparoscope is equipped with a miniature video camera and illumination, and transmits images to the outside.

lie (fetal)
The angle of a fetus in the uterus in relation to the mother's main body axis. Most commonly, the fetus is positioned with its backbone roughly parallel to the mother's backbone.

linea nigra
A pigmented vertical line that often develops on the skin of a woman's abdomen during pregnancy.

lobule
A small lobe or section of an organ, for example, in a mammary gland.

lochia
The fluids expelled from the uterus via the vagina in the days after birth.

lumen
The inside space of a tubular structure, such as a blood vessel or glandular duct.

luteal
Relating to the corpus luteum.

luteinizing hormone (LH)
A hormone secreted by the pituitary gland that acts on both ovaries and testes. Increasing levels of LH are necessary for puberty to occur in both sexes. LH stimulates testosterone production in men and plays various roles in the female menstrual cycle.

M
mammary gland
The milk-producing gland of a mammal. In women, much of the substance of the breasts is made up of mammary-gland tissue.

meconium
The greenish brown material expelled by a baby as its first bowel movement.

meiosis
A specialized type of cell division (strictly speaking, of nuclear division) in which haploid sex cells are created from diploid precursor cells. It is more complicated than normal cell division (mitosis) and takes place in two stages. *See also* haploid, mitosis.

menarche (pronounced "menarkey")
A girl's first menstrual period, which indicates that she is reaching sexual maturity.

menopause
The time in a woman's life when menstruation permanently stops (usually between 45 and 55 years of age).

menstrual cycle
The monthly changes that take place in a nonpregnant woman's reproductive system during her years of fertility. The cycle (of approximately 28 days) starts on the first day of menstruation and centers upon events in the ovary, where several egg-containing follicles begin to ripen. This is the follicular phase of the cycle. Usually, just one follicle develops fully each month, releasing its egg from the ovary (a process known as ovulation) halfway through the cycle, after which the empty follicle transforms into a corpus luteum, which marks the beginning of the luteal phase of the cycle. The lining of the uterus (the endometrium) also thickens, in preparation for possible pregnancy. If no pregnancy occurs in the days after ovulation, the corpus luteum breaks down; the lack of the hormone it produces (progesterone) causes the endometrium to break down, too, leading to menstruation, and the cycle begins again.
See also corpus luteum, endometrium, follicle.

menstruation
The expulsion of endometrial blood and tissue as part of the menstrual cycle, its occurrence each month being known as a menstrual period.
See also endometrium.

mesoderm
The middle of the three layers of tissue into which the embryonic disk becomes divided. The mesoderm later gives rise to many body tissues, including muscle, bone, and blood vessels.
See also ectoderm, embryonic disk, endoderm.

milk duct
A duct conveying milk from the milk-producing tissues of the mammary gland to the nipple of the breast. Around 15–20 milk ducts open separately on the surface of each nipple.
See also mammary gland.

miscarriage
The spontaneous loss of an embryo or fetus from the mother's body at a time that is too early for it to survive, which is usually taken to be any time earlier than 24 weeks of pregnancy. Later than this, the event is termed a premature or preterm birth. Miscarriage may be complete, or it may be incomplete (some material is left in the uterus, a situation that requires medical intervention). There are a variety of causes of miscarriage, but the cause may not be obvious.

mitosis
The process by which chromosomes are split and shared out during normal cell division. The two cells that are produced have the same number of chromosomes as the original cell.
See also meiosis.

morula
An early stage in the transformation of the fertilized egg into an embryo. It consists of a solid ball of cells. This stage proceeds that of the blastocyst.
See also blastocyst.

MRI
Short for magnetic resonance imaging, used to obtain images of internal organs and structures by causing atoms to absorb and emit radiofrequency waves while the body is held within a strong magnetic field. Compared with ultrasound scanning, MRI needs more time, greater precautions, and much more elaborate equipment. It is often used to investigate problems indicated by ultrasound scanning and is particularly useful for imaging the central nervous system.
See also ultrasound.

mucus plug
The protective plug of sticky material that seals the canal of the cervix during pregnancy. Its shedding from the vagina (known as "the show") indicates that labor is soon to begin.

mutation
A change to the genetic makeup of a cell, caused, for example, by mistakes in copying DNA before a cell divides. Mutations in sex cells, or in cells of the early embryo, may cause offspring to have unusual genetic features not present in their parents.

myelin
An insulating layer along the outside of many nerve cells. Its presence allows nerve impulses to travel faster.

myometrium
The muscular tissue forming the bulk of the uterus.
See also endometrium, perimetrium.

N

neonatal
Of or relating to a newborn baby (neonate).

neural tube
The hollow tube of cells formed in the early embryo from which the brain and spinal cord develop.

neuron
A nerve cell.

notochord
A rod of stiff material that develops along the back of the early embryo. It largely disappears later but marks the position of the future vertebral column (backbone).

nuchal translucency screening
A technique involving the use of ultrasound scanning to check the thickness of the fluid layer found in a young fetus beneath the skin of the back of the neck. A thicker than normal layer may indicate a chromosomal abnormality such as Down syndrome.

nucleus
The structure within a cell that contains the chromosomes
See also chromosome.

O

oocyte
An immature egg cell. Oocytes occur within follicles in the ovary
See also follicle.

ovary
One of the pair of structures in a woman's body in which unfertilized egg cells are matured and from which they are periodically released. The ovaries also produce important hormones, including estrogen and progesterone.

ovulation
The release of an unfertilized egg (ovum) from an ovary.

ovum (plural: ova)
An egg cell, especially one that has been released from an ovary and is ready for fertilization. The term can be applied to the fertilized egg as well.
See also gamete.

P

pelvic floor
A combination of muscles that supports the abdominal organs from below.

perimetrium
The outer covering of the uterus.
See also endometrium, myometrium.

perinatal
Relating to the period around birth, including the few weeks beforehand and afterward.

perineum
The area of skin and underlying tissues lying between the external genital organs and the anus. The mother's perineum stretches considerably during childbirth.

pituitary gland
A complex, pea-sized structure at the base of the brain, sometimes described as the body's "master gland." Its direct roles in reproduction include the secretion of luteinizing hormone and follicle stimulating hormone. It also produces oxytocin.

placenta
A disk-shaped organ formed on the wall of the pregnant uterus by the combined growth of tissues from the mother and the early fetus. The fetus's blood circulation flows close by the mother's in the placenta, allowing for the exchange of nutrients, dissolved gases, and waste products. The placenta also produces hormones.
See also umbilical cord.

placenta previa
A condition in which the placenta forms in the lower part of the uterus, sometimes blocking the entrance to the cervix. It may require delivery by cesarean section.

postnatal
For the baby, denoting the period after birth.

postpartum
For the mother, denoting the period after birth.

preeclampsia
A medical condition that some women experience during late pregnancy, which is characterized by a combination of high blood pressure and protein in the urine. It requires urgent medical attention (often including induction of labor) in case it develops into eclampsia, a life-threatening condition.

prenatal
The term used to describe the period before childbirth.

primitive streak
A linear arrangement of cells on the developing embryonic disk that indicates the future head- and tail-ends of the embryo.

progesterone
A hormone produced mainly by the corpus luteum of the ovary. Its action causes the lining of the uterus to be in a suitable condition to support pregnancy.

prostaglandins
Hormonelike substances produced by many tissues. They cause altered activity in neighboring tissues. Some prostaglandins promote uterine contractions and are used artificially to induce labor.

prostate gland
A gland that surrounds the urethra in men at the point at which it is joined by the ducts (vasa deferentia) leading from the testes. Its secretions contribute to semen.

puberty
The sum total of the bodily events involved in achieving sexual maturity and adult sexual characteristics, occurring over several years in both boys and girls.

R

relaxin
A hormone produced by the ovary and other tissues, and by the placenta, during pregnancy. Its functions include softening and relaxing tissues and ligaments in preparation for childbirth.

Rhesus factors
Molecules found on the surfaces of blood cells in most people (who are thus termed Rhesus positive or Rh-positive) but which are missing from a minority (Rhesus negative or Rh-negative). If a Rhesus-negative mother carries a Rhesus-positive fetus in a second or later pregnancy, her immune system may attack the fetus.

S

semen
The sperm-containing fluid released through the penis when a man ejaculates. Its fluid component derives from several glands, including the prostate gland.

seminiferous tubule
Any of the coiled tubes in the testis within whose tissues sperm cells are formed.

septa (singular: septum)
The membranes separating tissues of the body. The decidual septa are the divisions between the cotyledons of the placenta
See also cotyledon.

somite
One of several paired structures that form in the mesoderm from the fifth week of pregnancy onward. Somites eventually differentiate into the spinal cord and vertebrae, the muscles of the trunk, and the skin.
See also mesoderm.

sperm
A male sex cell, also called a sperm cell or spermatozoon. Each cell has a long, mobile tail that allows it to swim toward and fertilize an egg in the body of the female. The word is also used in nontechnical contexts to refer to semen.
See also gamete.

spermatids
The immediate precursors of sperm cells. When a secondary spermatocyte completes meiosis, it becomes an early spermatid. This small, round cell elongates and changes form to become a late spermatid as part of its transformation into a mature sperm cell.
See also spermatocytes.

spermatocytes
Cells that are at an intermediate stage in the creation of sperm cells. Spermatocytes undergoing the first stage of meiosis are called primary spermatocytes; those that have progressed to the second stage of meiosis are called secondary spermatocytes.
See also meiosis.

spermatogenesis
The overall process of sperm formation, from spermatogonia to mature sperm.

spermatogonia (singular: spermatogonium)
Cells that represent an early stage in the production of sperm cells. They derive from stem cells in the testis and, in turn, give rise to spermatocytes.

spiral artery
One of many small, spiral-shaped arteries that supply the endometrium of the uterus. During pregnancy, these arteries grow in size to supply blood from the mother's circulation to the placenta.
See also endometrium.

stem cell
A cell capable of dividing and differentiating into more specialized types of cells. The stem cells of the earliest embryo are capable of turning into any of the body's cell types, while later stem cells, including those in adults, can give rise to a more limited range of specialized cells.

surfactant
A substance that lowers the surface tension of water, allowing wetted surfaces to "unstick" from each other more easily. Surfactant in the air sacs of the lungs (the alveoli) plays a vital role in breathing, because it allows the sacs to inflate and collapse easily.

syncytiotrophoblast
The outer cells of the trophoblast whose contents are linked to form a continuum (a syncytium). They contribute to implantation. *See also* blastocyst, cytotrophoblast, implantation.

T

testis (plural: testes)
Either of the two sperm-producing organs in men, located in the scrotum outside the main body cavity. Also called testicles, the testes also secrete hormones, especially testosterone. *See also* testosterone.

testosterone
The main sex hormone in men, which also occurs in lower concentrations in women. In male fetuses, testosterone produced by the testes promotes the development of male genitals, while increased concentrations at puberty induce characteristics such as beard growth and are essential for sperm production.

transition
The final part of the first stage of labor, involving strong contractions and the completion of cervical dilation. *See also* labor.

trimester
Any of the three periods, of approximately three months each, into which pregnancy is divided. The first trimester is counted as starting from the woman's final menstrual period before pregnancy.

trophoblast
See blastocyst.

twins
The term used to describe two individuals who have developed in the same uterus at the same time. Nonidentical or fraternal twins occur when two separately fertilized ova implant in the uterus together. Identical twins (which are identical genetically) arise when a single fertilized ovum separates into two parts shortly after cleavage begins, each part going on to become a separate embryo.

U

ultrasound
The sound frequencies that are too high for the human ear to hear. They form the basis of ultrasonography, an imaging technique in which high-frequency sound waves bouncing off tissues in the body can be interpreted electronically to yield still or moving images. A similar technology called Doppler ultrasonography can visualize the speed of moving fluids, such as blood in an artery. Ultrasonography is convenient and lacks side-effects, so is commonly used to check fetal progress and sometimes to aid surgical operations.

umbilical cord
The flexible cord that attaches the developing fetus to the placenta. Fetal blood circulates to and from the placenta via blood vessels in the cord, allowing an exchange of nutrients and other materials with the mother. *See also* placenta.

ureter
Either of the two tubes that convey urine from the kidneys to the bladder.

urethra
The tube that conveys urine from the bladder to the outside of the body. In men it also conveys semen during ejaculation.

uterus (adjective: uterine)
The womb—the hollow muscular organ where the fetus develops during pregnancy. *See also* endometrium, myometrium, perimetrium.

V

vas deferens (plural: vasa deferentia)
Either of the two narrow muscular tubes in men that connect the epididymides to the urethra. They store and transport sperm in preparation for ejaculation. *See also* epididymis, urethra.

ventouse
Also called a vacuum extractor, a form of suction cap that is sometimes applied to an emerging baby's head during labor: pulling on the ventouse aids the baby's passage through the birth canal.

vernix
A greasy substance that coats and protects the skin of the unborn baby.

villi (singular: villus/adjective: villous)
The folded projections that form on the surface of some tissues. The placenta produces villi that are branched structures, comprising stem, secondary, and tertiary villi. The villi contain fetal blood vessels that allow efficient exchange of materials with the mother's blood supply.

Y

yolk sac
A membrane-bound cavity on the underside of the early embryo that is the location of the embryo's first blood-cell production. (It is not used for yolk storage in humans.)

Z

zona pellucida
The transparent protective layer around the ovum. It is shed by the blastocyst before implantation. *See also* blastocyst, ovum.

zygote
The diploid fertilized cell formed by the union of two gametes. *See also* diploid, gamete.

INDEX

ACKNOWLEDGMENTS

For their work on this revised second edition, **Dorling Kindersley** would like to thank Dr. Melissa Whitten and Dr. Paul Moran for consultancy, and Katie John for additional writing.

For their work on the previous first edition, Dorling Kindersley would like to thank Dr. Paul Moran of the Royal Victoria Infirmary, Newcastle, for providing ultrasound scans, as well as the women who gave permission for their scans to be used—Emma Barnett, Paula Binney, Sophie Lomax, and Katie Marshall. Sarah Smithies and Jenny Baskaya carried out additional picture research, and Laura Wheadon provided editorial assistance.

Picture credits

The publisher would like to thank the following for their kind permission to reproduce their photographs:

(Key: a-above; b-below/bottom; c-center; f-far; l-left; r-right; t-top)

4–5 Science Photo Library: Susumu Nishinaga (b). **6 Alamy Images:** Steve Bloom Images (bl). **FLPA:** Ingo Arndt/Minden Pictures (bc). **naturepl.com:** Doug Perrine (br). **Science Photo Library:** Dr. Yorgos Nikas (tl); Edelmann (tc, tr). **7 Ardea:** John Cancalosi (bc). **Auscape:** Shinji Kusano (bl). **Getty Images:** Photolibrary/Derek Bromhall (tl). **naturepl.com:** Yukihiro Fukuda (br). **Science Photo Library:** Custom Medical Stock Photo (tr); Dr. Najeeb Layyous (tc). **8 Science Photo Library:** Simon Fraser (tl). **8–9 Science Photo Library:** Susumu Nishinaga (t). **9 Science Photo Library:** Miriam Maslo (tr). **10 Science Photo Library:** Ian Hooton (tl); Zephyr (tc); Aubert (tr). **11 Alamy Images:** Janine Wiedel Photolibrary (tc); David R. Gee (tr). **Getty Images:** David Joel (tl). **12 Courtesy of the British Medical Ultrasound Society Historical Collection:** (bl). **Photograph courtesy of Doncaster & Bassetlaw Hospitals NHS Foundation Trust. :** (tc). **13 Science Photo Library:** ISM (fbr); CNRI (bc); Edelmann (br); Bernard Benoit (cb). **14–15 Dept of Fetal Medicine, Royal Victoria Infirmary**. **15 Science Photo Library:** Dr. Najeeb Layyous (br). **16 Dept of Fetal Medicine, Royal Victoria Infirmary:** (cl, br). **Science Photo Library:** Dr. Najeeb Layyous (bl); Thierry Berrod, Mona Lisa Production (tl). **17 Science Photo Library:** Tissuepix (t); Dr. Najeeb Layyous (bl, br). **18 Dept of Fetal Medicine, Royal Victoria Infirmary:** (bl). **Science Photo Library:** Edelmann (t, br). **19 Science Photo Library:** Edelmann (cl); GE Medical Systems (bl); Dr. Najeeb Layyous (tr, br, cr, tl). **20 Dept of Fetal Medicine, Royal Victoria Infirmary:** (bc, br). **Science Photo Library**. **21 Dept. of Fetal Medicine, Royal Victoria Infirmary:** (b/all). **Science Photo Library**. **22 Dept of Fetal Medicine, Royal Victoria Infirmary:** (c, cr, bl). **Science Photo Library:** Dr. Najeeb Layyous (cl); CIMN, ISM (bc, br). **23 Dept. of Fetal Medicine, Royal Victoria Infirmary:** (l). **Science Photo Library:** BSIP, Kretz Technik (cr). **24–25 Science Photo Library:** Susumu Nishinaga. **25 Science Photo Library:** Susumu Nishinaga (r). **26–45 Science Photo Library:** Susumu Nishinaga (sidebars). **28 Corbis:** Dennis Kunkel Microscopy, Inc./Visuals Unlimited (cr). **Science Photo Library:** Pasieka (bl). **30 Boston University School of Medicine:** Deborah W. Vaughan, PhD (cl). **Corbis:** Steve Gschmeissner/Science Photo Library (bc). **31 Getty Images:** Stephen Mallon (bl). **32 Science Photo Library:** Susumu Nishinaga (bl). **34 Corbis:** Image Source (cr). **Science Photo Library:** Pasieka (bl). **36 Science Photo Library:** (tl). **37 Science Photo Library:** Professor P.M. Motta & E. Vizza (tr); Steve Gschmeissner (br). **38–39 Lennart Nilsson Image Bank**. **41 Alamy Images:** Biodisc/Visuals Unlimited (c). **The Beautiful Cervix Project, www.beautifulcervix.com:** (tr). **Science Photo Library:** Steve Gschmeissner (bc). **43 Fertility and Sterility, Reprinted from:** Vol 90, No 3, September 2008, (doi:10.1016/j.fertnstert.2007.12.049) Jean-Christophe Lousse, MD, and Jacques Donnez, MD, PhD, Department of Gynecology, Université Catholique de Louvain, 1200 Brussels, Belgium, Laparoscopic observation of spontaneous human ovulation; © 2008 American Society for Reproductive Medicine, Published by Elsevier Inc with permission from Elsevier. (bl). **46–47 Science Photo Library:** Pasieka. **47 Science Photo Library:** Pasieka (cr). **48 Science Photo Library:** JJP / Philippe Plailly / Eurelios (ca). **48–55 Science Photo Library:** Pasieka (sidebars). **49 Science Photo Library:** Dr. Tony Brain (cr). **52–53 Getty Images:** Marc Romanelli (tc); Vladimir Godnik (c); Emma Thaler (ca). **52 Alamy Images:** Custom Medical Stock Photo (clb). **Corbis:** Photosindia (cr). **Getty Images:** Paul Vozdic (tr); Karen Moskowitz (cra). **53 Corbis:** Bernd Vogel (cla). **Getty Images:** JGI (cl); Steve Allen (cra); IMAGEMORE Co.,Ltd. (tl). **Science Photo Library:** Richard Hutchings (crb). **55 Press Association Images:** John Giles/PA Archive (br). **Science Photo Library:** BSIP, Laurent H.americain (bl). **56–57 Science Photo Library:** Susumu Nishinaga. **57 Science Photo Library:** Susumu Nishinaga (cr). **58 Getty Images:** Priscilla Gragg (cl). **Wellcome Images:** BSIP (b). **58–59 Getty Images:** DEA / G. Dagli Orti. **58–69 Science Photo Library:** Susumu Nishinaga. **59 Getty Images:** Darrell Gulin (bl). **Science Photo Library:** Ken M. Highfill (cra). **60–61 Getty Images:** Yorgos Nikas. **62 Getty Images:** Jupiterimages, Brand X Pictures (cr); PHOTO 24 (c); Beth Davidow (cl). **Science Photo Library:** Professors P.M. Motta & J. Van Blerkom (bl); Gustoimages (tl). **63 Getty Images:** Image Source (bl). **© 2008 Little et al. This is an open-access article distributed under the terms of the Creative Commons Attribution License, which permits unrestricted use, distribution, and reproduction in any medium, provided the original author and source are credited (see http://creativecommons.org/licenses/by/2.5/)** : Little AC, Jones BC, Waitt C, Tiddeman BP, Feinberg DR, et al. (2008) Symmetry Is Related to Sexual Dimorphism in Faces: Data Across Culture and Species. PLoS ONE 3(5): e2106. doi:10.1371/journal.pone.0002106 (cr). **Science Photo Library:** Steve Gschmeissner (tr). **64 Corbis:** Marco Cristofori (bc). **Science Photo Library:** Manfred Kage (tl). **66 Science Photo Library:** Zephyr (br); W. W. Schultz / British Medical Journal (cl). **67 Science Photo Library:** Professors P.M. Motta & J. Van Blerkom (br). **68 Getty Images:** Dimitri Vervitsiotis (cla). **68–69 Science Photo Library:** ISM (t). **69 Science Photo Library:** Pasieka (ca). **70–71 Science Photo Library:** Hybrid Medical Animation. **71 Science Photo Library:** Hybrid Medical Animation (r). **72–185 Science Photo Library:** Hybrid Medical Animation (sidebars). **72 Science Photo Library:** Cavallini James (tc); Science Pictures Ltd (tl); Dopamine (tr). **74 Science Photo Library:** Steve Gschmeissner (bl); Dr. Isabelle Cartier, ISM (cra); Gustoimages (cla). **75 Science Photo Library:** Anatomical Travelogue (br); Dr. Yorgos Nikas (bl). **78 Alamy Images:** Dick Makin (bl). **Science Photo Library:** Professor P.M. Motta & E. Vizza (br); Steve Gschmeissner (cl). **79 Wikipedia, The Free Encyclopedia:** Acaparadora (bl). **82–83 PhototakeUSA.com:** Last Refuge, Ltd.. **88 Alamy Images:** PHOTOTAKE Inc. (cb); MG photo studio (ca). **Corbis:** Jean-Pierre Lescourret (br). **Science Photo Library:** Lowell Georgia (cla). **89 Alamy Images:** Elizabeth Czitronyi (clb); Bubbles Photolibrary (tr). **Corbis:** Mango Productions (cr). **Getty Images:** Image Source (cla). **Science Photo Library:** Gustoimages (bc). **92 Science Photo Library:** Anatomical Travelogue (bl); Edelmann (br). **93 Science Photo Library:** Steve Gschmeissner (br). **96 Getty Images. Science Photo Library:** Edelmann (bl). **97 Getty Images:** B2M Productions (cra). **98 Prof. J.E. Jirásek MD, DSc.:** (bl). **99 Rex Features:** Quirky China News (br). **Science Photo Library:** Professor Miodrag Stojkovic (cl); Anatomical Travelogue (c). **100–101 Science Photo Library:** Edelmann. **102 Science Photo Library:** Edelmann (cl). **103 Science Photo Library:** Steve Gschmeissner (crb); Edelmann (tr). **104 Ed Uthman, MD:** (cl). **106 Getty Images:** Jim Craigmyle (cla). **Science Photo Library:** Edelmann (bl). **107 Getty Images:** Katrina Wittkamp (cr); Jerome Tisne (bl). **Science Photo Library:** Dr. Najeeb Layyous (cl). **110 Alamy Images:** MBI (cla). **Getty Images:** Stockbyte (bl). **111 Science Photo Library:** Dr. Klaus Boller (cl); Susumu Nishinaga (bc). **112–113 Science Photo Library:** Zephyr. **114 Science Photo Library:** Edelmann (tc). **115 Science Photo Library:** Dr. G. Moscoso (tr). **116–117 Prof. J.E. Jirásek MD, DSc.**. **117 Science Photo Library:** Steve Gschmeissner (tr). **119 Virginia M. Diewert:** (tc). **120 Virginia M. Diewert**. **121 Virginia M. Diewert**. **122 Corbis:** Frans Lanting (tr). **124 Getty Images:** Chad Ehlers - Stock Connection (tc). **Dept of Fetal Medicine, Royal Victoria Infirmary:** (tl). **Science Photo Library:** Neil Bromhall (tr). **126 Dept of Fetal Medicine, Royal Victoria Infirmary:** (cl). **Science Photo Library:** Edelmann (cr); Tissuepix (bc); Sovereign, ISM (bl). **127 Science Photo Library:** Saturn Stills (br); Astier (cr); Susumu Nishinaga (bl); Innerspace Imaging (cl). **130 Alamy Images:** Picture Partners (tl). **131 Science Photo Library:** Mendil (tl). **132 Science Photo Library:** Sovereign, ISM (cl); Ph. Saada / Eurelios (bl). **133 Getty Images:** Steve Allen (tl). **Science Photo Library:** Edelmann (cr). **134 Science Photo Library:** Neil Bromhall (cr); BSIP, Margaux (cl); Edelmann (bl). **135 Alamy Images:** Oleksiy Maksymenko Photography (bl). **Science Photo Library:** P. Saada / Eurelios (br). **138 Alamy Images:** Science Photo Library (fcr); Chris Rout (tr); Picture Partners (bc, br). **Science Photo Library:** Gustoimages (tc); (cr). **139 Corbis:** Ian Hooton/Science Photo Library (bc). **Science Photo Library:** (cl); Living Art Enterprises, Llc (ca). **140–141 Science Photo Library:** Neil Bromhall. **142 Alamy Images:** Nic Cleave Photography (b). **Science Photo Library:** Edelmann (tl). **143 Getty Images:** Photolibrary/Derek Bromhall (tr). **Science Photo Library:** (tl); Thomas Deerinck, NCMIR (br). **144 Getty Images:** Tom Grill (bl). **Science Photo Library:** Steve Gschmeissner (cra); Edelmann (cla, br). **145 Dept of Fetal Medicine, Royal Victoria Infirmary:** (br). **Science Photo Library:** Steve Gschmeissner (bl); CIMN, ISM (cl). **148 Science Photo Library:** Dr. P. Marazzi (ca); BSIP, Cavallini James (crb). **149 Science Photo Library:** Edelmann (tl); Ralph Hutchings, Visuals Unlimited (cra); Astier (crb); Anatomical Travelogue (br). **150 Corbis:** (tr). **Science Photo Library:** Edelmann (t); CIMN, ISM (bl). **151 Science Photo Library:** Penny Tweedie (b). **152 PhototakeUSA.com:** LookatSciences (tr). **154 Dept. of Fetal Medicine, Royal Victoria Infirmary:** (tr). **Science Photo Library:** ISM (tc); Ramare (tl). **156 Getty Images:** Ian Hooton (cr). **Science Photo Library:** Simon Fraser / Royal Victoria Infirmary, Newcastle Upon Tyne (bl); Dr. Najeeb Layyous (cl). **157 Alamy Images:** Glow Wellness (cr). **Getty Images:** Science Photo Library RF (bl). **Science Photo Library:** (cl); Dr. Najeeb Layyous (br/correct). **160 Getty Images:** Jose Luis Pelaez Inc. (cra); Science Photo Library RF (crb). **161 Science Photo Library:** Neil Bromhall (cla); Dr. Najeeb Layyous (bl). **162–163 Science Photo Library:** Simon Fraser / Royal Victoria Infirmary, Newcastle Upon Tyne. **162 PhototakeUSA.com:** Medicimage (br). **Science Photo Library:** Steve Gschmeissner (cl). **164 Getty Images:** Buena Vista Images (cr). **Science Photo Library:** Simon Fraser (cl); Dr. Najeeb Layyous (br); GE Medical Systems (bc). **165 PhototakeUSA.com:** LookatSciences (br). **Science Photo Library:** P. Saada / Eurelios (cr); Susumu Nishinaga (cl). **168 Getty Images:** Jose Luis Pelaez Inc (br). **169 Science Photo Library:** Thierry Berrod, Mona Lisa Production (cl); BSIP, Marigaux (br). **170 Science Photo Library:** AJ Photo (cl); Du Cane Medical Imaging Ltd (bl); Steve Gschmeissner (cr). **171 Dept of Fetal Medicine, Royal Victoria Infirmary:** (cl). **Science Photo Library:** Ian Hooton (br); Matt Meadows (cr). **174 Getty Images:** David Clerihew (bl). **Science Photo Library:** CNRI (cl). **176 Science Photo Library:** Steve Gschmeissner (tr); Sovereign, ISM (br). **178 Science Photo Library:** Sovereign, ISM. **179 Science Photo Library:** Sovereign, ISM. **180 Alamy Images:** Oleksiy Maksymenko (br). **Science Photo Library:** Dr. Najeeb Layyous (cl); Steve Gschmeissner (bl). **181 Science Photo Library:** Thierry Berrod, Mona Lisa Production (b/left & right); Du Cane Medical Imaging Ltd (tr). **186–187 Science Photo Library:** Pasieka. **188–189 Science Photo Library:** Simon Fraser. **188–203 Science Photo Library:** Pasieka (sidebars). **190 Corbis:** Radius Images (tr). **191 Science Photo Library:** BSIP, Laurent (cr). **194–195 Science Photo Library:** Custom Medical Stock Photo. **196 Alamy Images:** Angela Hampton Picture Library (b). **Science Photo Library:** Eddie Lawrence (cl). **198 Alamy Images:** Peter Noyce (cl). **198–199 Corbis:** Floris Leeuwenberg/The Cover Story. **200 Corbis:** Juergen Effner/dpa (cl); Rune Hellestad (bc). **Science Photo Library:** Professor P.M. Motta & E. Vizza (br). **201 Corbis:** Jennie Woodcock; Reflections Photolibrary (bl). **202 Alamy Images:** Chloe Johnson (br). **Science Photo Library:** Pasieka (cla). **203 Getty Images:** Vince Michaels (tl). **204–205 Science Photo Library:** Innerspace Imaging. **205 Science Photo Library:** Innerspace Imaging (cr). **206–207 Corbis:** Douglas Kirkland. **206 Getty Images:** Marcy Maloy (br). **206–213 Science Photo Library:** Innerspace Imaging (sidebars). **208 Science Photo Library:** Edelmann (bc). **209 Getty Images:** Lisa Spindler Photography Inc. (tl). **Photolibrary:** Comstock (br). **210 Corbis:** Howard Sochurek (cl). **211 Getty Images:** Jose Luis Pelaez Inc (clb); National Geographic (br). **Science Photo Library:** Ian Hooton (bc). **213 Alamy Images:** Christina Kennedy (tr). **214–215 Science Photo Library:** Professors P.M. Motta & S. Makabe. **215 Science Photo Library:** Professors P.M. Motta & S. Makabe (tr). **216 Getty Images:** Mike Powell (bl). **Science Photo Library:** (ca). **217 Corbis:** MedicalRF.com (crb). **Science Photo Library:** Dr. Arthur Tucker (cl). **218 Science Photo Library:** CNRI (tr, cl); Sovereign, ISM (b). **219 Photolibrary:** Medicimage (ca). **Science Photo Library:** Gustoimages (crb); Dr. Najeeb Layyous (bl/photo); John Radcliffe Hospital (br). **220 Science Photo Library:** Eye of Science (tc); Moredun Scientific Ltd (cl); Pasieka (cr). **221 Alamy Images:** Gabe Palmer (tr). **Science Photo Library:** Michael W. Davidson (br). **222 eMedicine.com:** Image reprinted with permission from eMedicine.com, 2010. Available at: http://emedicine.medscape.com/article/382288-overview (bl). **Science Photo Library:** Pasieka (tr). **223 Science Photo Library:** CNRI (cr); Dr. P. Marazzi (tl). **225 Science Photo Library:** Dr. Linda Stannard, UCT (cla); (cr). **227 Science Photo Library:** Zephyr (tl). **229 Science Photo Library:** BSIP DR LR (cl). **231 Science Photo Library:** Dr. Najeeb Layyous (br); Dr. P. Marazzi (cr); Professor P.M. Motta et al (t). **232 Corbis:** Nicole Hill/Rubberball (cl). **233 Science Photo Library:** Eye of Science (cr). **234 Children's Memorial Hospital, Chicago:** (bl). **Jamie Lusch / Mail Tribune photo:** (cr). **Science Photo Library:** Penny Tweedie (cl). **235 Science Photo Library:** Astier (tl); Du Cane Medical Imaging Ltd. (cr). **236 Science Photo Library:** Zephyr (tr); BSIP VEM (bc). **237 Corbis:** Leah Warkentin/Design Pics (cl). **Wellcome Images.** **238 Dept of Fetal Medicine, Royal Victoria Infirmary:** (cr). **Science Photo Library:** Dr. P. Marazzi (br). **239 CLAPA:** Martin & Claire Bostock (cb/before & after). **Science Photo Library:** Saturn Stills (tr); (br). **240 Science Photo Library:** Biodisc (cl); BSIP, Boucharlat (tr). **241 Science Photo Library:** (bc); BSIP VEM (cr); Sovereign, ISM (br). **242 Alamy Images:** Roger Bamber (cra). **Getty Images:** Alexandra Grablewski (tr). **243 Fotolia:** Lars Christensen (bc). **Science Photo Library:** Dr. P. Marazzi (tc); Ian Hooton (bl); Severine Humbert (br). *Endpapers:* **Getty Images:** Yorgos Nikas

All other images © Dorling Kindersley
For further information see: www.dkimages.com